CLIMATE CHANGE

THE FACTS 2020

CLIMATE CHANGE

THE FACTS 2020

EDITED BY
JENNIFER MAROHASY

CONTRIBUTORS

John Abbot Howard Brady Susan Crockford Bella d'Abrera
Arthur Day Geoffrey Duffy Scott Hargreaves Aynsley Kellow
Donna Laframboise Richard Lindzen Jennifer Marohasy
Paul McFadyen Jo Nova Peter Ridd Roy Spencer
Jim Steele Ken Stewart Henrik Svensmark
Marchant van der Walt Jaco Vlok

Institute of Public Affairs

First published 2020. Reprinted 2021

Institute of Public Affairs
Level 2, 410 Collins Street
Melbourne, Victoria 3000
Phone: 03 9600 4744
www.ipa.org.au

National & International Distribution
by Australian Scholarly Publishing
scholarly.info

ISBN: PB 978-1-925984-94-1

Cover photograph: Cumulonimbus cloud photographed near Lizard Island
in the Coral Sea on 19 January 2020 by Jennifer Marohasy
Cover design by Luke Harris
Typeset by Midland Typesetters, Australia

This book is dedicated to the
memory of meteorologist
Joanne Simpson

Contents

Contributors

John Abbot
John is a Senior Fellow at the Institute of Public Affairs with more than 120 peer-reviewed publications in international scientific journals, and more than a dozen recent publications in climate science. John is interested in understanding how natural systems change, and the application of artificial intelligence methods of forecasting rainfall. He is a graduate in chemistry from Imperial College (London), with a PhD from McGill University (Montreal), and also has a law degree from the University of Queensland (Brisbane).

Howard Brady
Howard is a geologist with a Distinguished Alumnus Scientist of the Year Award for contributions to Antarctic research (Northern Illinois University, 2011). Howard has published in *Nature, Science*, the *Journal of Glaciology*, and the *Antarctic Journal of the United States*. He is a member of the Australian Academy of Forensic Sciences and an emeritus member of the Explorers Club of New York. In 2018, Howard published the book *Mirrors and Mazes – A Guide Through the Climate Debate*.

Susan J. Crockford
Susan is an evolutionary biologist and a former adjunct professor at the University of Victoria, British Columbia. She is the author of several books, including *Rhythms of Life: Thyroid Hormone and the Origin of Species, Eaten: A Novel* (a polar bear attack thriller) and *Polar Bear Facts*

and Myths (also available in French, German, Dutch, and Norwegian). Susan is the principal at Pacific Identifications Inc. and blogs at www.polarbearscience.com.

Bella d'Abrera

Bella is the Director of the Foundations of Western Civilisation Program at the Institute of Public Affairs. She has a Bachelor of Arts in history and Spanish from Monash University, a Master of Arts in Spanish from the University of St Andrews, and a PhD in history from the University of Cambridge. She is frequently on Australian television and is a regular contributor to *The Australian*, the *Daily Telegraph*, *The Herald Sun* and *The Spectator Australia*.

(Robert) Arthur Day

Arthur is an earth scientist with a PhD from Monash University in volcanology. He worked at Australian Nuclear Science and Technology Organisation (ANSTO) developing methods for the safe immobilisation of nuclear wastes in collaboration with other scientists around the world. Arthur wants to see government policies based on evidence and science, untainted by partisan political or ideological agendas. He has no professional affiliations, and no career-related vested interests.

Geoffrey Duffy

Geoff is an emeritus professor at the University of Auckland, New Zealand. He has a PhD in chemical engineering and an earned Doctor of Engineering degree. He was the first chemical engineer to be elected a Fellow of the Royal Society in New Zealand. His research has focused on fluid mechanics and specialised rheology, heat transfer (especially radiant energy), drying and humidification. He has had more than 100 articles published in international journals, more than 130 published conference papers, and holds several patents. He has consulted widely in Scandinavia, Europe, Australasia and the USA.

Scott Hargreaves

Scott is the Executive General Manager at the Institute of Public Affairs, hosts the weekly *Looking Forward Podcast*, and is also the editor of the *IPA*

Review. Scott was once a policy adviser to government and private organisations, where he worked in corporate affairs including as a manager of sustainability. He has a Bachelor of Arts in politics and economics, a Postgraduate Diploma in public policy, a Master of Commercial Law in energy, environment and resources, all from the University of Melbourne, and an MBA from the Melbourne Business School.

Aynsley Kellow

Aynsley holds a PhD in political studies from the University of Otago. He is an emeritus professor at the University of Tasmania, where he taught from 1981–1984 and 1999–2018. He was foundation Professor of Social Sciences in the Australian School of Environmental Studies at Griffith University 1992–1998, a member of the Joint Academies Committee on Sustainability, and an expert reviewer for the United Nation's WGII Intergovernmental Panel on Climate Change's Fourth Assessment Report. His books include *Transforming Power* and *International Toxic Risk Management* (both Cambridge), and *Science and Public Policy: The Virtuous Corruption of Virtual Environmental Science* and *Negotiating Climate Change* (both Edward Elgar).

Donna Laframboise

Former *National Post* and *Toronto Star* columnist, and past vice president of the Canadian Civil Liberties Association, Donna is an investigative journalist whose work can be found at BigPicNews.com. Donna's book *The Delinquent Teenager Who Was Mistaken for the World's Top Climate Expert* (Connor Court Publishing) investigates the internal workings of the United Nation's Intergovernmental Panel on Climate Change (IPCC).

Richard Lindzen

Richard is an atmospheric physicist, Alfred P. Sloan Professor of Meteorology, Emeritus, at the Massachusetts Institute of Technology (MIT), and formerly the Burden Professor of Meteorology at Harvard. He was a lead author of the Intergovernmental Panel on Climate Change's (IPCC) Third Assessment Report on the science of climate change. An author of more than 230 peer-reviewed papers, he holds degrees in physics and applied mathematics from Harvard, though his doctoral thesis was on the inter-action of radiation and chemistry with the dynamics of the stratosphere. He has continued to work on the physics of the atmosphere since then.

Jennifer Marohasy

Jennifer is a Senior Fellow at the Institute of Public Affairs, with publi-cations in the international climate science journals *Atmospheric Research* and *Advances in Atmospheric Research*, as well as *Wetlands Ecology* and *Management, Human and Ecological Risk Assessment, Public Law Review* and *Environmental Law and Management*. Jennifer has a Bachelor of Science and PhD from the University of Queensland (Brisbane) and a life-long interest in natural history and long-range weather forecasting. Jennifer blogs at jennifermarohasy.com.

Paul McFadyen

Paul has a Bachelor of Science, Master of Science and PhD from the University of Queensland (Brisbane). He has first-hand experience of the interaction between science and public policy from his twenty years as a senior economic adviser in Queensland Treasury, mostly in budget and then two years as an adviser in cybersecurity and international tele-communications policy for the Australian government in Canberra. Paul worked as a scientist for three years in South America, mostly in Brazil, undertaking research.

Jo Nova

Jo is perhaps best known as a prolific and popular blogger on science-related issues. Her blog won Best Topical Blog of 2015, the Lifetime Achievement Award in the 2014 Bloggies and Best Australian and New Zealand Blog in 2012. She has a Bachelor of Science from the University of Western Australia (Perth) where she won the F.H. Faulding and the Swan Brewery prizes. A former associate lecturer of science communi-cation at the Australian National University (Canberra), Jo wrote *The Skeptic's Handbook* – translated into French, German (twice), Swedish, Norwegian, Finnish, Turkish, Japanese, Danish, Czech, Portuguese, Italian, Balkan, Spanish, Lao and Thai.

Peter Ridd

Peter is a former professor and Head of Physics (2009–2016) at James Cook University (Townsville). His consultancy work while at the university resulted in the development of specialist instrumentation for measuring aspects of water quality, with the profits used to fund student scholarships and research projects. He has published more than 100 papers in international science journals and supervised many successful PhD students.

Jim Steele

Jim has a Master's degree in ecology from San Francisco State University where he was appointed the Director of its Sierra Nevada Field Campus. He served in that capacity for 25 years, building its environmental education program and researching the effects of regional climate change on bird populations as part of the Sierra Nevada Neotropical Migratory Bird Riparian Habitat Monitoring project for the United States Forest Service. Jim's research culminated in the successful restoration of the Carman Creek watershed and its wildlife. In addition to lecturing at the university, he also taught science at inner-city schools.

Roy Spencer

Before becoming a Principal Research Scientist at the University of Alabama in Huntsville in 2001, Roy was a senior scientist for Climate Studies at NASA's Marshall Space Flight Center, where he and John Christy received NASA's Exceptional Scientific Achievement Medal for their global temperature monitoring work with satellites. Roy is the United States' Science Team leader for the Advanced Microwave Scanning Radiometer flying on NASA's Aqua satellite. Roy received his PhD in meteorology at the University of Wisconsin–Madison in 1981.

Ken Stewart

Ken is a retired teacher and school principal. With no formal qualifications in climate science but a long interest in climate and meteorology, he started the blog kenskingdom.wordpress.com in 2010. He has analysed and critiqued the Australian Bureau of Meteorology's adjusted datasets, weather station siting, one-second temperature data, heatwave detection, and cyclone reports – because he can and because he cares.

Henrik Svensmark

Henrik Svensmark is a physicist in the Astrophysics and Atmospheric Physics division at the Danish National Space Institute (DTU Space). He has held postdoctoral positions in physics at three other organisations: the University of California (Berkeley), the Nordic Institute for Theoretical Physics in Stockholm (Sweden), and the Niels Bohr Institute, University of Copenhagen (Denmark).

Jaco Vlok

Jaco is a Senior Fellow at the Institute of Public Affairs with degrees in electronic engineering from the University of Pretoria, South Africa, and a PhD from the University of Tasmania. He has undertaken research in radar and electronic warfare, including computer simulations, laboratory tests and field trials with South Africa's Council for Scientific and Industrial Research (CSIR). He has a particular interest in historical temperature reconstructions and developing alternative techniques using artificial neural networks, a form of artificial intelligence.

Introduction

Dr Jennifer Marohasy

Many people prefer their 'facts' to be short and to the point. But climate science is complicated and, despite the considerable expenditure on research, our understanding of the atmosphere is far from complete. Climate science is also important. I have no doubt that if we had a properly constructed theory of climate, scientists would be able to better forecast droughts, floods and bushfires. There would be less fear of catastrophic sea-level rise, and more curiosity rather than despair when it comes to polar bears and penguins.

In this latest book in the *Climate Change: The Facts* series, we give the scientific facts to date, whether or not they fit the catastrophic human-caused global warming paradigm. We also revisit the pioneering work of the late John Daly (1943–2004) and Joanne Simpson (1923–2010).

Simpson studied cloud formation and tropical thunderstorms, and how they could result in tremendous amounts of energy transfer from the Earth's surface to the top of the troposphere – where it can be radiated to space. According to Peter Ridd, writing in Chapter 12, these clouds are the heat engine of the atmosphere driving global circulation and mitigating any increase in temperatures from greenhouse gases. Ridd builds on the mathematics laid out by Simpson and Herbert Riehl back in 1958.

Simpson received numerous formal accolades and awards throughout her working life, including the American Meteorological Society's 1983 Carl-Gustaf Rossby Research Medal – the highest award in atmospheric sciences. She is remembered within this community as having a powerful combination of intellect, determination, leadership and drive. She also insisted that as a scientist it was always important to be sceptical, particularly of models that purport to represent reality while not being very good at forecasting rainfall. This book is dedicated to her memory.

The importance of an accurate weather forecast

Weather forecasting at the moment is woeful, as IPA Senior Fellow Dr John Abbot explains in Chapter 14. Abbot demonstrates an innovative new method for forecasting rainfall based on the latest advances in artificial intelligence. But this new, more skilful, method is shunned by leading climate scientists, perhaps because it would make general circulation models (GCMs) obsolete. GCMs, of course, underpin most assumptions about catastrophic human-caused global warming.

Skilful weather forecasts would not only be useful for agriculture, and management of dams for hydroelectricity, but more generally for scheduling activities in a range of industries, or for planning weddings. Skilful weather forecasts can also provide a military advantage. Some say it was a weather forecast that won World War II for the Allies. The D-Day landing was postponed by two days because British meteorologist James Stagg – who was trusted by the supreme commander of allied forces in Western Europe, Dwight D. Eisenhower – knew how to forecast the weather, enabling the Allies to take the Germans by surprise. Stagg understood the key variables that affect weather (including lunar cycles) and he had access to a network of weather stations, including one at a post office at Blacksod Point in the far west of Ireland. That weather station proved crucial in detecting the arrival of a lull in the storms that Stagg calculated would allow for an invasion on 6 June. Even as rain and high winds lashed Portsmouth on the night of 4 June, Stagg informed Eisenhower of the forecast for a temporary break. This forecast proved correct.

The scientific–technological elite versus the facts

Eisenhower went on to become president of the United States. In his farewell address, in a television broadcast on 17 January 1961, he warned about the dominance, or the 'capturing', of science-based public policy by what he called a 'scientific–technological elite'. It is this same elite who have dictated climate-change policies for some decades now, with wide-ranging economic implications. The Paris Climate Accord, for example, sets out to change the mix of energy types that we use. While these so-called climate policies have the veneer of science, it is unclear whether they really are fact-based. It could be, as Ansley Kellow explains in Chapter 18, that climate science falls short of the criteria for good science. Far too frequently climate science has demonstrated noble cause corruption – where the ends justify the means, or where there is a belief in a moral commitment – in producing evidence, any evidence, intended to have an influence on the political landscape.

A fact is something that is known, or proven, to be true. A scientific fact cannot be established by a consensus of opinion, or by the popular vote, or because it is morally good. A fact may contain offensive information, but may nevertheless still be true.

This book is prefaced by the assumption that we are all entitled to our own opinions, but not to our own version of the facts. For something to have the status of being 'a fact' in climate science it needs to be supported by objective evidence in the form of repeatable observations or transparent experimentation. It cannot simply be the output from a politically driven United Nation's working group, associated with what Eisenhower may have called 'the climate–industrial complex'. Yet this is increasingly the case, with significant implications, as Paul McFadyen, Scott Hargreaves and Bella d'Abrera explain in Chapter 19 with particular reference to university funding. In Chapter 15, I explain how less ideology and more active management of our eucalyptus forests could have saved human lives, property and so much wildlife over the 2019–2020 summer in south-eastern Australia.

The climate changes, but how does it change

We live at a time when climate change is deemed a morally important issue. It is claimed that greenhouse gases from the burning of fossil fuels are causing unprecedented and potentially catastrophic warming of the Earth's atmosphere. Those who disagree with this claim, or who ask for more evidence, are generally labelled 'deniers' of climate change.

In reality, the dispute is not whether the climate changes; it is what causes the change (could it be mostly natural rather than human-caused), and whether the current rate and magnitude of change is unusual. Then there is the issue of how this change is perceived and described – this has a philosophical dimension.

In climate science, as in life, we can perceive change as something that will tend to occur as a cycle that will one day come to an end. Alternatively, we may perceive change as more-or-less linear, as something that will continue in the same direction for a long time, perhaps forever. My mother lived through World War II as an adolescent in London, and she says that at the time, no one in her family had any idea that the war would one day end. Mum remembers D-Day. She says they woke up to clear skies on the morning of 6 June, and to the droning sound of aircraft overhead and what looked like thousands of aircraft in the sky. That was the beginning of the end of the war, but who really knew that at that time?

Mainstream climate scientists generally describe the rise in global temperatures since the late 1800s as linear – as consistently increasing in intensity and showing no sign of abatement. In contrast, sceptics generally perceive the current warming as part of a natural cycle that will one day end.

The first Intergovernmental Panel on Climate Change (IPCC) report, published in 1991, suggested that the current warm period is not quite as warm as the Medieval Warm Period, which extended from about 985 to 1200 AD. This was when the cathedrals with their very tall spires were built across Europe and the Vikings settled south-west Greenland. The Medieval Warm Period was followed by a cold period known as the Little Ice Age that included the Maunder Minimum when Greenland became too cold for human habitation. So, over just the last 1000 years or so, there is evidence of a climate cycle; it was warm and then cold and now it is warm again.

Howard Brady, who once worked as a scientist in Antarctica, explains in Chapter 2 that there is compelling evidence that south-west Greenland was warmer 1000 years ago than it is today, and even warmer between 5000 and 8000 years ago, during a period known as the

Mid-Holocene Thermal Maximum. Brady makes the case that climate change at both the North Pole and Antarctica has been occurring for at least 10,000 years – in cycles. Therefore, it might seem reasonable to assume that the warm period we are currently experiencing will eventually end.

This cyclical history of temperature change, however, has been arbitrarily altered; it has been remodelled in more recent IPCC reports. The Medieval Warm Period, for example, was 'flattened' by the hockey stick in the third assessment report of the IPCC, which was published in 2001. (This was discussed in some detail in the previous book in this series *Climate Change: The Facts 2017*, in Chapter 16 by Simon Breheny.) Flattening the Medieval Warm Period gives the false impression that global temperature change is linear, somewhat like the handle and shaft of a hockey stick, rather than cyclical, like waves at a beach.

Climate change, sea level change and volcanism in Antarctica

Antarctica is a focus of this book because it is so important in terms of global atmospheric circulation. Did you know that the Antarctic Circumpolar Current is the strongest ocean current on Earth and the only ocean current linking all major oceans – the Atlantic, Indian, and Pacific Oceans? Without the Antarctic Circumpolar Current, and its impact on planetary heat redistribution, the global climate would be very different.

According to mainstream climate science, global warming should be most pronounced at the North and South Poles. And you might also be misled by mainstream climate scientists on numbers of polar bears at the North Pole, and numbers of penguins at the South Pole, as scientists Susan Crockford and Jim Steele explain in Chapters 1 and 3, respectively. As journalist Donna Laframboise explains in Chapter 1, this has implications not just for science, but for our ability to freely communicate information.

Ken Stewart was once a primary school teacher. Now he is retired, and spends his days analysing temperature data and blogging. I asked him if he would have a look at the publicly available historical temperature measurements for Antarctica and write a chapter for this book. An advantage with Stewart's approach, detailed in Chapter 7, is that it is accessible to people without a science PhD. He presents the numbers in a straightforward way, as a schoolteacher might.

IPA research fellow Jaco Vlok follows up with a slightly more sophisticated analysis in Chapter 8, while John Abbot and I include some signal analysis and artificial neural network forecasting in Chapter 9. What we each find (Stewart, Vlok, Marohasy and Abbot) accords with Roy Spencer's analysis, that there is no 'global warming' at Antarctica. Spencer's analysis is detailed in Chapter 6. Spencer is a meteorologist, a principal research scientist at the University of Alabama in Huntsville, and the US Science Team leader for the Advanced Microwave Scanning Radiometer on NASA's Aqua satellite. He has served as senior scientist for climate studies at NASA's Marshall Space Flight Center and was awarded the NASA Exceptional Scientific Achievement Medal.

Something as basic as easily accessible, publicly available, unadulterated historical temperature measurements, as analysed by someone as experienced as Spencer, should be the building blocks for any theory of climate.

Then there is the issue of volcanoes and the possible risk they pose to the stability of the West Antarctica ice sheet. In Chapter 4, Arthur Day – who has a PhD from Monash University on modelling of volcanic processes – focuses on the implications of Antarctic volcanism for global sea-level rise. Specifically, he writes about volcanism in West Antarctica where there are at least 138 volcanoes comprising one of the world's largest active continental volcanic fields, 91 of them only newly discovered and hidden kilometres beneath the ice sheet.

I asked Day to write this chapter to help us better understand the situation, because both sides in the 'climate wars' often claim volcanism to advance their position. Alarmists believe that ice-sheet melting and sea-level rise are being accelerated because of 'climate change', and that the volcanoes magnify the risk to the ice sheet. Conversely, some climate sceptics claim that heat linked to volcanism is the main threat, contributing to melting and temperature rise, *instead* of carbon dioxide

(CO_2)-linked climate change. Day concludes that on any time frame relevant to human experience, there is no evidence that the intensity of volcanism will increase, future volcanism is unlikely to change *overall* ice-sheet melting rates; and, because the ice sheet is already in long-term balance with volcanic effects, future volcanism is extremely unlikely to destabilise the ice sheet and accelerate global sea-level rise.

It is the case that at some places on this Earth there has been recent sea-level rise, and yet for other coastlines, sea levels appear to be falling. In Chapter 17, Arthur Day extends his analysis of sea-level change at Antarctica to a world tour based on tidal gauge data, building on the work of the late John Daly. Daly's work remains relevant because it was mostly a compilation of particularly important historical sea-level records – placed in context. Daly was an inspiration to many concerned about the direction of climate science going back more than forty years. Daly was described by Professor Emeritus John Brignell of the University of Southampton on this death in 2004 as:

> Daly was the epitome of a new phenomenon of the post-scientific age, a lone scholar with all the traditions of meticulous attention to detail and truth that the word implies, with limited means upholding the principles of the scientific method in the face of adversaries with vast resources.

Arthur Day continues this tradition, updating John Daly's charts and providing more context.

The potential of clouds

In the early 1980s, Richard Lindzen was working as a meteorologist at the Massachusetts Institute of Technology, and he accepted the basic tenets of global warming theory. In Chapter 13, Lindzen explains that, like most people working on climate, he assumed relative humidity remained fixed as climate changed. He tried to measure this by measuring water vapour at different altitudes. At first Lindzen assumed a relationship between high cloud and concentrations of water vapour. He was surprised to find that the data suggested they were unrelated. Further, he went on to realise that clouds themselves were a major factor in Earth's radiative budget.

Low clouds, usually below 2000 metres (6500 feet), have a cooling effect on the Earth by reflecting incoming solar radiation and in this way casting a shadow. High clouds, above 6000 metres and possibly as high as 12,000 metres (20,000 to 40,000 feet), can have a warming effect by reducing the rate at which the Earth's surface and atmosphere radiate energy to space, and also by reradiating infrared radiation back towards Earth. In this way, clouds can contribute to what is known as the 'greenhouse effect'.

There are many different types of clouds, but one type of cloud dwarfs all others in size; it is cumulonimbus. The lower portion is composed of water-droplets, then there is a tower with an upper portion that is a roiling mix of ice, snow, hail and super-cooled water that has not yet frozen. At the very top of these clouds, the ice is ejected and spreads to form a characteristic wide, flat-top anvil shape. Multiple cumulonimbi towers can share a sprawling anvil-shaped top.

At any one time there may be more than 1000 of these towering cloud formations in a band across the Earth's tropical oceans. We can see them in satellite imagery as large storm cells that extend from the Earth's surface to the top of the troposphere, with vast quantities of wispy white cirrus cloud streaming away from them.

In Chapter 12, former head of physics at James Cook University, Peter Ridd, with Marchant van der Walt, make the analogy between these rising towers of cumulonimbi – that move vast quantities of heat from the surface of the Earth to the top of the troposphere – and the pistons in a car engine. This huge atmospheric engine helps cool the surface atmosphere. Building on the pioneering work of Joanne Simpson – who was the first woman in the world to receive a doctorate in meteorology in 1949 – Ridd describes how the resulting tropical convection drives atmospheric circulation. Convection causes warm air, which is less dense than cold air, to rise up.

The analogy of tropical convection as the heat engine of the atmosphere is a clever one. Applying some mathematics, Ridd shows that more greenhouse gases in the lower atmosphere can have the effect of making this heat engine more powerful. In short, Ridd shows that with increasing greenhouse gas concentrations in the lower atmosphere, air temperatures can increase and thus raise the water vapour content of the air if this occurs over tropical oceans. Water vapour, in turn, is the fuel driving this deep tropical convection – the giant pistons.

To summarise, according to Ridd, more greenhouse gases will have the effect of increasing the efficiency of the heat engine that drives global atmospheric circulation. Ridd calculates that for every 1 °C rise in tropical temperature, the heat transfer by the convection pathway will increase by 10%. This increase will mostly be within the towering cumulonimbi clouds that are transporting the additional heat to the top of the troposphere, where it can be lost to space through infrared radiation. This suggests a strong negative feedback to rising temperature, from any cause, including greenhouse gas concentrations.

A completely different type of cloud, called cirrus, forms from the dissipating tops of these storm cells. Wispy and white, cirrus clouds are composed only of ice crystals. Whipped about by the strong winds at very high altitudes of 7300 to 13,700 metres (24,000 to 45,000 feet) they form strands. In fact, the name 'cirrus' is derived from the Latin term for a lock of hair.

Richard Lindzen's work has been primarily focused on these cirrus clouds and their heating effect on the environment, because they reflect infrared radiation back to Earth. According to Lindzen, beneath cirrus clouds temperatures are known to rise by up to 10 °C because of the greenhouse effect. According to Lindzen's theory called the 'iris effect', as temperatures rise because of the increasing atmospheric concentrations of greenhouse gases, the concentration of upper-level cirrus cloud decreases relative to the area of cumulonimbus.

Lindzen makes the analogy with the pupils in our eyes changing size relative to how bright or dim the light is. Specifically, Lindzen has hypothesised that as the atmosphere warms from increasing concentrations of greenhouse gases, the area of cirrus cloud decreases, providing a negative feedback as more infrared radiation is able to escape into space.

So, Lindzen and Ridd concern themselves with different types of cloud and different types of processes. Both hypothesise that there are cloud-related negative feedback loops in place that will mitigate the potential effects of increasing concentrations of greenhouse gases on Earth's temperature. Neither of them deny the potential for greenhouse gases, especially water vapour and CO_2, to warm the Earth. Rather they explain that because of the complexity of the physical processes at work, in particular, and the role clouds play in facilitating negative (cooling) feedbacks, the Earth is unlikely to overheat.

Lindzen did not set out to disprove human-caused global warming theory. His seminal paper, with Ming-Dah Chou and Arthur Hou, simply recommended that mainstream models of global atmospheric circulation incorporate the iris effect in order to achieve a better match with observational data. However, rather than testing the iris effect as an hypothesis, an attack was launched by the climate science establishment against Lindzen's research – that was twenty years ago and continues today, as explained in Chapter 13.

Ridd's hypothesis, building on the work of Joanne Simpson from the 1950s to the 1970s, is outlined for the very first time in this book. In Chapter 12, he concludes that the complicated GCMs used by the IPCC will never be able to adequately simulate the role of convection and clouds because thunderstorms (the pistons in the heat engine) are far smaller than the grid-scale of their models. Ridd's hypothesis is detailed here in the hope that others may find a way to test it.

Before reading either Lindzen's or Ridd's chapters, it is worth taking the time to understand the facts in Chapter 11 by Geoffrey Duffy, former Professor and Head of the Department of Chemical and Materials Engineering, University of Auckland. This chapter also explains some of the basics of short-wave versus long-wave (infrared) radiation, which is fundamental to understanding something of the Earth's energy balance, and thus climate change.

Duffy also explains the nuts and bolts of the water cycle – water evaporation from land and sea, condensation as clouds, and precipitation as rain. Duffy does not spend much time on tropical convection, but then he is from New Zealand – the land of the long white cloud. (Ridd is from the tropics, from northern Australia.) Instead, Duffy focuses on incoming solar radiation and its re-radiation as infrared radiation, principally by water vapour and CO_2 – the basis of the greenhouse effect – and on popular concern about catastrophic human-caused global warming. Duffy explains in just enough detail why, if radiation is the key issue, the focus should be on water vapour rather than CO_2. He doesn't formulate an alternative hypothesis, as such, but he does explain that water vapour is twelve times more effective than CO_2 in long-wave (IR) radiation absorption and re-radiation. He also explains, tangentially, how water vapour can drive tropical convection because humid air is less dense than dry air. Water vapour is critical to our understanding of Earth's radiative energy balance and the operation of convection, as Duffy explains in Chapter 11.

Like Richard Lindzen, Henrick Svensmark is a university professor who has spent much of his career somewhat obsessed with clouds. But while Lindzen has focused on high-altitude cirrus clouds, Svensmark's theories concern low-altitude clouds.

Svensmark is a physicist in the Astrophysics and Atmospheric Physics division at the Danish National Space Institute (DTU Space) at the Technical University of Denmark.

Small and puffy white cumuli are perhaps the best known of the low-altitude cloud types. They are common over the tropical oceans. Strati clouds are uniform grey and featureless blankets: 'stratus' is from the Latin 'to spread out'. Stratocumuli are like fluffy blankets, more common in the subtropics and often thick enough to block the Sun. These are the main low-altitude cloud types. Altocumuli, altostrati and nimbostrati occur at higher altitudes, above 2000 metres (6500 feet).

According to Svensmark, since low clouds are very important for the radiative energy balance of the Earth, any systematic change in cloud cover will affect the temperature of the atmosphere and, ultimately, the climate. But according to Svensmark, it is changes in cosmic-ray flux, not water vapour, that drives changes in cloud cover. Remember, Lindzen, early in his career, assumed a relationship between high cloud and concentrations of water vapour but couldn't find one. This is perhaps because a key to the puzzle lies beyond Earth, even beyond our solar system.

In Chapter 10, Svensmark explains that cosmic rays are electrically charged particles (ions) ejected from exploding stars (some in far-away galaxies), and that cosmic rays drive the ion-nucleation of aerosols. Aerosols of many types are critical for cloud formation on Earth.

The cosmic-ray flux reaching the Earth's surface is modulated by changes in the Sun's magnetic field. Svensmark explains that the flux is also affected by our solar system's proximity to exploding stars. Our solar system travels around the Galactic Centre of the Milky Way in a journey that takes 240 million years. When our solar system enters those regions where there are open-star clusters (e.g. the Pleiades), our planet is showered with more energetic galactic cosmic rays that will affect cloudiness and thus climate.

In summary, understanding the clouds is critical if we want to understand climate change because clouds both influence, and are influenced by, atmospheric circulation. They affect the radiative energy budget of the Earth. The physical processes are complex and may be intrinsically self-regulating, but they may also amplify other processes. For example, any increase in greenhouse gases may increase the efficiency of tropical convection, which may, in turn, reduce the relative area of high-level cirrus cloud. This is my hypothesis, not necessarily supported by Ridd or Lindzen, although it is derived directly from their research. Increases in the cosmic-ray flux may increase the area of low-level cumulus and stratus cloud, also potentially cooling the Earth. In this way, changes in the cosmic-ray flux, through its modulating effect on cloud cover, could amplify by a factor of perhaps ten, changes in solar irradiation that are considered too small to affect climate change, according to the popular consensus.

If we want to truly understand climate change, then we need to place all the facts in context. To repeat what was stated at the outset of this introduction, a fact is something that is known, or proven, to be true. We are all entitled to our own opinions, but not our own facts. Climate change then becomes a puzzle that we are clearly still elucidating. I am hoping that Chapters 10, 11, 12 and 13 move us forwards in our understanding, perhaps by some significant margin.

Quality assurance

Some argue that the idea there are negative feedback mechanisms compensating for any increase in the concentrations of greenhouse gases – as proposed by Lindzen and Ridd in the previous section – is laughable, because we have temperature charts that clearly show a dramatic increase in global temperatures since at least 1880. But these charts are only 'constructs' of historical temperatures. There is nowhere on Earth where its actual temperature can be measured. Rather the statistics that show catastrophic global warming are weighted averages from thousands of weather stations situated at different latitudes and altitudes, each remodelled according to some particular formula. The issue of how temperatures are remodelled was detailed across six chapters in the last book in this *Climate Change: The Facts* series – published in 2017. In Chapter 16 of this book, I show how historic (already measured) temperatures for Australia were warmed by a further 23% by remodelling undertaken by the Bureau in 2018 – since the last book was published. The Bureau claims that this remodelling of temperatures is justified because of changes to the equipment used to record temperatures; and because of the relocation of weather stations. However, there have been no changes to equipment and no relocations since the release of ACORN-SAT Version 1 for either Darwin or Rutherglen, both of which are my case studies.

Vlok explains in Chapter 8 how the Australian Bureau of Meteorology artificially remodels a perfectly good temperature series recorded at a weather station at Mawson in the Australian Antarctica Territory by mimicking trends at a distant Russian weather station called Molodeznaya using complex software. At the same time, the Bureau appears to omit to check from thirteen months of overlapping data, that there is nearly half a degree difference in measurements between the old and new location for its Casey weather station, also at Antarctica. So, the original Casey values (which are a mishmash from different locations) are entered into official databases unchanged. While the trend in mean temperatures as recorded at Mawson (which has never moved) is changed from cooling at a rate of 0.284 °C to warming at 0.396 °C per century.

Any sceptic who raises such issues in polite society, however innocently, is likely to be branded a conspiracist as well as a science denier and a shill for oil companies. This sad state of affairs is explained by Scott Hargreaves in Chapter 20 with reference to Dante's great work about how we can find ourselves, most innocently, dragged down into hell. Perhaps, Chapter 20 should be read first, to serve as a warning to those of us who seek the truth through this book. The path is not without its dangers. Yet, I am reminded of the words of Thomas Huxley (1825–1895) that have always guided my approach to facts:

> Sit down before fact as a little child, be prepared to give up every preconceived notion, follow humbly wherever and to whatever abysses nature leads, or you shall learn nothing. I have only begun to learn content and peace of mind since I have resolved at all risks to do this.

Some years ago, wanting to better understand how CO_2 levels are determined from ice cores, I began reading some of the first papers that discussed the technique. These only date to the early 1980s – just 40 years ago. One of the first papers is by a Swiss scientist Werner Berner, published in the journal *Radiocarbon* (volume 22), it shows CO_2 levels much higher than at present (up to 500 ppm) just a few thousand years ago. This does not accord with more recently published ice-core studies that almost universally conclude that over the last 10,000 years atmospheric CO_2 levels have not exceeded 300 ppm – but the result does better accord with the known temperature history of the Holocene, specifically that there was a warmer Holocene optima some thousands of years ago when sea levels were also higher. I have discussed the need for a proper review of this literature with Jo Nova for some years, and she has done a version of this with a focus on plant stomata as an alternative measure of atmospheric CO_2 levels: a measure that has a 400 million year history. As Nova explains, one of the big mysteries of paleoclimate studies is why stomata are accepted as one of the main proxies of atmospheric CO_2 going back hundreds of millions of years, but not used to compare with results from ice cores for more recent periods.

The previous book in this series *Climate Change: The Facts 2017*, began with a chapter by Peter Ridd about the Great Barrier Reef in which he challenged the orthodoxy on Great Barrier Reef science and in particular reporting on coral calcification rates. In media promoting that book, Ridd argued for better quality assurance of Great Barrier Reef science and questioned the veracity of claims by some of his colleagues about the effects of climate change on the reef. This contributed directly to his subsequent sacking. Undeterred, Ridd continues to call for the release of fifteen years of missing coral growth data by the Australian Institute of Marine Science.

Ridd's fight not only for this data, but also for his job back, has attracted significant attention worldwide. It confirms what many people have suspected for a long time: Australia's universities are no longer institutions encouraging the rigorous exercise of intellectual freedom and the scientific method in pursuit of truth. Instead, they are now more like corporatist bureaucracies that rigidly enforce an unquestioning orthodoxy and are capable of hounding out anyone who strays outside their rigid groupthink.

The need for a new paradigm

Confirmation bias is a tendency for people to treat data selectively and favour information that confirms their beliefs. In the case of 'climate change' there is a trillion-dollar climate–industrial complex supporting a sophisticated research program generating 'facts' that cannot be questioned. So, we are told there is 'the science', which is 'settled', as though science is a fact, when science is a method of finding out about the world. As we learn more, our understanding should improve. But sometimes scientists are asking the wrong questions, based on politically motivated assumptions and using tools that are not useful. It is important to understand, for example, that there is nothing equivalent about the scientific methods that have established that Newton's law of universal gravitation is fact, and the theory of human-caused global warming. The law of gravity has been tested over and over and found to be true. Meanwhile, when Richard Lindzen tested the hard core of the greenhouse effect back in the 1980s and found it to be wanting, his findings were misrepresented and attacked, as explained in Chapter 13. He was told there were some aspects of climate science that could not be questioned.

It is possible to accept that high-level clouds absorb and reradiate long-wave (IR) radiation thus contributing to the greenhouse effect as Lindzen does, while also understanding that water vapour may be the most important greenhouse gas that fuels tropical convection as I do. This alternative 'science' of climate change provides at least two possible natural mechanisms in operation that could mitigate the future impacts of increasing atmospheric CO_2 levels from the burning of fossil fuels. In short, the climate system could be intrinsically self-stabilising, thus limiting the impacts of increasing CO_2. This would explain why vastly higher natural atmospheric CO_2 concentrations throughout much of geological history showed no relationship with past climate. This was discussed in the last book in this series, *Climate Change: The Facts* series – published in 2017, in Chapter 19 by Ian Plimer. Yet to suggest such an alternative explanation is to risk being labelled a denier.

This is because science, as a human endeavour, has always been more complicated that the simple assembling of facts through observation and testing. As the famous physicist and historian Thomas Kuhn explained in his seminal book back in 1962 titled *The Structure of Scientific Revolutions* (University of Chicago Press), a majority of scientists tend to operate within a particular research program (also known as a paradigm) that dictates not only the research method, but also which questions are permissible. The history of science would also suggest that such research programs will dominate and attempt to exclude all others. Furthermore, the established research programs, for example that more CO_2 through the greenhouse effect will inevitably lead to dangerous global warming, are rarely disproven until they are replaced.

Accepting the history of science would suggest that those of us who would like to see the current obsession with CO_2 discarded need to take an interest in possible replacement theories and actively support new competing research programs based on different methods.

For almost a generation, mainstream climate scientists across the Western world have defended the greenhouse effect. This has precluded the development of an understanding of the truly major climate changes that have characterised the Earth's history. The failed 'global warming' paradigm has also contributed to worse, not better, weather and climate forecasts. This has real implications for the efficiency of food production and also each nation's capacity to skilfully forecast weather fronts for battles that may be critical to winning future world wars.

This is the fourth book in the *Climate Change the Facts* series, a series which was conceived, and continues to be supported, by the Executive Director of the Institute of Public Affairs, John Roskam. Roskam is fond of the Galileo quote:

> In questions of science, the authority of a thousand is not worth the humble reasoning of a single individual.

Roskam says that Galileo's statement applies not only to climate science but to other fields of human endeavour and enquiry.

This book is a compilation of chapters by individuals, or at most three authors. Not all chapters come to the same conclusions about particular issues. For example, whether temperature change at Antarctica is the best measure of global climate change – or not. What is most important is that reasons are provided, and that the reasoning is logical and valid and supported by the available facts.

We have not been able to include everything or everyone in this book. I sincerely thank those scientists who submitted chapters that did not make it all the way to print. In the end judgements must be made about the extent to which a conclusion is adequately supported by the available facts – and there can only be so many pages in each book in this *Climate Change the Facts* series.

As you read through the chapters in this book, I hope you will find:

1. some rebuttal of questionable 'facts' derived from current research into the greenhouse effect
2. gain a better understanding of the importance of data integrity, and how some temperature, sea-level and ice-core series are constructs generated through the remodelling of actual measurements

3. realise that there are methods for researching climate change, and forecasting the weather, additional to GCMs, and
4. start to ask new questions about everything from water vapour to cosmic rays to emperor penguins.

Climate science necessarily touches on many different scientific disciplines, and it is complicated. It can also be so much fun! Treat it as a puzzle, that is my advice. Also, remember, it is much better to have questions that cannot be answered, than living and doing science according to answers that cannot be questioned.

SECTION I

FROZEN KNOWLEDGE ABOUT CLIMATE CHANGE

1 Walruses, Polar Bears and the Fired Professor

Donna Laframboise & Dr Susan J. Crockford

In April 2019, Netflix launched a new nature documentary series titled *Our Planet*. Narrated by cultural icon Sir David Attenborough, the second episode includes harrowing footage of walruses plunging over a cliff to their death. Viewers are told human-caused climate change has melted the sea ice, leaving these animals no choice but to gather on land. In a desperate attempt to escape overcrowding, we're told, hundreds climbed a cliff and fell 'from heights they should never have scaled'.

The Times of London promptly dubbed this footage a 'new symbol of climate change'. The producers, it said, regarded the plight of these walruses as 'the most powerful story they found during four years of filming' (Whitworth 2019).

Yahoo News described the reaction of viewers. Some sobbed. Others had trouble sleeping. One woman declared the falling walrus clip 'the saddest thing I've seen in my life', before adding: 'humans really suck'. A second person said the show had been 'Extremely tough to watch but necessary because things need to change' (West 2019).

Humans suck. Things need to change. No doubt these were precisely the reactions the *Our Planet* crew had hoped to elicit. This series is what happens when the world's leading internet entertainment service (Netflix) teams up with the world's largest conservation organisation (World Wildlife Fund Inc, WWF).

The WWF is run by professional lobbyists who know fundraising is more successful if the public makes an emotional connection, especially

Figure 1.1 Walruses and a polar bear in the Chukotka Autonomous Okrug, Russia

Source: Russian scientist via Shutterstock.

if we can be persuaded it is humanity's fault that walruses are plunging off cliffs.

Rather than being an accurate portrayal of reality, *Our Planet* is, as James Delingpole has aptly observed, 'green propaganda masquerading as entertainment' (Delingpole 2019). The script was written by professional spin doctors. Much of Attenborough's walrus narration is therefore pure fiction.

It turns out walruses have always gathered on land. They have a boom-and-bust population cycle and are currently particularly abundant, which accounts for more than 100,000 massing on beaches in recent years (Lowrey 1985; USFWS 2017). Being social animals, they crowd together – a behaviour that can lead to some of them being trampled, particularly if the herd is spooked. Walruses sometimes traipse about near cliffs. Sometimes they tumble off. In other words, the *Our Planet* clip shows perfectly natural events.

WALRUSES, POLAR BEARS AND THE FIRED PROFESSOR

Specialists refer to a walrus gathering on land as a 'hauling-out' or 'haulout'. In the words of United States (US) biologist Jim Steele, the 'notion that walruses only haul-out on land when deprived of ice is a story that would have been laughed at just 30 years ago' (Steele 2014).

The US Geological Survey in partnership with the US Fish and Wildlife Service have compiled a Pacific Walrus Coastal Haulout Database 1852–2016 that extends back to 1852 (Fischbach et al. 2016). An accompanying report explains that walruses 'spend most of their time at sea where they forage' on the seabed. 'Between foraging bouts, walruses rest out of water', returning 'repeatedly across seasons and years' to the same beaches.

Writing in the *Journal of Mammalogy* in 1923, Joseph Bernard described events near a village on the northeast coast of Siberia:

> … last year the beach was so crowded when the walruses hauled there, that many walruses were crushed to death just from overcrowding … Thirty years ago the walruses came in great numbers to Point Hope – by the thousands. The beach where the walruses hauled was near [Alaska's] Cape Lisburne … (Bernard 1923)

It so happens that walruses have been falling off a cliff in Alaska for some time. In 1996, the death of 60 animals in this manner at the Togiak National Wildlife Refuge made the *New York Times* (1996). A US Fish and Wildlife Service report from 2005 explains:

> … this activity seems to be a result of naturally occurring landscape changes. Prior to 1994, high grass-covered dunes confined walrus activity to a traditional haul-out site on the beach. But over time, wind and walrus activity eroded those dunes which opened a pathway that allowed access higher up … (USFWS 2005)

Over the years, refuge staff have hung plastic bags that flap in the wind on parachute cord in an attempt to discourage these 1360 kg animals from climbing near the cliff. Similarly, they have attached tarps to metal fence posts, and have physically herded walruses away from the precipice. In 2006, a 1 m-high, 76 m-long fence was erected. Why?

Because humans find these cliff deaths upsetting. In a 2006 news article, a refuge employee explained, 'The local people don't like to see walrus falling over. It's such a waste, but it's nature's way of doing things' (deMarban 2006).

In fairness, fake news about walruses and climate change didn't begin with *Our Planet*. After 130 young walruses were trampled to death in 2009, an Associated Press news story blamed 'a loss of sea ice' for a gathering of walruses on a beach elsewhere in Alaska (Joling 2009).

Trampling has long been recognised as a leading cause of walrus mortality. Famous US walrus researcher Francis Fay reported that more than 500 animals had died in this manner on the beaches of Alaska's St Lawrence Island in 1978 when sea ice was high (Fay & Kelly 1980). Yet in the Associated Press news coverage, Geoff York, a biologist on the WWF payroll, spun matters in a different direction: 'Were it not for the dramatic decline in the sea ice,' he declared, 'the young walruses at Icy Cape most likely would be alive on the ice and not dead on a beach.'

In sharp contrast, Canadian zoologist Susan Crockford has been trying to disperse the activist fog that currently envelops Arctic mammals. In 2014, she authored a briefing paper for the United Kingdom (UK)-based Global Warming Policy Foundation titled *On the Beach: Walrus haulouts are nothing new* (Crockford 2014). That paper was a direct response to media coverage of a massive gathering of walruses at still another Alaskan locale that same year.

An estimated three dozen trampling deaths had occurred, and officials from US government agencies had discussed declining sea ice at a press conference. Two quotes from WWF managing director Margaret Williams appeared frequently in that batch of news stories. The first declared that walruses 'should be scattered broadly in ice-covered waters'. The fact that they were gathering on beaches, said Williams, was 'just one example of the impacts of climate change' (*9 News* 2014).

The second Williams quote made the WWF's broader agenda plain: 'The walruses are telling us what the polar bears have told us ... the Arctic environment is changing extremely rapidly and it is time for the rest of the world to take notice and also to take action to address the root causes of climate change' (Linshi 2014).

For the activist WWF walruses and polar bears are convenient props, a means of cajoling the public to take notice and take action. But not just any action. The WWF lobbyists promote a specific agenda. They call carbon dioxide (CO_2), which is exhaled by every human being and is essential to all life on Earth, 'carbon pollution'.

On her own blog, *PolarBearScience.com*, Crockford has spent years providing calm counterpoints to the misinformation being spread by the WWF and others. Soon after the *Our Planet* series launched, she began challenging its falling walrus' narrative. Those particular walruses, she insisted, 'were almost certainly driven over the cliff by polar bears during a well-publicized incident' that took place in Siberia in 2017 (Crockford 2019a; *Siberian Times* 2017).

Those events were described by an online news outlet this way:

> 5,000 walruses recently hauled out on a shoreline near the village [of Ryrkaypiy]. The walruses were followed by about 20 polar bears, no doubt drawn by the stench of thousands of blubbery, flippered meals. The arrival of the bears caused the walruses to panic, and many attempted to flee. Per the *Siberian Times*, 'several hundred' fell to their deaths off the cliffs of the nearby Kozhevnikova Cape. The bears, naturally, went to town on the carcasses. (Stone 2017)

Under the bracing headline 'Netflix is lying about those falling walruses. It's another "tragedy porn" climate hoax', Crockford argued her case in Canada's *National Post* newspaper. Pointing to a WWF tweet declaring walruses 'the new symbol of climate change', she said the sole intention of the walrus clip was to 'shock viewers into taking climate change seriously' (Crockford 2019c; WWF 2019).

The same kind of emotional manipulation, she said, had been on display when *National Geographic* publicised a video of an emaciated polar bear two years earlier. The individuals who filmed the bear knew nothing about its medical condition or its history. Yet the public was told, 'This is what climate change looks like'. The video became 'the most viewed on *National Geographic's* website – ever' (Mittermeier 2018).

Initially, *Our Planet* producer Sophie Lanfear and cameraman Jamie McPherson denied that polar bears had been in the vicinity when the Netflix film footage was being shot. Several months later, it became clear this was untrue. In fact, the Netflix production crew sold additional footage to the BBC, which was incorporated into a new Attenborough TV series, *Seven Worlds, One Planet* (Asia) launched in November 2019 (BBC 2019a).

While there was some overlap between these two productions, the BBC program contained a critical new sequence. It showed hundreds of walruses being driven off a Ryrkaypiy cliff by more than a dozen polar bears – precisely the incident reported in the Siberian media in 2017.

The BBC footage made it clear that the falling walruses in the Netflix sequence had been filmed afterwards when the herd was almost certainly still agitated – in part because some bears were still in the vicinity. An associated BBC editorial further revealed that overhead drones were

Figure 1.2 Chukotka coastline

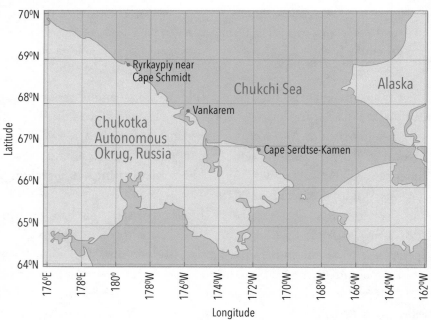

The walrus events of 2017 were filmed by Netflix crews at the Cape Schmidt cliffs and Cape Serdtse-Kamen beach, where super-herds of more than 100,000 walrus are known to occur. Vankarem is another known haul-out location.

Source: Susan J. Crockford.

used extensively in the filming of these events (BBC 2019b). Drones can send walruses into a panic, and for that reason their use is prohibited in US national refuge areas and discouraged near haul-outs on the American side of the Chukchi Sea (USFWS 2018, 2019).

While both productions claimed otherwise, scarce sea ice did not precipitate the walrus cliff deaths at Ryrkaypiy or the massive walrus gathering at Cape Serdtse-Kamen. In 2017, the US Fish and Wildlife Service concluded that walruses haven't been harmed by recent declines in sea ice, and that such harm was unlikely in the foreseeable future. This species, it says, 'has demonstrated an ability to adapt to changing conditions' (USFWS 2017).

Just as Pacific walruses are thriving, polar bears are also flourishing. The worldwide population has risen from about 10,000 in the 1960s (Crockford 2019b; USFWS 2008) to an official estimate of 26,000 in 2015 (Wiig et al. 2015). Surveys conducted since then suggest a total of about 28,500 (Crockford 2018b). When decades-old estimates for subpopulations are updated by extrapolation from regions that have been more recently surveyed, an estimate of about 39,000 becomes both plausible and scientifically defensible (Crockford 2017, 2019b).

This is surprisingly good news. Activists and some biologists had predicted that, as summer sea ice declined sharply by mid-century, two-thirds of the world's polar bears would vanish (Amstrup et al. 2007). When these summer sea-ice conditions arrived decades earlier than expected in 2007, polar bears appeared doomed (Amstrup et al. 2007; Stroeve et al. 2007; Wang & Overland 2009).

The health and population of the bears and their primary prey species (ringed and bearded seals) continues to be monitored in the field. By 2016, it was abundantly clear that no catastrophic decline in polar bears had materialised. Summer sea ice had stabilised at the predicted 2050 level (Crockford 2017, 2019a; NSIDC 2019), but polar bear numbers had not plummeted as expected (Crockford 2017; Wiig et al. 2015). Indeed, by 2018 it appeared polar bears were continuing the same steady rate of increase that began after a 1973 international treaty protected them from overhunting (Crockford 2019c).

Figure 1.3 Global increase in polar bear numbers

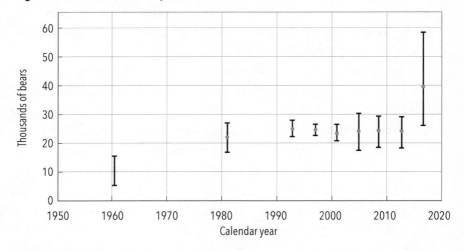

This new estimate for 2018 (39,000 with a range of 22,000-58,000) is a modest 4-6-fold increase over the 10,000 or so bears that existed in the 1960s and after 25 years, a credible increase over the estimate of 25,000 that the Polar Bear Specialist Group (PBSG) offered in 1993 (Wiig et al. 1995).

Source: Crockford 2019a.

Contrary to expectations, ringed and bearded seals in the Chukchi Sea are also doing better with less summer ice than in the 1980s when the ice was more extensive. This is because most of their feeding happens during the ice-free season.

Healthy seals lead to abundant, healthy polar bears. In 2016, the first Chukchi Sea population survey undertaken estimated the population to be about 3000 – about 1000 more bears than previously thought (Regehr et al. 2018; Wiig et al. 2015). Although bears that spend the summer on land in this region have to wait a month longer for the return of sea ice than was the case during the 1980s, researchers found no negative impact on their health or survival (Atwood et al. 2016; Rode et al. 2015). They were fat, healthy, and reproducing well (Rode et al. 2018).

Polar bears in the Barents Sea have experienced the most extreme reduction of summer sea ice in the entire Arctic (Regehr et al. 2016). Yet, in 2016, they too were reported to be in fine condition and

reproducing well (Aars 2018; Crockford 2019a). The same is t
many other recently surveyed Canadian subpopulations (York et al.
2016).

Bears in Western Hudson Bay and Southern Hudson Bay appear to
have suffered small, statistically insignificant declines in 2016 (Crockford 2019a; Dyck et al. 2017; Obbard et al. 2018). While this is blamed
on sea-ice loss (Polar Bears International 2019), several studies have
shown no trend in breakup or freeze-up dates since 2001 (Castro de
la Guardia et al. 2017; Cherry et al. 2013; Lunn et al. 2016). If these
populations are truly declining, something other than climate change
must be to blame.

The Southern Beaufort Sea is the only region where a dramatic
decline in polar bear numbers appears to be taking place. A 2001–2010
study documents a 25–50% drop in total population (Bromaghin et al.
2015). But rather than being caused by a shortage of summer sea ice,
thick spring ice kept pregnant ringed seals away for three years in a
row. Young bears didn't die in 2004–2006 because they were forced to
swim too much during the summer, but because there weren't enough
ringed seal pups for them to eat during the spring (Harwood et al. 2012;
Pilfold et al. 2016). This is a recurring phenomenon in that part of the
world (Crockford 2015; Stirling 2002). It also happened with similar
severity in 1974–1976.

Misrepresenting the naturally occurring spring ice events in the
Southern Beaufort as effects of climate change is part of the political
weaponisation of global warming rhetoric that some polar bear scientists actively promote (Amstrup 2019; Crockford 2019c). But the bulk
of field observations confirm that polar bears, like Pacific walrus, are
thriving despite almost 50% less summer sea ice than there was in 1980.
Neither species has experienced a climate emergency.

Amid cinematic exaggeration and widespread climate-change hyper-
bole, zoologist Crockford has been a calm, lonely, voice of reason. Much
like Australian physicist Peter Ridd, she has quietly challenged rhetorical
excesses, striving to ensure that the public hears scientifically accurate
information. Like Ridd, she has paid a price.

Figure 1.4 Sea ice extent

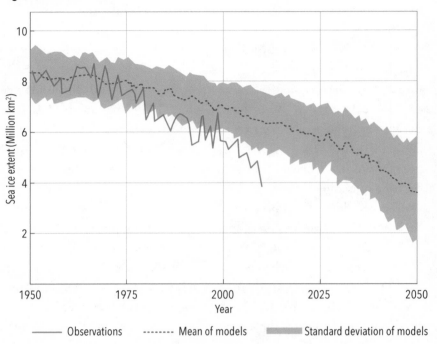

Observations ········ Mean of models ▬▬▬ Standard deviation of models

Sea ice reached an even lower extent in 2012 and all years since then have been below predicted levels (about 3–5 mkm²).

Source: Modified from Stroeve et al. 2007.

This world-renowned expert in evolution and animal bone identification, and role model for young women, is routinely hired by biologists and archaeologists in Canada and abroad to identify the remains of mammals, birds, and fish. She has helped catalogue museum collections, assisted police with forensic analyses, and produced a ground-breaking theory of how new species arise. But students at Canada's University of Victoria (UVic) will no longer benefit from her expertise (Laframboise 2019a).

After serving fifteen years as an Adjunct Professor, UVic's Anthropology Department has now withdrawn Crockford's Adjunct Professor status. In 2016, it had been renewed unanimously for a three-year term. The next time around was a different story. Crockford was advised in May 2019 that an internal Appointment, Reappointment, Promotion, and Tenure (ARPT) committee had 'voted not to renew your Adjunct Status'. No reasons were provided.

Crockford describes her expulsion as 'an academic hanging without a trial, conducted behind closed doors'. The position of Adjunct Professor is unpaid. In exchange for mentoring students, sitting on thesis committees, and delivering occasional lectures, adjuncts gain official academic standing and full access to library research services. Without a university affiliation, says Crockford, her ability to apply for research grants comes to a screeching halt.

When contacted by Canada's *National Post* newspaper, UVic spokesperson Paul Marck refused to say how many people were on the ARPT committee, how many voted against Crockford, or how many were zoologists in a position to make an informed decision about her expertise. When asked what safeguards ensure that adjuncts can't be excommunicated merely for expressing politically incorrect ideas, spokesperson Marck declined to respond, citing provincial privacy legislation. In his words, the university doesn't disclose 'information about internal processes. We must respect the privacy rights of all members of our campus community'.

Academia is a 'publish or perish' workplace, and Crockford is an accomplished scholar. In 2018, she was co-author of a paper published in *Science*, one of the world's most prestigious scientific journals. On any campus, the number of professors whose recent work appears in that journal is small. Once again citing privacy concerns, Marck declined to say how many other UVic professors have met this high standard.

Crockford isn't entirely surprised by her expulsion, given her previous ban from the UVic Speakers Bureau.

For the better part of a decade, that entity had arranged for her to deliver unpaid lectures to elementary and high school students, as well as to adult community groups. One talk concerned the early origins of domestic dogs and what it reveals about the evolutionary process of speciation. The other was titled *Polar Bears: Outstanding Survivors of Climate Change*.

There is every indication she was a popular speaker. But in 2017, Speakers Bureau coordinator Mandy Crocker advised her of a policy change. The chair of the Anthropology Department now needed to confirm that Crockford was 'able to represent the university' when discussing these topics.

Crockford's 2004 dissertation broke new ground with regard to the mechanisms by which wolves evolved into domestic dogs and how new species in general arise. UVic awarded her a PhD for that research. Yet thirteen years later, Chair of the Anthropology Department Dr Ann Stahl absurdly banned Crockford from giving public lectures about these matters.

In April 2017, Stahl advised, 'I will not be endorsing your request to be included in the Speakers Bureau roster for 2017–2018'. Admitting that she couldn't prevent Crockford from speaking elsewhere as a private citizen, Stahl drew the line at her doing so 'as a representative of UVic'.

Stahl said she respected 'issues of academic freedom', but Crockford's talks at schools had 'generated concern among parents regarding balance' and that this concern had 'been shared with various levels of the university'. Stahl did not respond to a request for an interview.

That was the first time Crockford, who is the author of five books about polar bears, was made aware of any problems. Because no one from the Speakers Bureau or the Anthropology Department ever advised her of any specific complaint, she was never given an opportunity to defend herself.

UVic's Speakers Bureau draws its volunteers from 'faculty, staff, graduate students and retirees'. Prospective speakers complete a form on its website, which says nothing about presentations needing to be balanced. In fact many appear to be overtly political.

Social Studies Associate Professor Jason Price, for example, currently delivers a lecture titled *Education and the Revolution: Climate Change and the Curriculum of Life* to students as young as kindergarten age. UVic graduate student Patrick Makokoro offers a presentation to audiences as young as ten about social justice.

Dwight Owens, an employee of Ocean Networks Canada, an entity affiliated with UVic, has no scientific training. His BA is in Chinese language and literature. His MA is in educational technology.

Nevertheless, under the auspices of the UVic Speakers Bureau, he has been giving talks about ocean chemistry and climate change for years.

When the *National Post* asked Marck how many people have been forbidden from participating in the Speakers Bureau, and what mechanisms are in place to vet presentations about controversial topics, Marck refused to address either of these matters. Speakers Bureau coordinator Crocker also declined to be interviewed.

While graduate students have apparently always required departmental oversight, no other category of speaker has ever needed permission 'to represent the university'. Until 2017, when one adjunct professor was targeted. In the absence of an alternative explanation, this policy change appears to have been invented solely as a means of silencing the eminently qualified, highly experienced Crockford (Laframboise 2019b).

Because her overall polar bear message that they're flourishing conflicts with activist rhetoric, and because activists apparently complained to administrators, her career as an academic researcher has come to an abrupt end. UVic Economics Professor Cornelis van Kooten, who holds a research chair in environmental studies, says he is 'appalled and distressed'. When, he asks, did 'universities turn against open debate? There's now a climate of fear on campus.'

Former chair of UVic's philosophy department Jeffrey Foss says Crockford has been punished for speaking her own mind about matters of fact, which means she has been denied academic freedom and free speech. 'I'm beginning to lose faith and hope in the university system,' he says.

During the time she delivered lectures to elementary school students, Crockford says she was continually 'astonished to learn that every single teacher believed that only a few hundred to a few thousand polar bears were left'. She feels duty bound as a scientist to speak up, to point out that the global population is officially estimated to be in the tens of thousands. 'I talk to groups about the adaptive features of polar bears that allow them to survive changes in their Arctic habitat,' she says.

In late 2019, Crockford spoke to audiences in Oslo, London, Paris, The Hague, Delft, Amsterdam, and Munich. High-school students in the Netherlands were given an opportunity that UVic chose to deny Canadian school kids. They heard Crockford's arguments firsthand.

At the Munich event, a two-day conference that featured more than a dozen speakers (EIKE 2019), protesters from a group called the Anti-Capitalist Climate Society invaded the lobby of the hotel in which the event was scheduled to take place. At the last minute, the hotel cancelled the months-in-advance booking, citing concerns about the safety of guests and staff (Crockford 2019d).

Alternative arrangements were made, and the conference went ahead elsewhere. But it was necessary to hire security personnel, and to keep the location secret. Scientists discussing scientific matters in Germany now need to fear persecution from violent mobs (Gosselin 2019; Taylor 2019; Williams 2019).

The motto of that protest group is 'system change, not climate change', which makes matters rather clear. Where global warming, polar bears, walruses, and related topics are concerned, political activism now overshadows rational and evidence-based factual discussion.

Figure 1.5 Walrus haul-out, Chukchi Sea

Walruses mass on beaches. Super-herds of more than 100,000 are known to occur at the Cape Schmidt cliffs and Cape Serdtse-Kamen beaches.

Source: Russian scientist via Shutterstock.

2 Climate Change in the Polar Regions: A Perspective

Dr Howard Brady

The polar quandary

The media's interest in the polar regions has not abated since the newspapers in London, Berlin, Oslo, Paris and even Cardiff helped fund expeditions in the 19th and early 20th centuries. But the storyline has shifted from describing heroic adventure to heralding dramatic climate change. The headlines are now about pack ice disappearing, ice sheets falling into the sea, and glaciers retreating at an alarming rate, all of which is causing the rise in sea level to accelerate, placing humans who live in coastal regions in danger. The ancient writers of mythology took poetic licence, condensing historical events that had taken thousands or millions of years to occur into a few days. So, is present-day climate science – fuelled by the media – creating a new mythology with talk of accelerating sea-level rise due to increased melting of polar ice sheets?

The key questions in polar regions are:

- What do we know of historical climate variability?
- What are the dynamics controlling the pack ice that forms in the polar seas?
- What are the dynamics controlling the volume of huge ice sheets that have formed over, and weighed down, the Greenland and Antarctic land masses?
- Is climate change in polar regions as dramatic as inferred in the media?

Sea-ice grows, forms, and melts in the ocean. It can take different forms, but it is always formed from seawater. When it is attached to the land it is fast ice, when it moves with the wind and currents it is drift ice. Drift ice that is closely packed together is known as pack ice. Ice sheets are not formed from seawater, but rather from snow. Their melt has the capacity to affect sea levels. The largest two ice sheets on Earth today cover most of Greenland and Antarctica. During the last ice age, ice sheets also covered much of North America and Scandinavia.

The pack ice

Our knowledge of the extent of the Arctic and Antarctic pack ice is limited to the past 200 years of exploration, while accurate satellite monitoring of the pack ice only began in 1979.

Pack ice forms as the surface waters of the polar oceans freeze, and since that ice comes from the ocean, it does not affect world sea level.

There is no land mass at the North Pole, only the Arctic Ocean. Some of this pack ice melts in summer months, therefore the winter pack ice is a mixture of new ice formed each winter and older ice formed in previous winters. There were various failed attempts in the 19th century to find a Northwest Passage, a route from the Atlantic Ocean north of Canada to the Pacific Ocean. The most famous was the ill-fated Franklin expedition in 1845–46.

The earliest annual Arctic pack-ice maps were Danish from 1920, but there were also Danish records of various ice-margin conditions north of Greenland since 1893 (Danish Meteorological Institute 2018). The Danish maps document:

- a very warm Arctic period in the 1920s and 1930s
- a cooling trend from 1938 onwards with the Arctic pack-ice extent increasing in the 1950s.

In Antarctica, a pathway through the pack ice was found by Sir James Ross between 1839 and 1843. His famous voyage paved the way for explorers such as Roald Amundsen and Robert Scott to reach the mainland and journey to the South Pole.

Since 2000, the extent of the Arctic winter pack ice has slightly decreased from 15 million km² to 14 million km². A dramatic decrease occurred in the summer pack ice from 5.5 million km² in 2000 to a record low of 3.41 million km² in 2012. The summer ice rebounded in 2014 to 5.02 million km² and is currently around 4.5 million km².

The extent of Arctic pack ice is influenced by ocean currents (Figure 2.1). There is a periodic switch in surface ocean temperatures in the North Pacific Ocean called the Pacific Decadal Oscillation (PDO). A 30-year phase of this 50- to 60-year cycle can explain the decrease in Arctic

Figure 2.1 Ocean current directions in the Arctic region

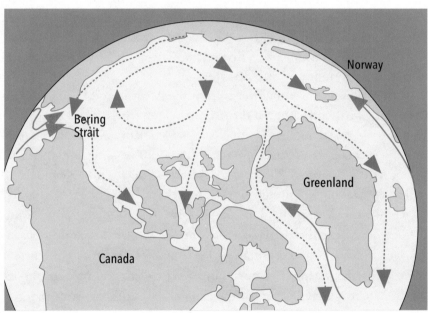

-------- Cold current ———— Warm current

Warm and cold currents, in periodic cycles, may be responsible for variations in the Arctic pack ice. Warm currents can come through Bering Strait near Alaska (in one phase of the PDO), and also along the west coast of Greenland and along the western coast of Norway in the Atlantic Ocean. The extent of the Arctic pack ice declined strongly in the 1920s and 1930s, then increased, and then declined between 1990 and 2012. It may be slightly increasing at the present time. Some villages in Greenland lie on the west coast because a warm current hugs that coast.

Source: College of Biological Science, University of Guelph, Canada.

ice during the 1920s and 1930s, as warm waters from the North Pacific Ocean flowed through Bering Strait into the Chukchi Sea in the western Arctic. In the other 30-year phase, no warm water will enter Bering Strait. There is a similar process in the North Atlantic Ocean. On different time scales, Atlantic currents along the coast of south-west Greenland or in the eastern North Atlantic can vary the amount of warm water entering the Arctic Ocean and can also affect the extent of the Arctic pack ice in the Barents and Kara Seas that lie along the northern Russian coasts.

Embarrassing statements about the dramatic decrease in Arctic pack ice have been made by popular climate catastrophists such as Al Gore, a previous vice-president of the United States. He promoted ideas developed in 2007 by Professor Wieslaw Maslowski of the Naval Postgraduate College of California, who had argued the Arctic could be ice-free by 2013 and that 'you can argue that our projection of 2013 could be too conservative' (Hollingsworth 2013).

While pulses of warm currents from the Atlantic and Pacific Oceans may cause further declines in the Arctic pack ice, there is still the danger that short-term trendology overlooks natural variability. After all, natural cooling events have occurred even in the last 400 years, such as the Little Ice Age from 1650 to 1705, and a cold period between 1790 and 1830 called the Dalton Minimum.

Unlike the Arctic pack ice, the pack ice in the Antarctic surrounds a huge continent that is separated from the rest of the world's oceans by the clockwise flowing Antarctic Circumpolar Current. This current forms a barrier, keeping warmer subantarctic waters away from the ice sheet. Where the cold waters of the Antarctic meet these warm waters is called the Antarctic Convergence (Figure 2.2).

Even though the surface of the Southern Ocean is cut off from the world's oceans by the huge Circumpolar Current, there is still a vast conveyor belt of ocean currents from the Atlantic, Pacific and Indian Oceans that flow at depth into the Southern Ocean. These deeper currents may, at times, be responsible for long-term erosion and under-cutting of polar ice shelves and glaciers, especially around the West Antarctic Peninsula that juts out towards South America.

Figure 2.2 Antarctica – the convergence, the ice shelves and the continent

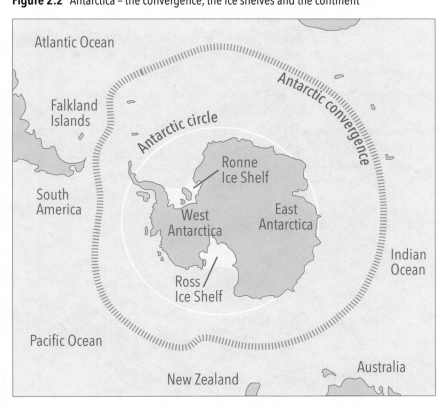

Where the Pacific, Indian and Atlantic Oceans meet the Southern Ocean is known as the Antarctic Convergence. This is the largest ecological barrier on Earth. Over a distance of only 50 km the water temperature drops from 5.5 °C to 2.8 °C. This completes the isolation of Antarctica by the Circumpolar current (the largest on Earth). The diagram also shows the Ronne Ice Shelf and the Ross Ice Shelf, each nearly the size of Spain. Both receive some ice from the West Antarctic Ice Sheet and the East Antarctic Ice Sheet.

Source: College of Biological Science, University of Guelph, Canada.

Unlike the Arctic, a greater percentage of the Antarctic pack ice melts in summer, so its winter pack ice is mainly new ice. As the summer ends, around late February, the Antarctic pack ice can form at rates well in excess of 100,000 km² a day until, by late October, it covers more than 18 million km² of ocean before stabilising and then decreasing in December and January to less than 3 million km².

Mapping indicates that since 1979 the average extent of the Antarctic pack ice increased while that of the Arctic decreased (Figure 2.3). This trend was sharply broken in 2018 when the Antarctic winter pack ice decreased by 2 million km². While some have dramatised this change, we have no idea of the long-term natural variability of the Antarctic pack ice. We only know that during the past 40 years it has not danced in unison with the Arctic pack ice. There are no obvious correlations between rising carbon dioxide (CO_2) levels, the decreasing trend in the Arctic pack ice and the gradual, almost linear, increase of the Antarctic pack ice to record levels over the past 40 years.

Figure 2.3 Arctic and Antarctic pack-ice extent 1979–2018

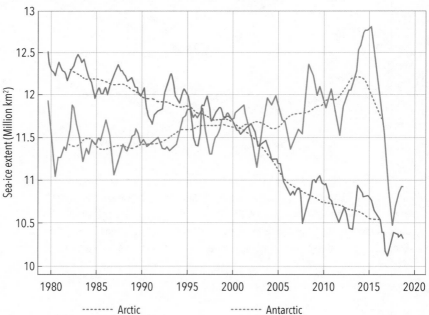

The average surface extent (summer and winter combined) of the Arctic and Antarctic pack are out of sync. The maximum extent of the Antarctic pack ice has been steadily increasing since 1981, except for the dramatic decrease in 2018. The maximum extent of the Arctic pack ice has been decreasing since 1979. A pattern of decrease may have occurred in the strong Arctic warming between 1920 and 1935 and one of increase during global cooling between 1945 and 1970.

Source: National Snow and Ice Center, USA.

The polar ice sheets and ice loss

Large ice sheets, 2–4 km thick, cover the island of Greenland in the North Atlantic Ocean, and the Antarctic continent in the South Pole region. Their weight depresses the landmass. In some places the West Antarctic Ice Sheet is below sea level. Inside these continents there are large areas of moving ice that behave much like slow rivers. These features are called ice streams because the moving ice is bounded by ice, not by rock nor mountains. These ice streams can be more than 100 km wide, 2 km deep and hundreds of kilometres long. They feed ice into coastal glaciers. In some areas of Antarctica these glaciers and ice streams contribute ice to form many areas of ice that rest on, or float over, the ocean floor. The largest such areas are the Ross Ice Shelf that covers 487,000 km^2 and the Filchner–Ronne Ice Shelf of 430,000 km^2 (Figure 2.2).

Ice loss in Greenland

In Greenland, there is evidence that ocean currents are causing the retreat of coastal glaciers by undercutting the glacial tongues that float in fjords along the Greenland coast. This undercutting is like a hidden cancer; sometimes not initially obvious from surface photographs.

Apart from ocean currents undercutting Greenland's glaciers, the surface of the Greenland ice sheet shows evidence of dynamic change. During the summer there are melt days where the surface is an icy mush and in many places small lakes form, sometimes draining quickly into rifts in the ice sheet. If these rifts, which can be more than 2 km deep, are filled with this lake water, the water pressure deep in the rift can create further cracks in its sides; this process is similar to the 'fracking' technique used in the oil industry to break open rocks at depth. If such rifts are near the coastal glaciers, this under-ice water can flow beneath glaciers, speeding their path to the sea. Dr Petr Chýlek, formerly of Dalhousie University in Halifax, Nova Scotia, has looked at the present relationship between coastal temperatures and mush days on the ice sheet and argues that these processes must have been also very active in the 1930s (Chýlek et al. 2007).

These losses should be balanced against ice gains due to increased snowfall near the coast when low-pressure atmospheric systems bring moisture over the plateau, where it is dumped as snow. Dramatic evidence of this snowfall comes from the crash of six P-38 fighters and two B-17 bombers on the Greenland ice sheet in 1942. One of the planes, now called Glacier Girl, was recovered from the ice sheet in 1992. It was buried beneath 81 m (or 268 ft) of snow – an ice-height gain in this region of 1.6 m a year over 50 years (Solly 2018).

The satellite data indicates that the annual ice loss from Greenland varies from year to year. The NASA/German Aerospace Centers' Twin Gravity and Climate Experiment (GRACE) indicates an average loss of 280 gigatonnes per year (Gt/year) between 2002 and 2016 (Caltech Jet Propulsion Laboratory 2016).[1] While these changes in the Greenland ice sheet look dramatic, they would cause an equivalent projected sea-level rise of about 8 cm per 100 years, if such losses continued annually. A constant trend is unlikely due to natural variability. Unfortunately, we have no estimates for ice loss during the warm 1930 period or of ice gain during the cooling period between 1945 and 1970. Such periodic climate variations in 50- to 60-year temperature cycles have been the norm during the overall warming trend of the last 150 years. It is reasonable to assume they will continue. These cycles have been independent of the continuing rise in greenhouse gas levels during the last 150 years.

Ice loss in Antarctica

The West Antarctic Ice Sheet is the smallest and deepest of two ice sheets at Anarctica, and includes the Ronne Ice Shelf and Ross Ice Shelf, as shown in Figure 2.2. It has so buckled the Earth's crust that the ice sheet effectively sits in a saucer with its coastal fringe rimmed by mountains. The most spectacular mountain chain is called the Transantarctic Mountains.

The average annual temperature over the polar plateau is between −50 °C and −60 °C. The surface is so cold that the draining of the Antarctic ice sheets by ice streams is not controlled by surface air temperatures

---

1 Ice mass is often described in gigatonnes (Gt), where one gigatonne is 1000 million tonnes or the weight of one cubic kilometre (1 km³) of freshwater. For perspective, it would take the melting of about 360 gigatonnes (Gt) of polar ice to raise world sea level by 1 mm and 360,000 Gt to raise the sea level by 1 m; in the latter case, an iceberg 3600 km by 100 km by 1 km deep!

or by recent global warming. Instead, it is the depth of the ice and plastic flow within the ice that are responsible for their movement. There are no 'mush' days, as there are on the Greenland ice sheet, nor are there surface lakes feeding water through rifts to its base.

Analyses of ice losses and gains on polar ice sheets use complex data from several satellites. The British Antarctic Survey has inspected ice core records and reported a dramatic increase in snowfall in West Antarctica during the 20th century (Wang et al. 2017). Such snowfall increase has to be offset against ice loss into the ocean from coastal glaciers and ice streams, or from ice loss due to the evaporation of ice by high winds (ablation). While there is little wind in the interior of the continent, the cold heavy air from the polar ice cap, which is 3 km above sea level, descends with fury when it nears the coast and falls to sea level. There are many 'blue ice' areas on the glaciers descending to the sea, where metres of ice are evaporated from the surface of glaciers by winds in excess of 150 km/hour.

There are many sub-surface freshwater lakes 3 km below the surface of the Antarctic ice sheets. The dynamic processes forming these sub-ice lakes are far from understood. Sometimes one sub-surface lake can drain and join another at a different pressure and the height of the ice sheet surface in that area will adjust. The relationship between these lakes and the flow of ice streams is also not understood.

The ice streams in the West Antarctic Ice Sheet flow towards the Ross Sea or in the opposite direction to the Weddell Sea and account for 80% or more of ice movement in that region. They have helped create the large Ross Ice Shelf in the Ross Sea and the Filchner–Ronne Ice Shelf in the Weddell Sea.

A paper from Professor Rignot and co-authors showed that the Mercer and Whillans Ice Streams decelerated slightly between 1997 and 2009, and that the Pope, Smith and Thwaites glaciers retreated 31 km, 35 km and 14 km respectively between 1992 and 2011 (Rignot et al. 2014). Another paper predicts the potential collapse of the ice sheet in the region of the Thwaites Basin that feeds these glaciers (Joughin et al. 2014).

In November 2015, Dr Jay Zwally, Chief Cryospheric Scientist at NASA's Goddard Space Flight Center and Project Scientist for the Ice

Cloud and Land Elevation Satellite reported in the *Journal of Glaciology* that the large Antarctic East Ice Sheet is growing, more than compensating for any ice loss from the West Antarctic Ice Sheet (Zwally 2015). This increase is due to the increased snowfalls. Dr Zwally reported that analysis of satellite data from 2003 to 2008 showed a net gain of 82 billion tonnes of ice per year during that period – a clear suggestion that the Antarctic ice sheets could contribute little to sea-level rise in the 21st century. Dr Zwally pointed out that his study was a report of what is actually happening now, and not a prediction for the future, and contradicts a 2013 Intergovernmental Panel on Climate Change (IPCC) report that indicated a net ice loss in the Antarctic of 147 billion tonnes a year between 2002 and 2011.

In January 2019, Eric Rignot and co-authors published in the *Proceedings of the US National Academy of Science* a detailed study on Antarctic Ice Sheet mass balance between 1979 and 2017 (Rignot et al. 2019). They analysed eighteen regions and 176 basins. Their findings disagree with Zwally in that they provide evidence for ice loss from numerous ice shelves in East Antarctica and contend that East Antarctica and West Antarctica have been contributing to overall mass ice loss. Their total ice loss is:

- 40 +/– 9 Gt/yr between 1979–1990
- 50 +/– 14 Gt/yr between 1989–2000
- 166 +/– 18 Gt/yr between 1999–2009
- 252 +/– 26 Gt/yr between 2009–2017.

In practical terms, Rignot and his fellow authors calculate Antarctic ice contribution to sea-level rise to be about one ninth of one millimetre a year between 1979 and 1990 – a figure too small to have much credibility. Their figure of 252 Gt/yr for a recent Antarctic contribution to sea-level rise translates forwards to around 7 cm per 100 years.

The difficulty of interpreting and then extrapolating such short-term changes in Antarctic ice-mass balance was highlighted in a recent article based on Australian research into the Totten Ice Shelf. Dr Gwyther and co-authors demonstrated that despite rapid thinning in the five-year period between 2003 and 2008, a longer time series over eighteen years showed high interannual variability without any apparent trend

(Gwyther et al. 2018). The authors pointed out that previous studies had indicated the Totten Glacier was the 'canary in the coalmine' holding back at least 3 m of sea-level rise. The authors noted in particular the variability in the temperature of Circumpolar Deep Water, which is one of the main mechanisms quoted for melting the base of ice shelves and glaciers – a mechanism not unlike the deep, warmer, ocean currents, already mentioned that undercut coastal glacial tongues in Greenland.

An historical perspective

Evidence is being uncovered that demonstrates considerable historical natural variability of the Greenland and Antarctic ice sheets since the beginning of the collapse of continental ice sheets 10,000 years ago.

It is generally accepted even in IPCC reports that the Mid-Holocene Thermal Maximum between 8000 to 5000 years ago was the warmest period in the last 10,000 years. The spectacular meltwaters and lakes that appear at present during summer on the Greenland ice sheet must have occurred during the Mid-Holocene Thermal Maximum – yet there was no ice sheet collapse. It would not be unreasonable to suggest that the fastest flowing glacier in the world today (the Greenland Jakobshavn) was at that time flowing even faster.

There is considerable information about the climate in the Arctic region during the Medieval Warm Period 1000 years ago. The Vikings had cattle and some crops in South Greenland, and they could easily bury their dead due to an absence of permafrost in that southern coastal region. If Southwest Greenland in the Medieval Warm Period was warmer than today, then how can terms such as 'unprecedented' be used to describe today's warming in the Arctic?

Dr Chýlek notes that in Greenland the strongest warming trend last century was in the 1930s, as shown by the temperature records of the settlements at Godthaab/Nuuk and Egedesminde (now known as Aasiaat) on the west coast (Chýlek et al. 2006). This is borne out in the local temperature records (Figure 2.4).

With respect to the West Antarctic region there is panic about the strong retreat of the Pine Island and Thwaites Glaciers in recent decades. However, a recent study by J. Johnson and others indicates a strong

Figure 2.4 Average summer temperatures at Godthaab/Nuuk and Egedesminde

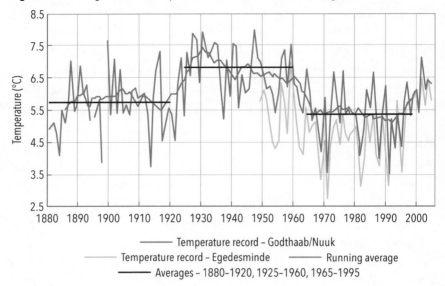

Local temperatures on the West Greenland coast in the 1930s were higher than those of today. This West Greenland data is consistent with the 1912-2010 temperature record of the island of Svalbard in the Greenland Sea where the warming between 1920 and 1925 was the fastest recorded anywhere in the 20th century. An analysis of the temperature records for 37 Arctic stations and seven sub-Arctic stations also showed the highest temperatures up to 2005 were in the 1930s (Przybylak 2000).

Source: Chýlek et al. 2006.

thinning of the Pine Island Glacier during the Mid-Holocene Warm period:

> We demonstrate, using glacial-geological and geochronological data, that Pine Island Glacier (PIG) also experienced rapid thinning during the early Holocene, around 8000 years ago. Cosmogenic ^{10}Be isotope concentrations in glacially transported rocks show that this thinning was sustained for decades to centuries at an average rate of more than 100 centimeters per year, which is comparable with contemporary thinning rates. The most likely mechanism was a reduction in ice shelf buttressing. Our findings reveal that PIG has experienced rapid thinning at least once in the past and that, once set in motion, rapid ice sheet changes in this region can persist for centuries. (Johnson et al. 2014)

There is also ice-core evidence from West Antarctica that some warming trends in the 18th and 19th century were stronger than those of recent times. Dr Thomas and her colleagues from the British Antarctic Survey analysed an ice core taken by them on the Bryan Coast, Ellsworth Land, adjacent to the Bellingshausen Sea in 2010–2011 (Thomas et al. 2013). This was one of a series of ice cores, and it provided a 308-year temperature record. There were various volcanic ash layers in the ice core that could be dated within an error margin of one year. These ash layers were from various eruptions of Chilean volcanoes in the 18th, 19th and 20th centuries.

> The record shows that this region has warmed since the late 1950s, at a similar magnitude to that observed in the Antarctic Peninsula and central West Antarctica; however, this warming trend is not unique.

> … the recent isotropic warming trend is not the largest in the 308 year record. Larger 50 year warming trends occurred in the middle to late eighteenth century [+4.1‰ dec^{-1}(1740–1789)] and the mid-nineteenth century [+3.8‰ dec^{-1}(1888–1839)] with several equally large cooling trends. Overall there is no significant trend in the average ^{2}H [deuterium] record since 1702 A.D. (Thomas et al. 2013)

Distinct layers in this core show 50-year temperature cycles. This is remarkably similar to the 60-year pause/warm cyclicity in global temperatures over the past 150 years that is evident in the IPCC historical temperature diagrams (Folland et al. 1990). This data links mid-18th century climate change to the present. It is clear that most computer climate models do not show the step-like pattern in temperature movements over the past 150 years, and consequently cannot possibly mimic the natural variability of climate change in polar regions.

The sea level problem

Since we are currently in an interglacial period the question of sea-level rise, especially within the next 100 years, is a critical one. While global warming can cause some expansion in the upper ocean, and the melting of mid-latitude glaciers, the largest sources of freshwater for sea-level rise are the ice sheets in Greenland, West Antarctica and East Antarctica. The general estimate is that they hold around 6 m, 6 m and 57 m of world sea level, respectively.

It is clear from articles already cited that there has been ice loss from the Greenland ice sheet, and opinions about the mass balance of the Antarctic Ice Sheet are divided. But is the ice loss accelerating or not? One could sensibly argue that present sea-level rise data should supply an indirect but credible answer to this problem. After all, if sea-level rise is not accelerating then the loss of ice from polar ice sheets into the oceans cannot be accelerating.

The technical committees on sea level at the IPCC meetings in 1990, 1995 and 2001 closely examined tide-gauge evidence from all over the world for the 20th century. They all concluded that no acceleration of sea-level rise had been detected, even though the general circulation computer models (GCMs) predicted that sea-level rise would accelerate as the world warmed. For example:

> No significant acceleration in the rate of sea-level rise during the 20th century has been detected. (Church et al. 2001)

In 2007, the IPCC technical committee on sea level noted that the evidence from the Topex/Poseidon and Jason-1 series of satellites since 1993 showed sea-level rise had accelerated to around 30 cm/100 years; this message was repeated in the IPCC assessment report in 2014. This would suggest a sudden dramatic acceleration, whereby the rate of sea-level rise nearly doubled in a decade.

> Sea level likely rose 1.7 ± 0.5 mm yr^{-1} during the 20th century, but the rate increased to 3.1 ± 0.7 mm yr^{-1} from 1993 through 2003, when confidence increases from global altimetry measurements. (Trenberth et al. 2007)

There are serious difficulties with the IPCC's acceptance of the satellite data in preference to a huge database of tide-gauge data spanning nearly 200 years. While tide gauges may be affected by local crustal movement, such movements are much slower than sea-level rise rates, and modern surveying techniques can easily calculate adjustments for such movement. In 2011, after noting no significant acceleration in well-kept tide gauges that have been installed for periods of 60 years or more, Professors Houston and Dean seriously questioned the accuracy of the Jason satellite system (Houston & Dean 2011). Their concerns were vindicated when the NASA engineers released a report admitting design and altimetry problems with the present Jason system in a report promoting a new system, the Geodetic Reference Antennae in Space (GRASP) (Bar-Sever et al. 2012).

Despite problems with satellite telemetry, the National Oceano-graphic and Atmospheric Administration of the USA (NOAA) holds the view that satellite data shows sea-level rise acceleration and correctly reflects what is happening on a global level. A NOAA website released in 2015, and since updated, shows sea-level rise data from tide gauge stations along the West Coast, East Coast and Gulf Coast of the USA, the Caribbean islands, as well as seven Pacific and six Atlantic islands (NOAA 2020). There are no claims that worldwide sea-level rise can be detected in this tide gauge database as '…The sea level trends measured by tide gauges that are presented here are local relative sea level (RSL) trends as opposed to the global sea level trends.'

But, a complete disjunct between the worldwide tide gauge database and the open ocean makes no sense and even the detection of sea-level rise acceleration in the satellite data has proved difficult. For example, in 2016 University of Colorado scientists who manage the Jason sea-level rise satellites reported sea level *deceleration* in the previous decade. It was proposed that a cooling event due to the 1991 volcanic eruption of Mt Pinatubo in the Philippines had delayed the sea-level acceleration that would have eventuated by 2016 but 'barring another major volcanic eruption, a detectable acceleration is likely to emerge from the noise of internal climate variability in the coming decade…'. (Fasullo et al. 2016). However, in 2018 an article examining the same data, and with some of the same scientists as authors, reported that sea-level rise accel-eration had now been detected but 'based only on the satellite -observed changes over the last 25 years.' (Nerem et al., 2018).

If proving sea-level rise acceleration is so elusive serious questions need to be asked, not only about the climate models, but also about articles claiming significant accelerating ice sheet losses in Greenland and Antarctica. If these articles are correct then any resulting huge volumes of extra meltwater pouring into the ocean should have caused an easily detectable acceleration in sea-level rise by now, in both the world-wide tide gauge database and the satellite sea-level rise database.

Some final observations

There is considerable natural variability of the climate system in polar regions that have always been sensitive to the smallest changes in climate. Many recent polar studies have concentrated on short-term trends and ignored the natural periodic variability that can be clearly seen in the last 10,000 years. The Mid-Holocene warming, the strong polar warming pulses in the 18th and 19th century, and the marked warming in the 1930s, place the warming of the last 40 years in perspective. Furthermore, the General Circulation Models, which are so often used to predict future trends, cannot mimic these periodic pulses of warming interspersed with periods of cooling, or periods with little change in temperature. Of particular note is the mid-18th century and 19th century warming in West Antarctica, the rates of which were equal to or greater than those in the much-heralded warming decades of the 20th century.

It is important to note that recent changes in ice volume of polar regions provides no simple indication of cause. Cause is a separate question. The fact that the relatively recent shift in polar ice sheet volume does not correlate well with changes in greenhouse gas levels indicates, for example, that other factors are continually at play. There has been polar ice loss, but the steady rise of sea level without acceleration, as documented in the worldwide tide-gauge data base, indicates there has not been any acceleration of ice loss to date.

Because a general warming trend exists at the present time, the monitoring of the Antarctic and Greenland ice sheets is important. These ice sheets are the only areas that could supply large volumes of fresh water to raise world sea levels significantly. However, it is clear that there has been much hype and exaggeration in polar studies.

3 Counting Emperor Penguins
Jim Steele

The Emperor Penguin (*Aptenodytes forsteri*) is the largest of all living penguin species. Many know them from the Academy Award-winning Best Animated Feature *Happy Feet* and the Oscar-winning Best Documentary Feature *March of the Penguins* that followed their 'march' at the end of summer across vast swathes of stable, winter sea-ice. Here, despite brutally cold winter conditions, each male Emperor Penguin incubates a single egg. Meanwhile, their mates journey back to the sea before returning two months later with food for the newborn chick. After breeding, Emperor Penguins can migrate 1200 km to dense pack-ice where they moult (Kooyman et al. 2004). Despite annual cycles involving such significant travel, many penguin researchers doubted breeding colonies were relocating to more suitable habitats: stable sea-ice with reasonable access to open water. Instead, the relocation of a colony has often been wrongly reported as a local extinction due to global warming (i.e. Trathan et al. 2011).

Counting penguins is not an exact science. A single colony's growing abundance may be the result of reproductive success or immigration or both. Likewise, a colony's shrinking numbers or a colony's total disappearance may be the result of poor survival, or emigration to a more sustaining location. The timing and scope of surveys have not accurately accounted for immigration and emigration and thus have not yet properly integrated relocations between colonies within a region. Therefore, significant uncertainty surrounds population estimates.

Figure 3.1 Emperor Penguins at Terre Adélie, Antarctica

Source: Adobe Stock, photograph by Fabrice Beauchene.

Some researchers mistakenly assumed changes in one colony exemplified a trend for all colonies. That erroneous assumption prompted predictions of approaching species extinctions. But improving technology is now allowing scientists to better account for emigration and immigration between all colonies, and thus provides more reliable estimates of change in the total number of Emperor Penguins.

Wrestling penguins at Point Géologie

Accurately counting the total number of penguins foraging at sea is impossible, therefore the number of non-breeding individuals is unknown. Monitoring Emperor Penguin's breeding populations on land is more doable, but still extremely difficult. In order for their large chicks to fully develop by December and fledge when summer conditions are most hospitable, mating begins in March–April when sea-ice begins to form and winter approaches. However, the harsh winter conditions endured by breeding populations also make surveys extremely difficult.

Emperor Penguins are awkward on land. So, breeding colonies are restricted to easily accessible coastal features, typically fast-ice. Fast-ice is usually flat and fastened to the land or to grounded icebergs. Fast-ice does not circulate like offshore pack-ice. It typically forms and melts each year, but storms and icebergs can destroy it prematurely. Prime Emperor Penguin breeding habitat requires stable fast-ice that can support a colony for nine months. Good breeding habitat must also be located relatively near open water. Researchers concerned with possible effects from global warming theorise that the Emperor Penguin's critical fast-ice breeding habitat could thin and more easily be lost as the Earth warms. That concern is the basis for catastrophic projections. But is this consistent with what we know about one of the best studied Emperor Penguin colonies?

The Point Géologie Emperor Penguin population, in the region of eastern Antarctica known as Terre Adélie, is immediately adjacent to the French research station Dumont d'Urville. Located just 500 m from the colony, the research station also provides reliable data from which researchers can estimate weather effects. And due to its accessibility, this colony was videoed in the documentary *March of the Penguins* (2005).

Unfortunately, the 'mark and recapture' method used by researchers to measure survival rates, population trends and individual movements confounded any interpretation of changes to the Point Géologie colony. That method required attaching a uniquely numbered band to a penguin's flipper. But flipper bands are now known to be detrimental to the penguins' survival (Dugger et al. 2006; Saraux et al. 2011). Furthermore, capturing and wrestling each penguin to attach a flipper band is a strenuous task. Emperor Penguins need to conserve all their energy, especially males who must fast for four months during extremely harsh winter conditions while balancing an egg on their feet. Tussles with researchers potentially deplete precious energy. Furthermore, when researchers drove penguins into lines of twos or threes in order to systematically read flipper bands, they disrupted the penguins' heat conserving huddling behaviour.

Researchers attached flipper bands from 1967 to 1980, coinciding with the period during which the Point Géologie population steadily plummeted to half of their pre-banding abundance. After banding ceased in 1980, the population quickly stabilised. The decade of decline had two possible explanations. One: banded penguins failed to return the following year because they did not survive, possibly due to starvation or

Figure 3.2 Known locations of Emperor Penguin colonies

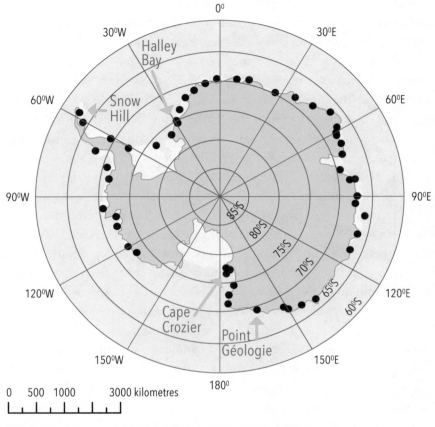

Black dots are the locations of known Emperor Penguin colonies in 2019

Source: Dr Barbara Wienecke (Wienecke 2011).

depredation. Or, two: banded Emperor Penguins were still alive but chose to move to a different breeding spot to avoid further human disturbance.

Without evidence, researchers had presumed Emperor Penguins are always loyal to their breeding location (philopatry). Thus, they assumed if a banded penguin did not return then it must have died. Furthermore, they dismissed the alternative possibility that Emperor Penguins had emigrated to a nearby colony, incorrectly believing the nearest colony was more than 1000 km away. Unfortunately, those assumptions took scientific analyses down a wrong path. Researchers then assumed the population decline must have been due to environmental disruption that lowered survival. Climate change was conveniently blamed because the shrinking population also coincided with a short-term warming spike (Barbraud & Weimerskirch 2001).

During the population decline, a short-term warming spike raised winter temperatures from −17 °C to −13 °C. But such warming should have increased penguin survival as they huddled against deadly cold. Emperor Penguin survival also depends on foraging in much warmer, above-freezing water. The warmer land temperatures were still far colder than the ocean and thus arguably irrelevant to the Emperor Penguins' survival. Furthermore, an observed contraction of pack-ice extent was also arguably irrelevant. Sea-ice extent determines the distance from the shore to the edge of the pack-ice. However, breeding Emperor Penguins are not dependent on the extent of offshore sea-ice.

Emperor Penguin colonies need to be within a reasonable distance of open-water access. Fast-ice breakouts provide ocean access that's far closer to a colony than the edge of the pack-ice. When fast-ice break-outs allowed quicker access to open water, Point Géologie's Emperor Penguins' reproductive success increased to more than 75%. If more extensive fast-ice remained in place and forced penguins to march greater distances to open water, reproductive success fell to 50% or less (Massom et al. 2009). Thus, the more parsimonious hypothesis

Figure 3.3 Annual mean winter temperatures at Point Géologie

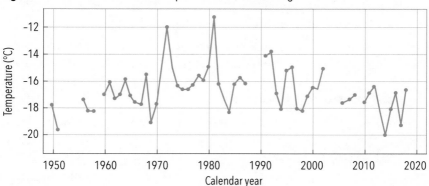

Source: Unadjusted mean annual temperatures as recorded at Dumont d'Urville, downloaded from the Global Historical Climatology Network (GHCN).

for Point Géologie's declining population was the Emperor Penguins had emigrated elsewhere to avoid researcher disruptions (Steele 2013, pp. 51–67). A newly discovered, nearby colony supports that hypothesis, as do additional observations of other Emperor Penguins rapidly relocating between colonies (LaRue et al. 2015).

Nonetheless, a paper was published claiming the decline of Point Géologie's Emperor Penguins was due to the 1970s 'warm event' (Jenouvrier et al. 2009). Because climate models predict more warming as carbon dioxide (CO_2) concentrations rise, researchers predicted a quasi-extinction by the year 2100, asserting at least 96% of the Point Géologie's Emperor Penguin population would be lost. To hype their conclusions, they issued a press release with the headline 'Emperor Penguins March Towards Extinction?'. Their later paper (Jenouvrier et al. 2014) unwisely extrapolated a ten-year snapshot of Point Géologie Emperor Penguin declines to suggest global warming will cause all Emperor Penguin colonies to decline by 2100.

However, observations of cooling winter temperatures at Point Géologie since the 1970s now suggest climate models are not reliably predicting changes on the eastern Antarctic coast. Furthermore, throughout eastern Antarctica, fast-ice has slightly but significantly increased (Fraser et al. 2012). For Antarctica overall, climate models incorrectly predicted CO_2-driven warming would reduce sea ice (Landrum et al. 2012), but sea-ice extent increased for three decades, reaching record highs in 2014. In 2016, sea-ice extent dramatically fell, but slightly rebounded in 2018. Researchers now suggest these sea-ice changes are likely due to natural climate variations. Although a human contribution cannot be ruled out, a clear anthropogenic effect has not yet emerged (Wang et al. 2019).

Improving survey methods

The advent of the satellite era has improved our ability to estimate emigration and immigration between breeding colonies. Instead of extrapolating the dynamics of one colony to estimate changes in the entire species abundance as Jenouvrier (2014) did, satellite estimates now allow researchers to integrate changes in all colonies within a defined region. Genetic analyses can define a region as containing closely related individuals, suggesting ongoing interbreeding between those colonies (Younger et al. 2017). Interbreeding requires frequent movements between colonies. Thus, if a decline at the Point Géologie colony was simultaneously offset by an increase at a nearby colony, most likely there was no change in the overall regional population, just a relocation of individuals. Such short-distance relocations are unlikely to be driven by global factors, but by local factors.

High-definition imaging from satellites can detect a colony's guano (excrement) stains and alert researchers to the presence of overlooked or relocated breeding sites. Researchers then calibrate and validate satellite estimates of abundance with on-the-ground counts to produce more accurate population estimates. With these observations, the assumption of breeding site loyalty has been increasingly questioned. After carefully examining satellite data, LaRue et al. (2015) 'documented six cases of movement of an entire breeding colony' as well as the detection of new colonies. LaRue et al. (2015, p. 115) concluded: 'We find compelling evidence that emperor penguins are not strictly philopatric [loyal to one breeding location], and that the location of emperor penguin colonies is more dynamic than previously thought.'

Some researchers have long suspected large changes in local population abundance were better explained by immigration and emigration forced by Antarctica's ever-changing icescape. For example, the Cape Crozier colony in the Ross Sea is one of the most southerly colonies, and global warming is not expected to affect its fast-ice breeding habitat for hundreds of years. Still, the colony's abundance has naturally fluctuated. In 1961, a high of 2000 adults was observed, but in 1976 only 127 adults were counted. In 2000, more than 1200 chicks were counted, but in 2001 zero chicks and zero adults were observed. By 2012 more than 1100 adults were again breeding (Kooyman & Ponganis 2017).

The wild swings in Cape Crozier's abundance were due to icebergs. An iceberg that broke off from the Ross Ice Shelf collided with Cape Crozier's fast-ice, destroying the Emperor Penguin's breeding habitat. Incubating adults were crushed or trapped in ravines of crumpled fast-ice. Survivors abandoned the colony, which totally failed in 2001 (Kooyman et al. 2007).

Seventy kilometres away at the Beaufort Island colony, fragments of that iceberg blocked the Emperor Penguins' easy access to the open waters of the Ross Sea. Icebergs created more extensive fast-ice by trapping ice normally blown out to sea during winter storms. The combination of more extensive fast-ice and insurmountable icebergs created a 150-km barrier blocking normal access to open water. Adults were forced to take longer routes in order to feed, causing chick production to plummet to 6% of pre-iceberg conditions (Kooyman et al. 2007). Given the dynamic nature of Antarctica's ever-changing icescapes, if the Emperor Penguins didn't have the good sense to move they would not have survived. Strict philopatry would be suicidal. With satellite observations covering the entire Ross Sea region, researchers will be more able to determine how many penguins die and how many emigrate to other colonies during such events.

A small breeding colony on the northern segment of the western Antarctic peninsula – Emperor Island, Dion Islands – decreased from 150 pairs in 1948, to fewer than 20 pairs by 1999. By 2009, this colony had been totally abandoned. This colony's 'extinction' was uncritically blamed on global warming and considered to be evidence of a climate-driven shrinking of the species' population (Trathan et al. 2011). However, temperatures along the peninsula that were once warming, are now cooling and sea-ice is slightly increasing (Turner et al. 2016). The existence of a new breeding colony, first detected in 2013 and located just 190 km from the abandoned colony (LaRue et al. 2015), suggests researchers, again, misleadingly labelled the Emperor Penguins' reloca-tion as a population extinction. Suitable fast-ice for a breeding colony largely depends on local micro-environments. Penguins often breed on one side of an island sheltered by the wind but not the other. Depending upon the direction of the prevailing winds and regional storm tracks, some islands and their inlets can provide shelter for more stable fast-ice for decades. When natural oscillations shift the direction and strength of local winds, ice stability and breeding penguins adjust accordingly.

Nonetheless, advocates of catastrophic climate change viewed the disappearance of the Dion Islands' colony as consistent with model predictions that colonies furthest north would be most vulnerable to global warming (Ainley et al. 2010). But a recently discovered colony disputes that prediction. The Snow Hill colony, located on the eastern side of the Antarctic peninsula, is the most northerly Emperor Penguin colony, and it is thriving. When the first census was conducted in 1997, only 2400 breeding adults were observed (Coria & Montalti 2000). In a 2004 census, more than 8000 breeding adults were counted (Todd et al. 2004), and recent surveys suggest the colony is still growing (Libertelli & Coria 2017). Apparently, the colony is not suffering any detrimental effects of climate change. But to what extent this colony is growing due to increased reproductive success as opposed to immigration of adults from shrinking colonies elsewhere remains to be determined.

An obvious solution to resolving the species' population uncertainty requires simultaneously surveying all Emperor Penguin colonies in a region, like the Ross Sea, to evaluate changes in that regional population (Kooyman & Ponganis 2017). Then, by integrating regional estimates, researchers can more reliably monitor changes in the species' total population.

The importance of a regional estimate is well illustrated in a recent study by Fretwell and Trathan (2019) concerning the Halley Bay colony located on the eastern side of the Weddell Sea. It had been considered the second largest Emperor Penguin colony in Antarctica. Sheltered fast-ice near a persistent stretch of open water (a polynya) had provided ideal breeding conditions. Nonetheless since 1956, the colony's population fluctuated between 14,000 and 23,000 breeding pairs.

In September 2016, in conjunction with a strong disruptive El Niño, the Halley Bay colony experienced its highest average wind speeds in 30 years, which destabilised the fast-ice. As a result, the colony's population crashed. Unfortunately, there are no long-term datasets to evaluate how other El Niños had affected the colony. But with the aid of satellite data, we now know that coincident with the Halley Bay population crash, a small colony just 55 km away experienced a population increase of more than 1000%, adding more than 13,000 breeding pairs in less than three years. Such an increase can only be due to immigration. The colony did not 'vanish'. It relocated.

Unfortunately, the public understanding of penguin adaptability is obscured when media outlets hyped the colony's probable relocation as a 'colony wiped-out' and a 'colony vanishes' (*New York Times* 2019). The researchers knew exactly where the colony had relocated. Instead of celebrating the penguins' resilience and ability to relocate, the media uncritically republished information suggesting a climate catastrophe.

Misleading official designation

Designations of a species' health by the International Union for the Conservation of Nature (IUCN) is considered the global gold standard for conservation. In 2009, the IUCN ranked this ice-dependent penguin as 'least concern'. 'Least concern' is the healthiest of all IUCN rankings. At that time the Emperor Penguin population consisted of an estimated 270,000 to 350,000 adults. The population was deemed stable in the absence of evidence for any declines or substantial threats. In 2012, their population was still considered stable, and the estimated number of known adults had increased by 36% to 476,000 adults. Although larger populations are considered more resilient, the IUCN downgraded their status to 'near threatened' because models projected a rapid population decline over the next three generations owing to the effects of projected climate change (Ainley et al. 2010).

By 2017, the total number of known Emperor Penguins increased again to about 595,000 adults and that number is likely still rising, as new colonies have been discovered but not yet included in the species' population estimate. However, the IUCN maintains the Emperor Penguin's status as 'near threatened' (BirdLife International 2018), because it remains uncertain to what degree increasing numbers of known Emperor Penguins are the result of a growing population versus newly discovered colonies. Nonetheless, other than large variability in single colonies, there is no evidence the species' overall numbers are diminishing.

4 Fire and Ice, Volcanoes at Antarctica

Dr Arthur Day

Volcanism has existed under the ice in Antarctica for millions of years, and, in fact, one of the world's largest active continental volcanic provinces is found there – a rare combination of fire and ice. More than 90 of the volcanoes (two-thirds of the total) have only recently been discovered because they are subglacial volcanoes that lie completely concealed beneath the kilometres-thick West Antarctic Ice Sheet (WAIS). The potential impact of these needs to be evaluated. In this chapter, I contemplate whether subglacial volcanism could accelerate melting and impact global sea levels and conclude there is no evidence the intensity of volcanism will increase; future volcanism is unlikely to change *overall* ice-sheet melting rates; and, because the ice sheet is already in long-term balance with volcanic effects, future volcanism is extremely unlikely to destabilise the ice sheet and accelerate global sea-level rise.

In Antarctica, large ice sheet-bound volcanoes were not known until 1940 when first seen from the air. Mount Takahe is so remote it was not visited until 1957. Mount Sidley, the highest volcano in Antarctica, was not climbed until 1990, and Mount Siple, another giant volcano, has never been climbed.

To put the Antarctic volcanism into perspective, it is useful to briefly describe the enormous range in destructive power of volcanic eruptions.

At one extreme, consider the Cerberean Cauldron volcanic eruption that took place about 374 million years ago in the Marysville region of Central Victoria, Australia, just 100 km east-northeast of the city of

Melbourne. The Cerberean Cauldron volcano was not a mountain, but rather an area of subsidence measuring 35 km by 27 km. This is arguably the most dangerous kind of volcano. It sat above a 30 km-wide shallow magma chamber (Birch 1978; Clemens & Birch 2012). Its eruption was triggered when the roof over the magma chamber sagged. As it sagged, the roof began to fracture. The entire magma chamber then exploded into the atmosphere, mostly through a circular eruptive fissure system that opened up as a ring fracture around the roof tapped into the magma chamber below. The fissure system had a circumference of about 100 km. It can still be seen today.

When the erupting column of thick volcanic ash subsequently collapsed after climbing tens of kilometres into the sky, hot gaseous ash-flows spread across the surrounding countryside at hundreds of kilometres per hour, smothering thousands of square kilometres under hot dust and rock fragments at temperatures as hot as 700 °C. These ash deposits then thermally re-consolidated into a series of thick rock units. The total volume of magma involved was about 900 km³.

Experimental high-pressure and high-temperature replication of the pre-eruptive conditions inside the magma chamber (Clemens & Birch 2012) indicates the magma had a dissolved water content of between 4.1 and 5.3 weight per cent. This water would have added considerably to the explosive power of the eruption upon release of the confining pressure over the magma chamber. The effect of the explosion would have been immediately catastrophic over a significant proportion of the state of Victoria. In addition, there would have been significant transient global climatic impacts. Some of the ash filled the hole that was created when the roof collapsed into the emptying magma chamber. This hole was about 27 km wide and 1 to 2 km deep. Today, despite 374 million years of erosion, the remaining solidly re-welded granite-like deposit of fragmented rock filling that hole is at least 1700 m thick.

At the other extreme, volcanoes such as Mauna Loa in Hawaii are far less explosive. Eruptions have had limited impact, being small and dominated by flowing lava. Thousands of people live on the flanks of Mauna Loa despite it being the world's largest active volcano.

These two examples demonstrate the wide range in destructive power of individual volcanic eruptions.

The volcanoes beneath the WAIS are relatively non-explosive and more like the Hawaiian volcanoes. The violence of a volcanic eruption is linked to the chemical composition and volume of magma. Very viscous magmas with higher silica content make for larger and much more explosive eruptions (Victoria), compared to low silica magmas with much lower viscosity (Hawaii). Thus, the destructive power and ice-sheet impact of the recently discovered subglacial volcanoes will be linked to magma chemistry and consequent volcano types. The geologic setting that gave rise to the highly destructive Victorian volcano does not exist beneath the ice sheet.

Fears about ice-sheet melting and the 'instability' of the WAIS are well documented. The additional impact of subglacial volcanism is not always considered but it can be complex and non-intuitive. It is not just about melting. In fact, a subglacial eruption could actually *hinder* ice-sheet loss, depending on the circumstances of the eruption. Adding to the challenge is that predicting the timing, power, and location of individual volcanic eruptions is impossible.

It is widely accepted current melting is influenced by warm ocean currents along the immediately adjacent coastline. To understand any potential impacts of volcanism it is necessary to understand both the ice sheet and the volcanoes, as well as the underlying geological rift structure that hosts them, namely the West Antarctic Rift System (WARS).

The West Antarctic Rift System

The WARS is a giant continental rift with high heat flow and volcanism. It hosts all the recently active volcanoes on the Antarctic continent, including those beneath the WAIS. Stretching for 3500 km from the Ross Ice Shelf to the Antarctic Peninsula, and about 1000 km wide, it is the only geological structure of its kind comprehensively buried beneath a thick continental ice sheet. It contains a series of tensional geological faults bounding giant valleys. These host large glaciers that drain the ice sheet and feed the Ross Ice Shelf. Volcanism has existed in the WARS since rifting began 175 million years ago, long before the appearance of ice sheets.

Subglacial volcanism in the West Antarctic Rift System

The West Antarctic volcanic province associated with the WARS is one of the world's largest continental volcanic provinces. It comprises at least 138 volcanoes. The most famous is Mount Erebus, the world's most southerly active volcano.

Van Wyk de Vries et al. (2017) analysed ice-penetrating radar echograms, as shown in Figure 4.1, to identify 91 previously unknown volcanoes. Some are very large. Another 47 had already been identified because they are not concealed. They are widely distributed, but with

Figure 4.1 Ice-penetrating radar echograms

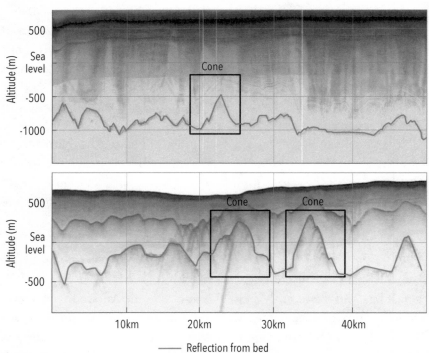

—— Reflection from bed

The upper panel shows an ice-penetrating radar echogram from NASA's Icebridge mission (NSIDC 2014). The lower panel shows an echogram from Corr and Vaughan (2008) with basal topography picking out two volcanic cones. The suspended dark layer is 2000-year-old volcanic ash believed to be from the Hudson Mountains subglacial volcano.

Source: Modified from Figure 3 in Van Wyk de Vries et al. (2017), https://sp.lyellcollection.org/content/461/1/231.

concentrations in Marie Byrd Land and along the central WARS axis, as shown in Figure 4.2. Volcanoes have a wide range of sizes with cones ranging in height from 100 m to 3850 m. Basal diameters range between 4.5 km and 58.5 km. Most have good basal symmetry. The largest has a volume of 2542 km³.

Figure 4.2 Map of WARS volcanoes

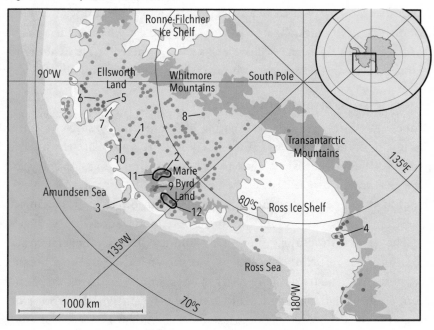

● Visible volcanoes ● Volcanoes hidden beneath ice

1. Mount Takahe 2. Mount Sidley 3. Mount Siple 4. Mount Erebus 5. Hudson Mountains 6. Hudson Mountains Subglacial Volcano 7. Pine Island Glacier 8. Mount Casertz 9. Mount Flint 10. Mount Murphy 11. Executive Committee Range, five volcanoes north to south: Mounts Hampton, Cumming, Hartigan, Sidley and Waesche 12. Flood Range, four volcanoes west to east: Mounts Berlin, Moulton, Kohler, and Bursey

Puce dots represent visible volcanoes that have been confirmed in other studies. These include large ice-sheet-bound volcanoes, such as those in Marie Byrd Land, which protrude through the ice. The cyan dots represent volcanoes that are completely hidden beneath the ice, and which, with a few exceptions, are newly discovered.

Source: Modified from Figure 2 in Van Wyk de Vries et al. (2017), https://sp.lyellcollection.org/content/461/1/231.

All of the 91 newly discovered volcanoes lie beneath a thick layer of ice. It remains unknown how many are dormant and likely to erupt again or how many are extinct.

Volcanology of the subglacial volcanoes

The range in sizes of the volcanoes corresponds with a range in eruptive styles and volcano types. The different eruptive styles determine how the volcanoes interact with the ice sheet. Volcanoes formed under subglacial conditions have unique features. These depend on whether the volcano breaks through the surface of the ice sheet to form explosive steam eruptions, the size and frequency of its eruptions, and how long lived it is.

Shorter-lived simple volcanoes

Cones up to several hundred metres high are usually single-eruption 'simple' volcanoes. They are most likely basaltic and relatively small in volume. Their active lives range from a few months to up to several years. Pedersen and Grosse (2014) carried out a detailed study of 33 subglacial basaltic volcanoes in Iceland that are now no longer enclosed in ice. These are good analogues for the smaller WAIS volcanoes.

The stages in eruption of small monogenetic subglacial volcanoes and the resulting landforms, are shown in Figure 4.3. Smaller examples are tindars while the larger ones are tuyas. Tuyas can either have conical or flat-topped profiles depending on whether the structure is capped by lava. Structures such as tindars and conical tuyas begin with the eruption of a core-forming cone of pillow lavas.[1] An entire eruption can be encased in ice without breaching the ice sheet's surface (Phase 1). Cones with deposits of hyaloclastite[2] have breached the ice sheet, resulting in a steam explosion vent and a meltwater lake (Phase 2). Tuyas that continue to erupt may develop flat tops of solid

---

1 Pillow lavas have erupted under water, or melted ice. They are frequently found in subglacial volcanoes. When a basaltic lava erupts into water its surface rapidly chills. Eruption proceeds via repeatedly bulging envelopes of lava encased in sacks of a thermally insulating skin of glass. The resulting structures are around a metre in size and resemble pillows.

2 A hyaloclastite is an accumulation of glassy fragments formed when lava erupts into water or ice, rapidly chills into glass, then shatters.

Figure 4.3 Development of short-lived simple volcanoes and resulting landforms

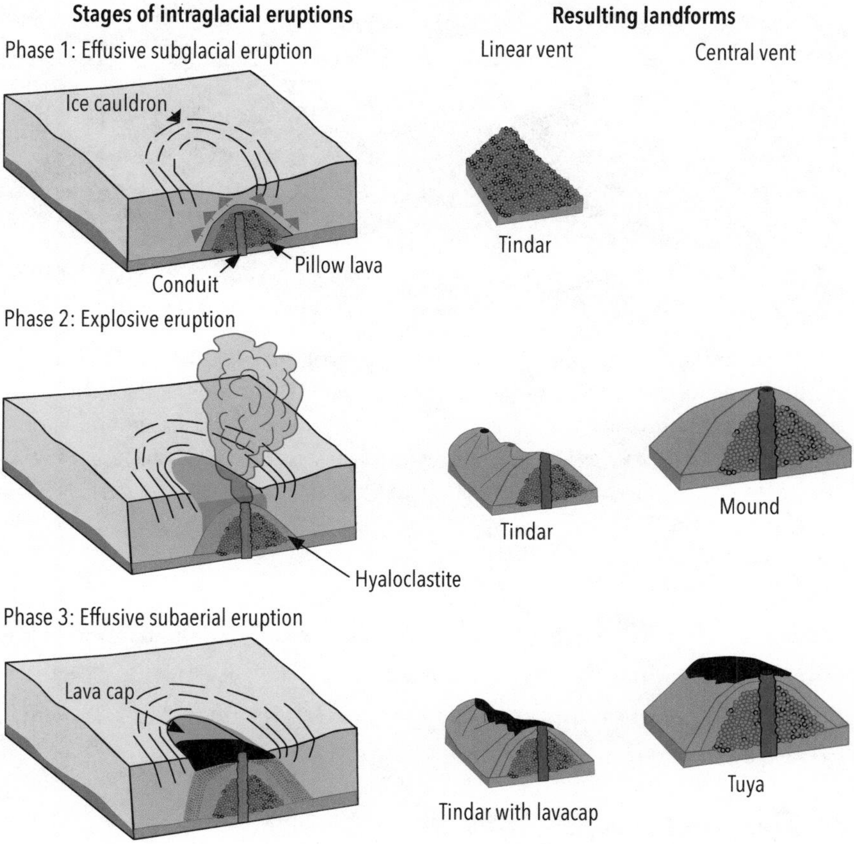

Stages of intraglacial eruptions

Resulting landforms

Phase 1: Effusive subglacial eruption

Linear vent

Central vent

Ice cauldron

Conduit

Pillow lava

Tindar

Phase 2: Explosive eruption

Hyaloclastite

Tindar

Mound

Phase 3: Effusive subaerial eruption

Lava cap

Tindar with lavacap

Tuya

Source: Redrawn from Figure 1 in Pedersen & Grosse (2014), https://ars.els-cdn.com/content/image/1-s2.0-S0377027314001838-gr1.jpg.

lava if the meltwater lake drains away (Phase 3). A good example of such a tuya is pictured in Figure 4.4.

Smaller eruptions deep beneath the WAIS might be too short-lived to breach the ice sheet. Volcanoes whose summits rise to within a few hundred metres of the surface could break through. Some may have already done so but any ice cauldron[3] depressions (Phase 1) or steam

---

3   An ice cauldron is a depression in an ice sheet that forms above an area of basal melting caused by a subglacial volcanic eruption (see Phase 1 in Figure 4.3).

Figure 4.4 Brown Bluff is a basaltic tuya on the Antarctic Peninsula

This volcano erupted in a single event a million years ago beneath the former Antarctic Peninsula Ice Sheet. When the ice retreated the tuya was exposed. It is about 12–15 km in diameter. All three of the eruptive phases illustrated in Figure 4.3 can be seen in this 800 m-high cliff face. There is an initial (dark) pillow lava at the base, an intervening light-brown layer of basaltic hyaloclastite, and then a capping of solid lava erupted after the meltwater lake drained away.

Source: Robert Wyatt, Alamy Stock https://www.alamy.com/brown-bluff-antarctic-peninsula-panorama-image237775919.html.

explosion vents (Phase 2) could have filled with ice again after eruptions ceased. Consequently any surface evidence of eruptions beneath the ice sheet could disappear from view, making it difficult to assess their frequency of occurrence over time.

Longer-lived complex volcanoes

Subglacial volcanoes up to thousands of metres high may be longer-lived, multi-eruption, shield volcanoes. They are likely to be similar to the large, exposed, shield volcanoes nearby and, therefore, may be dominantly basaltic and trachytic. Figure 4.5 shows one-on-one cross-sections of three of the recently discovered large shield volcanoes concealed beneath the WAIS and compares their profiles with those of three other well-known large shield volcanoes from around the world. Their volume–height characteristics and basal diameters closely match.

A shield volcano is built up by flow after flow of fluid lava. The non-explosive style of eruption means 90% of the volcano is made of lava. Volcanoes this size can have an active lifespan of well over a million

Figure 4.5 Three concealed WAIS cones compared with three well-known shield volcanoes

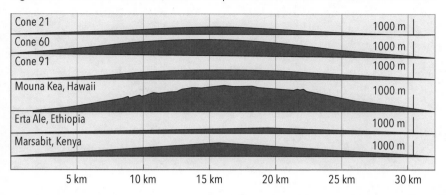

One-on-one cross-sections of three cones from Van Wyk de Vries et al. (2017) (Cones 21, 60 and 91) compared with three well-known shield volcanoes, namely Mauna Kea (Hawaii), Erta Ale, and Marsabit (Africa).

Source: Redrawn from Figure 4 in Van Wyk de Vries et al. (2017), https://sp.lyellcollection.org/content/461/1/231.

years. Individual eruptions can occur thousands of years apart, decreasing in frequency as the volcano ages. An eruption could last several years but produce only small volumes of lava. Volcanoes beginning under water or ice can eventually emerge if they grow large enough. For example, Mauna Loa began underwater at least 700,000 years ago and broke through the surface after about 300,000 years.

Evidence for recent subglacial volcanism

There are multiple lines of indirect evidence for recent volcanic eruptions beneath the ice. In one case the activity may be ongoing. It is highly likely eruptions have been frequent for as long as there has been an ice sheet. The province should be classified volcanically active because future eruptions are a certainty.

The Hudson Mountains subglacial volcano

Compelling evidence for subglacial volcanism lies in ice-penetrating radar data from the Hudson Mountains in the Ellsworth Land area of Antarctica (Corr & Vaughan 2008). This shows a single volcanic ash layer over an elliptical 23,000 km² area, as shown in Figure 4.1. Its depth dates the eruption to about 207 BC, ±240 years. In terms of geological time this is an extremely recent event. The ash layer seems thickest where a large topographic feature, more than 20 km wide and 1000 m high, rises to a point where its summit is less than 100 m beneath the surface of the ice sheet. The shape and size of this feature suggests it might be a medium-sized shield volcano. It has been named the Hudson Mountains subglacial volcano. Its summit is close enough to the surface to allow explosive steam eruptions to burst through. If this is a long-lived shield volcano, which last erupted only 2000 years ago, it is potentially active.

There is more evidence. The rare isotope helium-3, carried up from the Earth's deep interior by volcanism, has been detected in seawater at the front of the Pine Island Glacier's ice shelf (Loose et al. 2018). This is evidence of upstream volcanic or geothermal activity. The location downstream from the Hudson Mountains subglacial volcano might be no coincidence. However, scientists carefully point out that the chief cause of melting beneath the front of the glacier remains currents in the Amundsen Sea, and not volcanism.

Mount Casertz

Further evidence of recently active, or possibly ongoing, subglacial volcanism lies with a concealed mountain peak north-west of the Whitmore Mountains, upstream of the glaciers feeding the Ross Ice Shelf. The peak is known as Mount Casertz. In an aerogeophysical survey, an approximately 6 km-wide and 48 m-deep depression was detected in the ice sheet. Beneath the depression, ice-penetrating radar revealed a steep-sided conical peak in the subglacial topography. The base of the peak is 6 km wide. It rises quickly to a height of 650 m above the surrounding topography to within 1600 m of the surface. It is associated with a magnetic anomaly signature.[4] Winberry and Anandakrishnan (2003) reported a tight cluster of fourteen small

---

4   A magnetic anomaly is a local variation in the Earth's magnetic field resulting from variations in the magnetism of rocks. The effect can be used to detect structures obscured by overlying material, such as volcanoes beneath ice, because volcanic rocks are more magnetic than ice.

shallow earthquakes between February and April 1999, all focused beneath Mount Casertz. Such a swarm is indicative of highly localised magmatic activity.

The surface depression may be an ice cauldron. Along with the earthquakes, its existence is very strong evidence that Mount Casertz is either quietly volcanically active or has erupted recently. The topographic feature is consistent with a medium-sized conical tuya. It might not have progressed beyond the Phase 1 pillow lava stage of Figure 4.3 because of its significant depth. The thickness of the overlying ice might prevent any direct signs of volcanism reaching the surface, except for the ice cauldron depression. This is quite feasible because thousands of kilometres of submarine rift volcanism is continually taking place deep beneath the world's oceans. The absence of visible eruptive activity at the surface is the norm rather than the exception.

Large protruding shield volcanoes standing in kilometres-deep ice

Eighteen large highly visible shield volcanoes have coexisted with the ice for millions of years. Some are illustrated in Figure 4.6. They have characteristic low shield-like profiles and large ice-filled calderas.[5] It is highly likely the recently discovered shield volcanoes completely concealed beneath the surrounding WAIS are geologic equivalents. Therefore they may share the same age and volcanology.

Mounts Cumming and Hartigan are almost completely buried in ice. Mounts Berlin and Siple are active. Mounts Waesche and Takahe have erupted very recently. Recent volcanic ash layers have been found in ice near Mounts Waesche, Berlin, Moulton, and Takahe. Lava compositions range from basalts to trachytes. Some volcanoes commenced with pillow lavas and hyaloclastites, indicating subglacial beginnings. Volcano ages range widely from the old heavily glacially dissected Mounts Flint and Murphy at 19.1 and 8.7 million years respectively, to the youngest well-preserved Takahe at 0.31 million years. Most volcanoes

---

5 These calderas are enlarged eruptive craters. They widen by inward wall collapse following withdrawal of lava lakes.

Figure 4.6 Shield volcanoes embedded in the WAIS

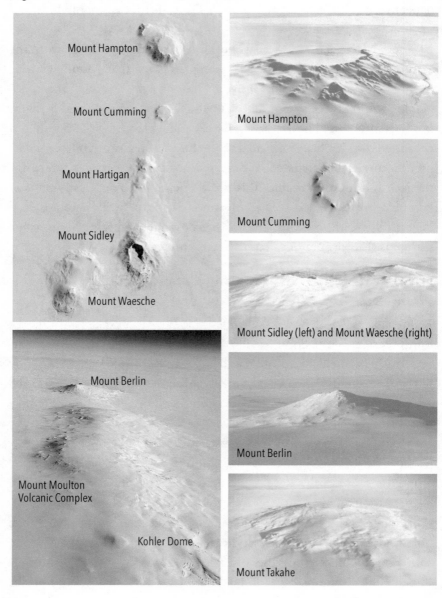

Clockwise, previous page: 2014 Landsat 8 image of the five volcanoes comprising the **Executive Committee Range**. These lie in an 80 km-long north–south line.

Source: NASA Earth Observatory image by Jesse Allen, using Landsat data from the US Geological Survey. Public domain, https://upload.wikimedia.org/wikipedia/commons/thumb/b/b5/Sidley_oli_2014324_lrg.jpg/800px-Sidley_oli_2014324_lrg.jpg.

Mount Hampton, with a summit dominated by a large ice-filled caldera.

Source: US Navy photo.
Public domain, https://upload.wikimedia.org/wikipedia/commons/d/de/MountHampton.jpg.

Mount Cumming is buried up to its caldera.

Source: Cropped from the adjacent image.

Mounts Sidley (left) and **Waesche** (right). At 4285 m Sidley is the highest volcano in Antarctica. It has a steep-sided 5 km-wide caldera breached by an outward flank collapse (see adjacent image).

Source: Courtesy of USGS. Public domain, https://sp-images.summitpost.org/144277.jpg?auto=format&fit=max&h=1000&ixlib=php-2.1.1&q=35&s=2732d54677389d5c82d9f5551f7b283c.

Mount Berlin has a summit caldera 2 km wide. It is considered active. Several fumarolic ice towers emitting steam have been observed on the caldera rim, making this the only volcano in Marie Byrd Land with current geothermal activity. (Photo Nelia Dunbar).

Source: New Mexico Bureau of Geology and Mineral Resources,
https://geoinfo.nmt.edu/staff/dunbar/images/berlin%20small.JPG.

Mount Takahe, a large shield volcano protruding over 2000 m through 1500 m of ice. The summit caldera is 8 km wide.

Source: US Navy photo. Public domain,
https://upload.wikimedia.org/wikipedia/commons/6/6e/MountTakahe.jpg.

Flood Range looking west from 12,500 m aboard NASA's DC-8 Airborne Science flying laboratory, showing large volcanoes extending in an east–west direction for 96 km.

Source: NASA Laser, Vegetation, and Ice Sensor project. Public domain,
https://lvis.gsfc.nasa.gov/images/MtBerlin_470pxW.jpg.

are 5–10 million years old. For Mount Sidley, detailed investigations imply a 1.5 million year eruptive lifespan between 5.7 and 4.2 million years ago.

The oldest ice in West Antarctica, at almost 500,000 years, lies trapped inside the Mount Moulton caldera. Because it can't drain away, it is hundreds of thousands of years older than the oldest ice in the surrounding WAIS. The presence of such old ice inside a caldera shows how long these big volcanoes can remain dormant.

Mount Takahe (Figure 4.6) began under ice with pillow lavas and hyaloclastites 310,000 years ago. Its total height is 3460 m, is roughly circular, and about 29 km across. Its volume is 780 km³. Takahe is considered potentially active because, compared to its long lifespan, the time since its last eruptions – about 17,700, 7900, and 5550 years ago – is relatively short. It produced a series of nine eruptions over a 200-year period. It is likely to erupt again.

Anatomy of the West Antarctic Ice Sheet and sea-level impacts of subglacial volcanism

To understand the risks that subglacial volcanism presents to sea level it is critical to understand the ice sheet. The volume of the WAIS is about 2.2 million km³, or just under 10% of the entire Antarctic ice volume. Its area is a few million square kilometres, around a fifth of the continent, as shown in Figure 4.7. Overall, it is about 2 km thick, although it can reach 4 km.

Ice-sheet dynamics and the paths taken as it flows into the sea are heavily influenced by the depressed basement, defined by the WARS, that it infills and flows through. The ice sheet drains into the Ross and Weddell Seas. Smaller outlet glaciers also descend into the Amundsen Sea. 'Grounding lines' define the boundary between the bedrock-supported ice sheet and the floating marine portion that makes up the ice shelves. The grounding lines lie several hundred metres below sea level. Inland, the bedrock supporting the ice sheet deepens, so that much of it is grounded on bedrock about 1 to 2 km below sea level. This depth is partly because the weight of the overlying ice sheet has depressed the bedrock by about 0.5 to 1 km. The WAIS is fringed by

Figure 4.7 Profile through the WAIS

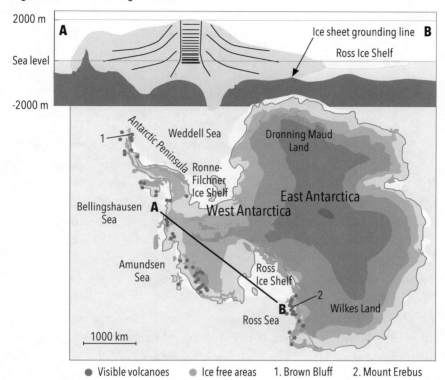

● Visible volcanoes   ● Ice free areas   1. Brown Bluff   2. Mount Erebus

(A) Bellingshausen Sea – WAIS – Ross Ice Shelf – Ross Sea (B). The profile shows most of the WAIS is grounded well below sea level. Beyond the grounding lines the ice sheet forms floating ice shelves. White areas hugging the coastlines in the plan view are floating ice shelves. Light brown areas are ice free – 2.8% of the continent. Grey areas are the West and East Antarctic ice sheets. The progressively grey shading in the plan view increments ice thickness by 1000 m. Puce dots are the locations of visible volcanoes.

Source: Redrawn from a cross-section by Hannes Grobe 21:51, 12 August 2006 (UTC), Alfred Wegener Institute for Polar and Marine Research, Bremerhaven, Germany, own work, CC BY-SA 2.5, https://commons.wikimedia.org/w/index.php?curid=1048117.

floating ice shelves. The two largest are the Ross and the Ronne-Filchner Ice Shelves. These are hundreds of metres thick, extend hundreds of kilometres seaward of their feed glaciers, and may play an important role as buttresses impeding glacier flow into the sea.

As mentioned in the introduction, one of the world's largest continental volcanic provinces lies beneath the ice, so the potential impact of volcanism needs to be evaluated. There are two opposing ways volcanism could impact an ice sheet:

- by increasing basal melting and flow lubrication, which could *accelerate* ice flow, or
- by building eruptive structures that could *impede* ice flow.

In terms of meltwater generation, the impacts on ice flow can be quite nuanced. This is because ice sheets do not flow uniformly. About 10% of the volume of the WAIS consists of fast-moving ice streams.[6] The effect of additional meltwater from a volcanic eruption could differ quite markedly depending on where it is generated. Outcomes could be quite different if meltwater from an eruption were to enter an existing fast-moving ice stream compared to if it took place under slow-moving ice at a topographic high where the ice-sheet base might be relatively 'dry' and stable.

On the other hand, if an eruption were to build a large volcanic cone in the path of an ice stream it could constitute an obstacle blocking its movement, at least temporarily, until subglacial erosion removed the cone long after eruption ceased. However, by then, the transient burst of lubricating meltwater would also have ceased. The interaction between glaciation, topography, and volcanism is therefore quite complex. Not only can the eruptive environment change the ice sheet, but the ice sheet could also alter the eruptive environment because of the impacts of ice overburden pressure on volcanism, and subglacial erosion.

Clearly, trying to evaluate the net impact of volcanism on an ice sheet is not trivial. The challenge is compounded by considerable historic variability in the extent of the ice. Theory has long suggested the WAIS may be inherently 'unstable'. There is abundant geologic evidence for large changes. These are recorded in sediment deposits and microfossils collected from beneath the WAIS itself (Scherer 1991),

---

6   Ice streams are a type of glacier within an ice sheet. Moving at more than 1000 m per year they flow much faster than the main ice sheet, moving at only 1–100 m per year. They account for most of the ice discharged into the sea.

and in sediment cores from the nearby Ross Sea (Naisch et al. 2009). Conditions fluctuated from ice-free freshwater lakes on land, and completely open water in the Ross Sea, to a thick ice sheet with extensive ice shelves. Simulations by Pollard and De Conto (2009) show total Antarctic ice volume fluctuated by up to 11 million km^3 as many as ten times during the last million years. At least twice, ~400,000 and ~125,000 years ago, parts of the WAIS may have disappeared almost entirely. Ten thousand years ago, the Marie Byrd Land coastal WAIS was more than 700 m thicker than today (Stone et al. 2003). These changes are in response to global glaciation cycles. In the face of this variability, understanding any volcanic impacts can only be enlightened by first examining the more conventional paths to instability that have already been contemplated.

The proposed instability of the WAIS is based on the marine ice-sheet instability (MISI) hypothesis. Its basis is that a self-sustaining retreat of the ice-sheet grounding lines, triggered by oceanic or atmospheric circulation changes, could lead to increased melting along the base of the floating ice shelves. This could result in 'rapid and irreversible inland migration of the submerged grounding lines', leading to 'runaway' retreat and 'collapse'.

Hypothetically, should the WAIS 'collapse', the global sea level would rise 5 to 6 m. However, an ice sheet of this size cannot simply 'collapse'. 'Collapse' is an emotive term. It is not an appropriate description of the situation. The ice sheet is simply too large. Pollard and De Conto (2009) show the transition between glacial, intermediate, and 'collapsed' states takes one to several thousand years.

Numerous *attempts* have been made to construct meaningful computer models of the impact of *future* ice-sheet 'collapse' on sea level. Hypothetical outcomes vary widely, depending on the underlying assumptions. This is because a lot of the critical *calibration adjustments* that need to be made to the computer models are 'only loosely constrained by present day changes' (Ritz et al. 2015). For example, a key unknown is the reliability of the calculated dependence of ice-sheet flow rates on basal friction.[7] Basal friction is heterogeneous and highly spatially variable beneath an ice sheet. It remains poorly characterised. To add to the computational difficulty, the relationship between basal friction and ice-sheet flow is highly complex, nonlinear, multivariate, and chaotic.

Any sea-level impacts from volcanic eruptions, of the kind expected under the WAIS, are highly unlikely to directly result from the tiny volumes of meltwater they'd produce, relative to the 2.2 million km^3 volume of the WAIS. Volcanism might impact locally by increasing geothermal heat flux and producing lubricating meltwater, but these can also both vary independently of volcanism. In any case, mapping their distribution kilometres beneath the ice is highly problematic.

Adding to the challenge is the style of volcanism. Based on observations of similar volcanism elsewhere, the most likely style of eruption consists of random small volume and relatively non-explosive short-lived events, lasting only a few years each, occurring sporadically from individual volcanoes over millions of years. Consequently, injections of lubricating meltwater beneath the ice sheet are also likely to be similarly random, transitory, and piecemeal, rather than permanent.

Ice sheets flow via complex processes that are far from understood. They can stop and start moving again for no apparent reason. Unsurprisingly, attempts to model the sea-level impacts of MISI have produced widely different results. In 2013, the Intergovernmental Panel on Climate Change (IPCC) concluded, based on modelling at the time, that MISI of the entire Antarctic ice sheet might increase sea levels by 0.5 to 1.0 m by 2100, and 1.4 m by 2200 under the climate change linked 'IPCC Special Report on Emission Scenarios (SRES) A1B' scenario simulation (IPCC 2013).

---

7   Basal friction beneath an ice sheet is the force resisting the motion of the ice sheet over the surface supporting it. Basal friction is subject to significant variability. It will differ considerably depending on whether the base of the ice sheet is in contact with bare bedrock or a lubricating mush of ground-up rock and meltwater. Basal friction is *lower* beneath fast-moving ice but *higher* at the base of slower-moving ice beneath the more stable portions of ice sheets near topographic divides.

In contrast, using the same IPCC SRES A1B emissions scenario in modelling ice-sheet melting, Ritz et al. (2015) established that the IPCC estimates were 'implausible under current understanding of physical mechanisms and potential triggers'. Their modelling gave drastically lower estimated sea-level increases of 30 cm by 2100 and 72 cm by 2200. That's a halving of the model-estimated future impacts of MISI. This reflects the sensitivity of the models to the inbuilt assumptions. Such uncertainty exists even before taking volcanism into account.

Ice-sheet 'collapse' modelling needs to be viewed in the context of the IPCC's admission that it is 'not yet clear' whether human-induced climate change has influenced the circulation of warm circumpolar deep water contributing to melting, or how this circulation might change in the future (IPCC 2013). So, if the cause of changes in circulation is unknown, it could simply be a random change in response to natural variability. In climate science, natural variability is well recognised. Its effect is most pronounced at the local scale within individual climatic regions, like Western Antarctica. There are many sources of uncertainty not simulated in complex computer models. This is just one of them.

Adding potential effects of volcanism to MISI models will have no impact on these sources of uncertainty, so trying to refine models of ice-sheet 'collapse' (and sea-level rise) by inserting the impacts of volcanism would make them even more speculative. Computer modelling has its limitations. 'The science is *not* in' on ice-sheet 'collapse', nor on consequent sea-level rise.

Conclusion

There are highly compelling signs volcanic eruptions have occurred beneath the ice sheet very recently. One small eruption might be ongoing. Such eruptions have occurred as long as the WAIS has existed and will continue in the future. However, eruption rates are extremely modest. For example, Mount Takahe took 310,000 years and countless small lava eruptions to grow to a volume of 780 km^3. In contrast, the Marysville volcano described in the introduction explosively ejected 900 km^3 of material in one eruption.

The WAIS has repeatedly advanced and retreated in response to global glaciation cycles. At times, parts of the WAIS may have disappeared almost entirely, but each time the ice returned. All the big WAIS volcanoes have witnessed these cycles. It is difficult to conceive how volcanism can be a threat if the WAIS has repeatedly grown and buried dozens of volcanoes in an active volcanic field with no discernable impacts on its viability.

Volcanism-linked destabilisation on any short timescale relevant to human experience would require the sustained eruption of all the more than 100 volcanoes associated with the ice sheet at once. The probability of such an event happening would be once in many millions of years because it is not the eruptive style of the types of volcanoes that exist beneath the WAIS. Thus, on a century-by-century basis, it is an incredibly low probability event that would still take millennia to unfold.

5 Sacred Bubbles in Ice Cores
Jo Nova

In the world of evidence there was nothing more compelling than ice-core bubbles. They filled the first graph in the first executive summary of the first Intergovernmental Panel on Climate Change (IPCC) report (IPCC 1990). And in all the years since, there is no other dataset like it. Nothing else shows carbon dioxide (CO_2) and temperature 'in lock step', moving up and down in parallel.

Yet buried in a subclause of the 'Summary for Policymakers' was the rather devastating caveat:

> Measurements from ice cores going back 160,000 years show that the Earth's temperature closely paralleled the amount of carbon dioxide and methane in the atmosphere.
>
> Although we do not know cause and effect ...

With a candour that disappeared in later reports, the IPCC even admitted they weren't sure whether CO_2 led the temperature or temperature led CO_2. The resolution of the early cores was not good enough to tell. But billions of dollars were about to be poured into solving a crisis and the ice cores were about to become a poster child. The world was captivated by the unnerving parallelism of temperature and CO_2 shifting together through the eons. But it was a very new and immature science. Only fifteen years earlier it had not even seemed realistic to estimate atmospheric CO_2 levels from bubbles (Berner et al. 1980).

Figure 5.1 *New Scientist* 1989 ice-core graph

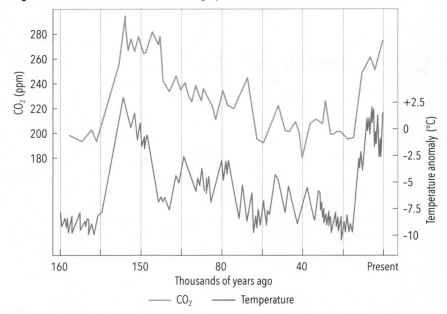

Redrawn here, the chart was captioned: 'Where carbon dioxide leads temperature follows', though the IPCC report acknowledged that 'we do not know cause and effect'.

Source: John Gribbin, 'The End of the Ice Ages', *New Scientist*, 17 June 1989.

Such was the runaway enthusiasm of the day that in 1989 *New Scientist* magazine prematurely ran the same early Vostok ice-core graph with the caption, 'Where carbon dioxide leads, temperature follows' (Gribbin 1989).

It would take another ten years after the first IPCC report before scientists realised that it was temperature that rose first, not CO_2. Between 1999 and 2003 it was discovered (quietly) that temperature was driving CO_2 levels, and not the other way around. Cause and effect were back to front and thereafter were barely spoken of again. In the end, it was just basic chemistry; the oceans contain 50 times the amount of CO_2 as all the air in the sky. As the oceans warmed, they degassed and released CO_2.

Whatever drives temperature, by default, indirectly drives CO_2.

That ought to have been a major blow to the theory of a man-made catastrophe, but the express train to a carbon-based scare was already

Figure 5.2 Temperature and CO_2 in the Vostok ice core

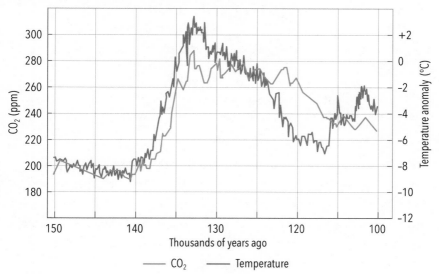

By 2003 it was obvious that temperature led CO_2 levels by hundreds of years. CO_2 followed temperatures up and, with a longer delay, back down. Sometimes there was a mismatch for several thousand years where temperatures fell, but CO_2 stayed constant.

Data from the Carbon Dioxide Information Analysis Center: http://cdiac.ornl.gov/ftp/trends/temp/vostok/vostok.1999.temp.dat; http://cdiac.ornl.gov/ftp/trends/co2/vostok.icecore.co2.

Source: Graph by Jo Nova 2008, from http://joannenova.com.au/global-warming-2/ice-core-graph.

well fuelled and running 'at speed'. Paleoclimatologists (and so many others) were implacably convinced that CO_2 controlled the climate. They argued that even though temperature initiated the rise in CO_2, after that, CO_2 amplified the warming. And perhaps it does to some small extent, but it's not large enough to measure. Teams of researchers have hunted ever since, with little success.

The silence was deafening

Having told the world that carbon led temperature, *New Scientist* magazine didn't rush to correct the record. The peer-reviewed papers detailing the 'lag' were published in the theoretically prestigious journals of *Science* and *Nature*, but even years later, in 2007, the 170,000 subscribers of *New Scientist* would have had no idea. I wrote to the editors in 2007 and asked if I had missed the news.

Michael Le Page, from *New Scientist* replied, acknowledging that they did not cover any of the papers, but bristled that it was entirely unnewsworthy, saying:

> Why do you think we should have reported a finding that no scientist found at all remarkable at the time, because it confirmed what they had always thought.

As if *New Scientist* wouldn't have done glorious headline-grabbing cover stories on those same papers had the results been the other way. This data is central to a topic that *New Scientist* calls 'the most important threat we face', and the magazine found the space to mention 'Vostok' eighteen other times during the same five-year period the papers were released, but somehow the 800-year lag was not even news?

I pointed out the incorrect caption on the 1989 graph, and Le Page stopped replying. Three months later, *New Scientist* mentioned the lag for the first time in a 'myth debunking series' with the snappy title 'Climate Myths: Ice cores show CO_2 increases lag behind temperature rises, disproving the link to global warming'. So the news that was too obvious to mention in 2003 was later worth debunking because punters were starting to think the 'lag' *New Scientist* had kept secret might matter?

It speaks volumes for how science journalists became activists and how powerful the Antarctic ice cores were as a form of advertising.

It's what they don't say that matters

The largest marketing campaign using the ice-core results was Al Gore's *An Inconvenient Truth*. Look at the extreme care he took to frame the graph:

> The relationship is actually very complicated but there is one relationship that is far more powerful than all the others and it is this – when there is more CO_2 the temperature gets warmer, because it traps more heat from the sun inside. (Gore 2006)

So he enlarged the graph to 20 square metres, and talked at length about it, was aware of how complicated it was, but didn't find the time to say the four words 'temperature rises before carbon'.

He won a Nobel Peace Prize for what was arguably stage-managed misinformation.

Slow formed bubbles completely smooth out spikes in CO_2

The ice cores were sold as time capsules from the past. But the bubbles form in slow motion – taking anything from ten years to 2000 years to completely seal off from the air above.

The snowfall piles up gradually year after year. Incrementally, it compresses into firn and finally, eventually, it compacts into ice, which is when the last channels to the atmosphere above are closed. Therefore, bubbles are more like a smoothed average of the air above. Automatically it means that spikes or dips in CO_2 will be flattened out. The coldest, driest locations with the longest records are also the slowest to seal. So if CO_2 levels rose rapidly for 100 years in 90,000 BC, the ice cores wouldn't detect it at all.

It is simply not possible for a scientist to use long-term ice-core graphs and claim that CO_2 levels have 'never risen as fast as they are rising today'. They don't have the resolution. Most of the data points in the Vostok core are between 200 and 2000 years apart.

The smoothing effect of ice-core bubble formation was well known by climate researchers and had been for years. Yet Al Gore explicitly says that the bubbles measure the CO_2 for 'the year that snow fell'. He also declares that in 650,000 years the CO_2 level has never gone above 300 parts per million (ppm).

All ice cores have much better resolution of temperature data – because that depends on isotopes in the ice, which is set down in layers, around the bubbles. But the bubbles within that ice are younger and smoothed out.

The chemistry is so extreme that these bubbles have been to hell and back

There are many ways the bubbles of air could be gradually changed even within the vaults of ice.

After 60,000 years at Vostok, and when a kilometre of ice presses from above, the pressures are so large that bubbles transform into clathrates – a sort of crystal lattice that surrounds individual molecules. The process takes 100 years for an individual bubble and in these 'transition' zones the forces on the molecules may go through a kind of fractionated sort and separation process (Stauffer & Tschumi 2000). CO_2 is less soluble in ice than in bubbles. It's all so much messier than we are led to expect. Indeed chemists use terms like 'gravitational sorting' and 'preferential adsorption', and describe microscopic liquid layers that wander through the ice (Stauffer & Tschumi 2000). Veins of liquid water may still be liquid even at –50 °C, shifting as the crystals grow and recrystallise (Mulvaney, Wolff & Oates 1988).

Among all the other changes, there is the potentially explosive release as the cores are extracted and the pressure falls.

Not to mention that ice cores may be contaminated during collection, handling, storage and transportation. Let's not underestimate the daunting challenge of collecting a bubble from two kilometres below the surface, on a continent where few humans live, and where none of them could survive long without outside assistance.

Where are all the other proxies?

Ice cores only contain information about 2% of the last 100 million years for a small part of the world. They dominate paleoclimatology but other proxies reach further back, and cover more continents or oceans.

The European Project for Ice Coring in Antarctica (EPICA) ice core reached 2.7 km down to get an 800,000-year record. Ice cores now stretch back to 2 million years (Nealon 2019) but can't go much further, though teams from Australia, America, Russia, China and Japan are all looking for long records (Amos 2019). But as they get closer to the bedrock, geothermal heat melts away the ice.

For most of the last 100 million years, the planet has been warmer, and had less ice. Which brings us to the odd incongruity that while there are many CO_2 proxies that potentially cover the last million years, the ice cores 'own' the Pleistocene and it's a rare graph that shows other proxies in the Holocene, the Eemian, or even for the last 1000 years. Like the dog that didn't bark, where are those comparisons?

There are many different ways to estimate ancient CO_2, all of them, of course, difficult. But there are plenty of other alternatives like plant stomata, boron isotopes, phytoplankton, paleosols, liverworts, and oceanic carbonate sediments – and some of these go back 400 million years (Beerling & Royer 2011). But these are judged by how well they agree with the ice cores. Paleoscience is stuck in an endless loop of circular reasoning and confirmation bias. When other proxies disagree it is a major mark against them. Despite the extreme chemistry and poor resolution, the ice cores are 'king'. All other proxies bend to ice cores.

> In many ways, ice cores are the 'rosetta stones' that allow development of a global network of accurately dated paleoclimatic records using the best ages determined anywhere on the planet. (Alley 2010)

But there are other proxies and they suggest there is much more natural variation in CO_2 levels, with some suggesting that CO_2 levels were often 20 or 30 ppm higher than is shown by the ice cores.

The puzzle of stomata versus ice cores

Plant stomata make sensitive CO_2 meters. They are the pores that exist on every leaf of nearly every plant, and their job is to capture CO_2. They were the great invention that 400 million years ago drove plant growth. Plants are finely tuned and hyper-evolved, to make the most of CO_2. When CO_2 is scarce, plants produce more stomata on every leaf in a race to catch more of their main building-block material. But there is a price to pay – the more stomata they have, the more water they lose. So each year plants grow new generations of leaves adapted to the level of CO_2 they find around them. As those ancient leaves fall, they capture that season in time. A fossilised leaf will hold the same pattern of stomata – there's no 'smoothing' – although there are other uncertainties, such as dating the fossil and calibrating it. Each species has its own stomata response 'curve'. Therefore, we can use living plants to calibrate the way they react. That gets difficult with older long-extinct

Figure 5.3 Gingko leaves and stomata

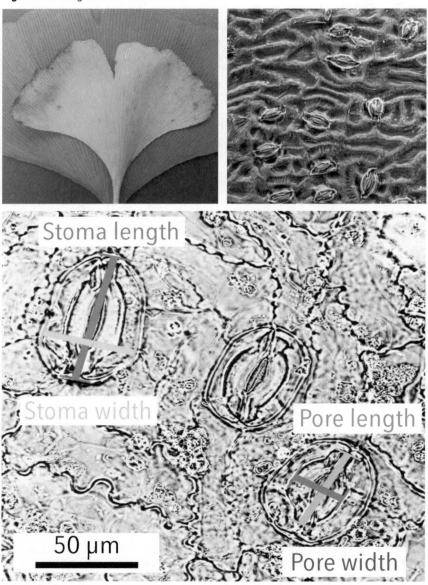

Clockwise: two gingko leaves, electron microscopy of gingko stomata, close-up of three stomata

species. Stomata researchers have had to concentrate on a few 'living fossils' such as the ginkgo family, which hasn't changed much in 200 million years.

The big mystery of paleoclimate studies is that plant stomata are accepted as one of the main proxies of atmospheric CO_2 levels right back to 400 million years ago, but they're largely ignored when it comes to studying the last million years – in what has become the era of 'sacred' ice-core readings. The paradox is all the more inexplicable because fossils from more recent times are from species that we are most familiar with. They ought to be the easiest to calibrate and most reliable, and yet they are forgotten, discounted, and rarely mentioned.

The technique was first reported in 1987, when it was discovered that plant stomata had changed markedly in the last 200 years as CO_2 levels rose. As CO_2 climbed from 280 – 340 ppm the density of stomata fell by 40% (Woodward 1987).

And stomata are used in analysis of the last 1000 years, so they top and tail the ice-core era. But here is the strange case of the last 1000 to 1 million years. The CO_2 levels in ice cores run right through that time frame at a flat 285 ppm, but the estimates from stomata vary wildly.

Having just been calibrated to ice core and flask CO_2, you might expect stomata to reliably report the accepted preindustrial levels of 280 ppm for at least a few hundred years – but they don't (Kouwenberg et al. 2005). Stomata proxies might be good 20 million years ago, but we apparently can't even get them to agree with ice cores a mere 250 years

Previous page: The Smithsonian Environmental Research Centre in Edgewater, Maryland, is part of an experiment called 'Fossil Atmosphere' designed to test how the concentration of carbon dioxide affects the number and size of stomata on gingko plants, often referred to as 'living fossils.' Because the frequency of the stomata varies with carbon dioxide concentration it is possible to estimate past levels of atmospheric carbon dioxide.

The close-up is of the abaxial leaf side of *Ginkgo biloba* showing stomatal parameters that are often measured. This is figure 1 from Šmarda P. *et al.* https://doi.org/10.1038/s41438-018-0055-9 republished under Creative Commons Attribution 4.0 International License.

Source: https://www.smithsonianmag.com/blogs/national-museum-of-natural-history/2017/06/08/can-you-help-us-clear-fossil-air/ and https://www.nature.com/articles/s41438-018-0055-9/figures/1.

ago. Most ice-core records show CO_2 barely varied by 12 ppm from the time of William the Conqueror through to the invention of the steam engine. But when stomata suggest the natural variation is common and much larger, some researchers dismiss it as less reliable because it doesn't agree with ice cores. The same researchers admitted in 2009 that few studies have been done to compare the two proxies (Finsinger & Wager-Cremer 2009). It's as if no one wants to look?

In the 1990s, several stomata studies showed that during the Holocene, CO_2 levels were often higher, and even as high as 350 ppm,

Figure 5.4 Stomatal frequency counts and ice-core measurements

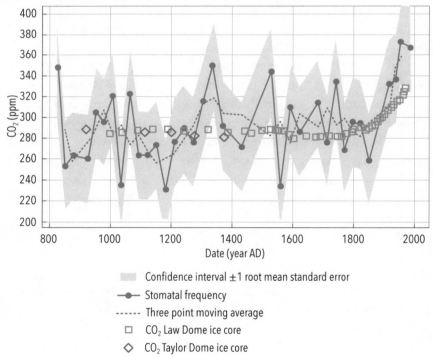

Changes in the frequency of stomata on Tsuga heterophylla suggest that CO_2 is higher and more variable than that calculated from ice core data – sometimes even 50 ppm higher. Yet climate models assume the flat ice-core graph is correct and preindustrial CO_2 was always 280 ppm.

Source: Kouwenberg et al. 2005.

a level human industrialisation only achieved again with a thousand coal plants or so in 1988. Yet preindustrial levels were defined as being 280 ppm, as if everything above that was de facto man-made.

As Wagner says:

> Our results falsify the concept of a relatively stabilized holocene CO_2 concentration of 270 to 280 ppm until the industrial revolution. SI based CO_2 reconstruction may even suggest that during the early Holocene atmospheric CO_2 concentrations that were >300 ppm could have been the rule rather than the exception. (Wagner et al. 1999)

The message from fossilised leaves is that CO_2 levels varied a lot more than is shown in the ice cores. Instead of being a constant 270 ppm during the Holocene, many stomata suggest that plants were responding to levels more like 300 ppm or 320 ppm.

A study from the Eemian period about 120,000 years ago, showed for thousands of years birch trees were consistently responding as if the air had 30 ppm more CO_2 than was apparently recorded in the Vostok ice cores (Rundgren et al. 2005).

Figure 5.5 Stomatal frequency counts and ice-core measurements

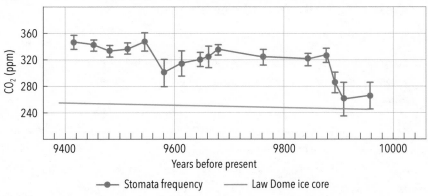

Stomata levels suggest CO_2 levels were often higher than ice cores indicated, which were close to 250 ppm at Law Dome 10,000 years ago.

Source: Wagner et al. 1999.

What looks like confirmation bias, smells like confirmation bias ...

CO_2 levels from stomata research just don't agree with the CO_2 levels from the ice cores whether from the last 10,000 or 120,000 years ago, but no one wants to hear that.

A string of papers came out from 1990 to 2005 with similar results, showing that stomata proxies produce higher and more varied estimates than ice cores, but everyone seemingly lost interest in pursuing this. In the fifteen years since, the stomata researchers shuffled sideways to estimating climate sensitivity from 2 million, 20 million or 50 million years ago, while definitely not trying to estimate CO_2 levels during the same period as the reign of the sacred ice cores.

Among paleoclimate researchers, it was well known that stomata results disagreed. As one researcher said, 'Stomatal studies face fierce "competition" from ice-core people, who do not like it at all' (H Birks 1999, pers. comm. 24 March).

Yet who knew? The public were never informed that ice-core bubbles were flattened averages and that other major classes of proxies showed that CO_2 levels might have been systematically higher by 20 or sometimes 50 sacred ppm.

A whole field built around one assumption

The paradigm is so encompassing, it is claimed:

> Atmospheric CO_2 is a critical component of the global carbon system and is considered to be the major control of Earth's past, present, and future climate.

Researchers go to great lengths to accommodate their results through the lens of CO_2 as a climate controller.

Paleoclimatology seems apparently to exist largely to estimate the climate sensitivity of CO_2. But ponder how many angels are dancing on this pinhead? If we are unable to calculate an accurate climate sensitivity from measurements on Earth today with global satellite coverage, thousands of thermometers, robot ocean buoys, and 28 million weather balloons, what are the odds of calculating climate sensitivity from air, rain or leaves that fell half a million years ago?

By unquestionably relying on ice cores, we have underestimated natural variations and subsequently overestimated human influence on CO_2 levels.

Does the climate even matter?

If life on Earth depended on unpacking the ancient climate to understand the future one, surely someone somewhere would have been calling for more stomata proxies? Surely governments would have poured money into a key technique – one of the few – that can map out CO_2 levels for 400 million years and all over the world?

For four hundred million years there have been falling leaves all over the globe. Presumably this means that there is no shortage of samples we haven't yet found, but oddly the IPCC and activists don't even appear to be looking? Could it be because the answers are not the ones they're looking for?

The only thing we know for sure about Antarctic ice cores is that climate scientists have repeatedly downplayed the problems, hidden the complexity, oversimplified the task, and not objected when ice-core bubbles are portrayed as perfect time capsules of air.

6 Monitoring Temperatures and Sea Ice with Satellites

Dr Roy W. Spencer

Since 1979 satellites have provided truly global measurements of temperature. This is also the period during which warming should have been strongest, according to global warming theory. Thus, the satellite record, specifically measurements of the lower troposphere, provides an opportunity to test climate models' predictions of warming. In fact, the global lower troposphere warming measured by satellites has been +0.13 °C per decade (1979 through 2019), which is only about 50% of that produced by the models forced by increasing greenhouse gases.

There is significant regional variability and very different trends in the two polar regions. The Arctic has warmed more than the global average, with a decadal trend of +0.26 °C. There has been significant decline in the area of sea ice, particularly late season sea ice. Considering the longer sea-ice record for this region, at least some of the recent warming is likely to be natural and an extension of the warming trend evident since the Little Ice Age. In contrast, at the Antarctic, satellite temperature measurements indicate no tropospheric warming, and an overall increase in the area of sea ice that covered nearly 16 million km² in September 2014.

Technical considerations

Why would we be interested in monitoring deep-layer atmospheric temperatures, when the surface of the Earth is where people live? Global coverage is one obvious answer. But at least as important is the way in which the climate system works. Global warming theory is fundamentally based upon how efficiently the climate system cools through infrared (heat) radiation to outer space. Most of that radiation comes from the atmosphere, not from the surface. The 'effective emitting temperature' of the climate system is rather close to the temperature that the satellites measure. Thus, to understand just how much of a threat anthropogenic global warming is, we need to monitor temperatures in the atmosphere as well as from the surface.

Monitoring of global temperature change with any technique is difficult because the signal is so small, generally 0.15 to 0.25 °C per decade. This is only a couple of hundredths of a degree per year. No environmental temperature measurement system was designed to measure such a signal accurately, which would require a couple of thousandths of a degree per year stability. Instead, we must use systems that were designed for weather monitoring of degrees of change, or tens of degrees of change, over hours or days. Corrections of such data for known influences are inevitable, whether from satellites or from surface-based thermometers. Some corrections, such as the urban heat island effect on surface thermometers, are difficult to distinguish from long-term global warming.

Surface-based thermometers are the most common temperature measurement, with reasonably good coverage of much of the Earth's land surface dating back to the late 1800s, with lesser coverage over the oceans. Radiosondes (weather balloons) have measured the vertical profile of atmospheric temperature, but only at a relatively small number of sites around the world since the 1950s. Finally, satellites have provided globally distributed atmospheric temperatures over fairly deep layers of the atmosphere, but only since 1979. Thus, depending upon the system, we are limited in either time sampling or geographic sampling.

In addition, weather forecast centres around the world have used all of these data sources and more: from surface thermometers, radiosondes, commercial aircraft, ocean buoys, and a wide variety of satellites to produce global 'reanalysis' datasets, where all of the data are interpreted with the help of known atmospheric physics and thermodynamics to arrive at a best-fit to all of the data. The reanalysis system is not perfect, but if done well it does give a good semi-independent check on some of the individual datasets.

The satellite technology for monitoring deep-layer temperatures uses microwave radiometers that measure the thermal emission of micro-wave energy by oxygen in the atmosphere (Spencer & Christy 1990). The temperature of the air layer being sensed is directly proportional to the intensity of the microwave radiation. The system is calibrated by observation of an on-board warm calibration target whose temperature is monitored with redundant high-accuracy platinum resistance ther-mometers, as well as by viewing the cosmic background radiation at a temperature of 2.7 Kelvin (K) (−270 °C).

Despite the relatively short data record (40+ years), satellites have the unique advantage of providing daily, near-global coverage, even in remote areas of the world that have no surface thermometers or radiosonde data.

John Christy and I used a total of fourteen different satellites operat-ing at different times over the more than 40 years' record. Each satellite measurement is a spatial average of one 25,000 km^3 volume of atmo-sphere (50 × 50 km horizontal resolution over a depth of about 10 km) every 50 milliseconds. Combined with the satellite movement and the scanning geometry of the instruments, most of the Earth is sampled every day, with small zones of no data centred at the poles.

Two independent research groups provide the most widely referenced global temperature datasets, our University of Alabama in Huntsville (UAH) (Earth System Science Center) group, and Remote Sensing Systems (RSS) (RSS n.d.), a private research firm in Santa Rosa, Califor-nia. The products produced by the two groups give very similar results in the tropics, but new revisions to the RSS data (Version 4.0) have resulted in greater global-average warming trends in the RSS data than in the UAH data.

A thorough comparison of the satellite datasets to radiosonde data was published in 2018 (Christy et al. 1990), where we argued that the UAH dataset likely provides the most accurate temperature trends due to the RSS inclusion of the full record of old NOAA-14 MSU data, which had substantial spurious warming in its calibration compared to the simultaneously operating (and better-calibrated) NOAA-15 AMSU instrument.

Satellite global temperature trends, 1979–2019

The monthly global-average temperature variations of the lower troposphere from 1979 through to April 2020 for the UAH dataset are shown in Figure 6.1.

When we average the monthly variations seen in Figure 6.1 into calendar years as shown in Figure 6.2, we find that 2016 was the warmest year in the satellite period, due to the record warm El Niño event of 2015–2016; 2019 was the third warmest.

So, the satellite data suggests that there has been a weak warming trend since 1979 of about +0.13 °C/decade. If warming continued at this pace, it would amount to 1.3 °C in 100 years, by which time increasing prices of dwindling fossil fuel resources combined with technological advances in energy generation might mean fossil fuels are no longer the primary source of humanity's energy. Adapting to that level of warming would be easy. It is not very meaningful to project

Figure 6.1 Monthly global-average temperature anomalies

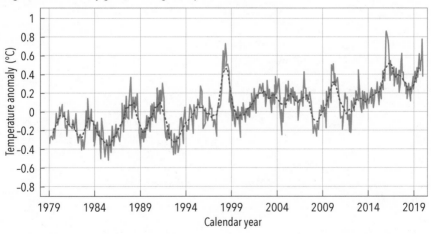

Monthly global-average temperature anomalies of the lower troposphere since 1979, as the departure from the 1981 to 2010 average. The Version 6.0 global average lower tropospheric temperature anomaly for April 2020 was +0.38 °C, down from the March value of +0.48 °C.

Source: Roy Spencer and John Christy, https://www.nsstc.uah.edu/data/msu/v6.0/tlt/uahncdc_lt_6.0.txt.

Figure 6.2 Ranking of years, warmest to coolest

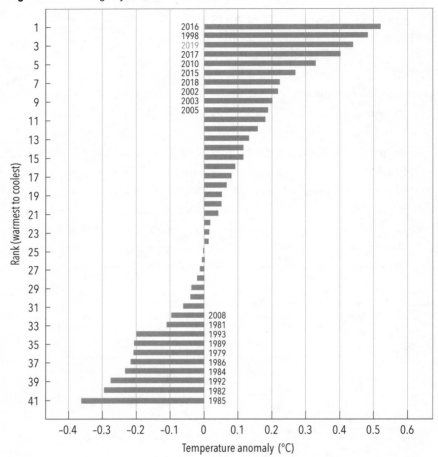

Ranking of the calendar years 1979–2019 from warmest to coolest in the UAH Lower Troposphere satellite data. The difference between 2016 and the previous record warm year (1998) is not statistically significant.

warming past 80 years into the future anyway, since we cannot know what technological advances might be made in the energy sector. But this slow rate of warming is not what is expected by many climate researchers. Significantly, the satellite-indicated warming rate is only about 50% of the warming rate predicted by the computerised climate model projections (KNMI Climate Explorer 2019a) tracked by the

United Nation's Intergovernmental Panel on Climate Change (IPCC). If we compare the warming predicted by 102 climate model runs, with the satellite-measured warming, we see that they have diverged since about 1998, as shown in Figure 6.3. Four of the reanalysis datasets (ERA-Interim, ERA-5, JRA, and MERRA), when averaged to match the same atmospheric layer measured by the satellites, also suggest a disagreement with the climate models.

The disagreement between the observations and models is even greater in the tropics (not shown). While an average of 102 climate models, as shown in Figure 6.3, indicates the average behaviour of climate models

Figure 6.3 Lower tropospheric temperatures, models versus observations

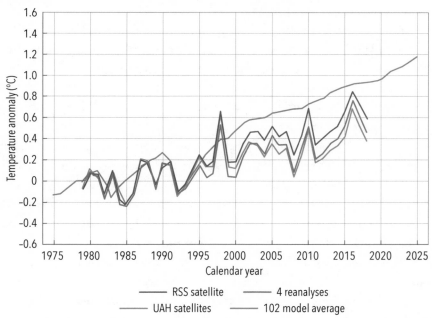

Output from the Coupled Model Intercomparison Project Phase 5 (CMIP5) versus satellite and radiosonde observations for yearly, global-average tropospheric temperature anomalies relative to their respective 1979–1984 averages. The reanalysis series is based on surface thermometers, radiosondes, commercial aircraft, ocean buoys, and a wide variety of satellites where all of the data are interpreted with the help of known atmospheric physics and thermodynamics to arrive at a best-fit.

Source: CMIP5 data available from Royal Netherlands Meteorological Institute at https://climexp. knmi.nl/start.cgi?id=someone@somewhere.

in general, virtually none of the individual models projected such a slow rate of warming as has been observed by satellites and diagnosed by the reanalysis datasets. Only one of the 102 models had a cooler trend than the UAH trend, while four of 102 models had a cooler trend than the RSS trend; 98% of the models had a warmer trend than the average of the four reanalysis datasets.

The discrepancy between models and observations is important because the climate models tracked by the IPCC provide the basis for changes in energy policy to reduce the emission of greenhouse gases. The satellite measurements (and reanalysis datasets, which include surface thermometer data, radiosondes, and many other *in situ* data sources) are critical to the discussion of just how necessary such energy policy changes really are.

There are several potential reasons for the discrepancy between models and observations. One possibility is that there has been a temporary, but unknown, cooling influence not contained in the models, which will eventually end, and rapid warming will resume in the observations at some point in the future. In other words, the models will eventually be proved correct, even though they do not contain the temporary cooling influence, and that we just haven't waited long enough for the full strength of global warming to be realised. This seems to be the attitude of the IPCC and the climate modelling community, as preliminary results from the new Coupled Model Intercomparison Project Phase 6 (CMIP6) climate models suggest the new models will produce at least as much warming as the CMIP5 models represented here.

An increase in heat storage by the deep ocean, which is very cold, might also explain the discrepancy. But a recent reanalysis of deep-ocean warming suggests that observed rates of warming are close to that predicted by the average of the climate models (Cheng et al. 2019).

The most likely explanation for the difference between models and observations is the models are tuned to be too sensitive to our greenhouse gas emissions. The direct warming effect of doubling carbon dioxide (CO_2) in the atmosphere is only about 1 °C, and it is positive feedbacks in the models that amplify this to about 3 °C. Yet, none of the global flows of energy in and out of the real climate system are known with sufficient accuracy to model the climate system from first physical principles. Instead, the models are 'fudged' to produce no natural climate change, and then the warming effect of increasing CO_2 is added to them. This means all of the warming produced by the models is due to CO_2, because that's what was assumed at the outset. Since the feedbacks which produce most of the warming in the models are rather poorly known, the models can be further adjusted to produce warming rates that roughly match the observations. But it is entirely possible that, say, only 50% of recent warming was due to humans, and that feedbacks in the climate system do not amplify the weak direct warming effect of CO_2.

Regional trends across the globe

The regional distribution of temperature trends are shown in Figure 6.4, where we can see a general increase in warming progressing northward from the Antarctic to the Arctic. This is what one would expect for any source of warming, since there is an increase in land coverage of Earth with latitude as well (except for the Antarctic continent and the Arctic Ocean). The difference is due to the huge heat capacity of the oceans versus land. Under an imposed energy imbalance, it takes longer to heat a deep body of water than it does a land surface, since a water body can readily mix warmed surface waters with cooler, deeper waters – something that does not happen on land.

If we do area averages over the polar regions, we find warming is indeed strongest in the Arctic, with a decadal warming trend of +0.26 °C per decade, as shown in Figure 6.5.

I have had to broaden the temperature scale in Figure 6.5 because the poles experience wide swings in temperature due to variations in heat inflow from the lower latitudes – an issue that does not exist in global averages. The poles are kept warm by import of heat from warmer regions, and this influx of warmth varies as large-scale weather systems come and go. There can also be natural variations in ocean heat transport from the Pacific and Atlantic towards the poles, but these are only poorly known due to a lack of sufficient ocean data.

While Arctic sea ice has indeed experienced melting since satellite observations began in 1979, newer proxy measurements of sea-ice

Figure 6.4 Regional linear temperature trends

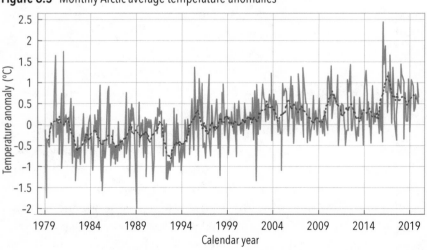

Temperature (°C/decade)

Regional linear temperature trends (°C/decade) in the UAH lower tropospheric satellite temperatures, January 1979 through to December 2019.

Source: University of Alabama in Huntsville.

Figure 6.5 Monthly Arctic average temperature anomalies

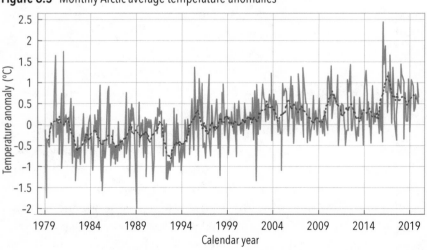

Monthly average temperature anomalies of the lower troposphere in the Arctic region since 1979, as the departure from the 1981 to 2010 average. The Arctic region covers the latitude 60° north to 83° north.

Source: Roy Spencer and John Christy, https://www.nsstc.uah.edu/nsstc/.

variations over the last 1200 years (Cabedo-Sanz et al. 2016), consistent with the historical Koch sea-ice index, suggest that there have been large variations in sea ice in the past, as shown in Figure 6.6.

The period of modern sea-ice monitoring is very short. During that 41-year period there has been a 3.8% per decade decrease in sea-ice area since 1979, as shown in Figure 6.7. Most of that decrease has been during the end of the summer melt season, as much as –11.4% per decade during September. This decline is consistent with the warming in the satellite temperature record.

Figure 6.6 Arctic sea ice before the 20th century

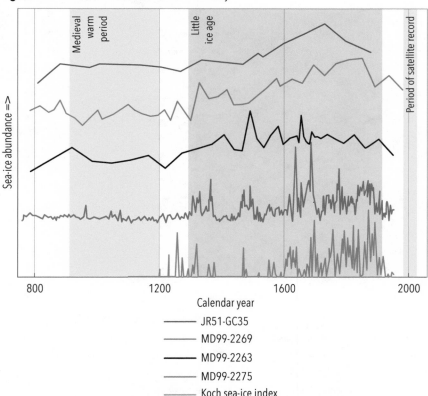

The top four series are proxy estimates of Arctic sea ice, which together with the Koch historical sea-ice index, show that Arctic sea ice was in a long-term decline well before the satellite record began.

Source: Cabedo-Sanz et al. 2016.

The situation is very different in the Antarctic, with no long-term warming trends from 1979 to 2019, as shown in Figure 6.8. This is in contrast to the average of 40 CMIP5 climate models, which produce an average warming rate of +0.21 °C/decade over the same period of time (KNMI Climate Explorer 2020b).

The sea ice around Antarctica shows a slight upward trend of +1.3% per decade during the period of the satellite record (Figure 6.9). The satellite temperature observations do not show a warming trend (Figure 6.8).

Figure 6.7 Arctic sea-ice area since 1979

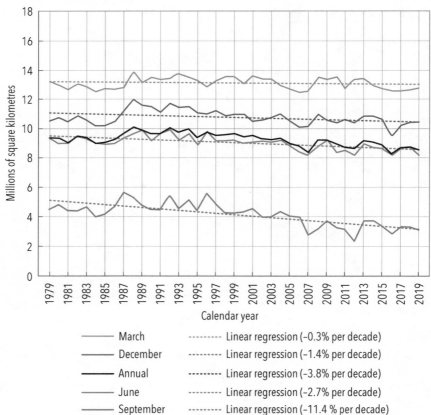

Satellite measurements of Arctic sea-ice area, 1979–2019, along with the decadal trends in four different calendar months.

Source: Data from the National Snow and Ice Data Center.

Figure 6.8 Monthly Antarctic average temperature anomalies

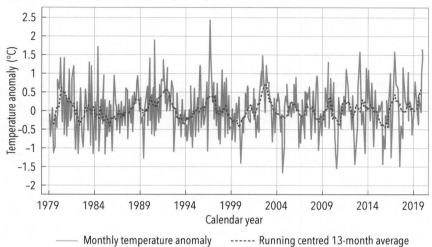

—— Monthly temperature anomaly   ------ Running centred 13-month average

Monthly average temperature anomalies of the lower troposphere in the Antarctic region since 1979, as the departure from the 1981 to 2010 average. The Antarctic region covers the latitude 60° south to 83° south.

Source: Roy Spencer and John Christy, https://www.nsstc.uah.edu/nsstc/.

Sea ice at the Antarctic covered an area of nearly 16 million km² in September 2014, as shown in Figure 6.9. There was a large decline in area of sea ice in February 2016 corresponding with the El Niño event of that Southern Hemisphere summer.

Conclusions

The satellite measurements of the lower troposphere have the advantage of being the only truly global measurements that are monitoring the portion of the climate system (the troposphere), which largely governs how the Earth cools to outer space. According to global warming theory, the lower troposphere is expected to warm faster than the surface (IPCC 2014) but the observations suggest that the opposite is happening. This gives us clues about how the atmosphere and surface are responding to greenhouse gas emissions. Indeed, monitoring of tropospheric temperatures is critical to understanding the response of the global climate system to increasing CO_2 concentrations.

Figure 6.9 Antarctic sea-ice area since 1979

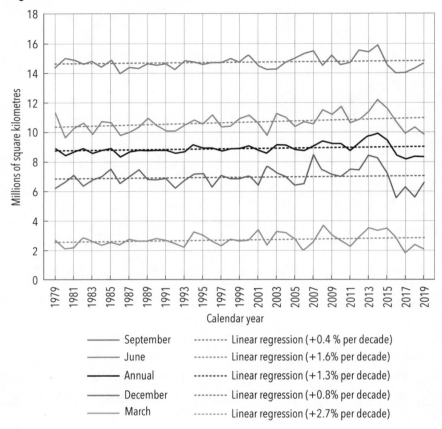

Satellite measurements of Antarctic sea-ice area, 1979–2019, along with the decadal trends in four different calendar months.

Source: Data from the National Snow and Ice Data Center.

To reiterate, since global warming theory is fundamentally about the role of CO_2 in reducing the ability of the Earth to cool itself, this makes tropospheric temperatures necessary for testing global warming theory.

Since 1979, the global-average lower troposphere has warmed at a rate only about 50% of what is expected by the CMIP5 climate models. The discrepancy also exists in the average of four popular reanalysis datasets, which do a synthesised physical and thermodynamic analysis of a variety of observational data from surface thermometers, radiosondes, aircraft, buoys, and many different satellite sensors.

While warming in the Arctic has been substantial over the satellite period, various proxy estimates of sea ice going back as far as 1200 years suggest that sea ice was declining since at least the early 1800s. Because the Arctic and Antarctic are sensitive to natural variations in equator-to-pole heat transport, they make poor indicators of human-caused climate change. In Antarctica, the satellite measurements suggest no tropospheric warming trend since 1979. Consistent with the satellite temperature data, the satellite sea-ice data shows no downward trend in sea ice, and in fact a slight upward trend of +1.3% per decade.

This observational evidence suggests a need to re-examine assumptions underpinning climate models. They cannot be designed from only first physical principles because various models produce a variety of global-average energy flows and global-average temperatures. Furthermore, we cannot even test which of them is closest to the truth because our measurements of the radiant flows of energy into and out of the global climate system are not accurate enough. Because of this uncertainty, climate modellers simply assume that the climate system is naturally in energy balance, and so any long-term changes must be the fault of humans. That they point to their models as evidence of this causation then becomes circular reasoning. It is time to start taking observations from satellites more seriously and make adjustments in climate models that bring them more in line with the observed behaviour of the global climate system.

7 Winter Temperature Trends in Antarctica

Ken Stewart

Antarctica is the coldest place on Earth. The lowest natural surface temperature ever directly recorded on Earth is –89.2 °C (–128.6 °F; 184.0 K), which was at the Soviet Vostok Station on 21 July 1983. The mean winter temperature at Vostok is –66.6 °C, while at the South Pole, Scott–Amundsen Base has a mean winter maximum of –55.3 °C, and mean winter minimum of –62.2 °C.

Even at the most northerly station in Antarctica, Esperanza, the mean maximum winter temperature is –5.7 °C and the mean winter minimum is –14.1 °C. At Australia's Mawson Station, the hottest-ever summer day reached 10.6 °C in January 1974, while the temperature in winter dropped to –36 °C more than once.

Winter temperature trends in Antarctica are a direct test of anthropogenic global warming theory for three main reasons. The first is that in winter Antarctica is surrounded by hundreds of kilometres of sea ice, therefore, any influence from a warming or cooling ocean is minimised. Second, the solar influence is also minimal because from June to August the noon solar angle varies from negative to 15°, giving just zero to nine hours of daylight. Third, because of the intense cold, the atmosphere is very dry, therefore minimising the influence of water vapour, which is a strong natural greenhouse gas. Any warming trend at Antarctica might, therefore, be attributable to the influence of anthropogenic greenhouse gases.

While there is data from 1945 for the first of the stations used in this study, Esperanza, and a few more stations commenced recording temperatures in the 1950s including Mawson, it was not until 1959 that the number of stations in Antarctica reached double figures.

There is regional variability in winter temperatures at Antarctica. Overall the trend has been one of warming from the beginning of the record to 1989, and cooling in many regions since 1989. The cooling is particularly evident at the five stations within the Australian Antarctic Territory. This can be contrasted with warming in winter on the Antarctic Peninsula in West Antarctica.

There is generally good agreement between the surface temperature record and the University of Alabama in Huntsville (UAH) satellite record. The overall winter temperature trends at Antarctica, however, are not in agreement with global warming theory, which claims that warming should be greater in the polar regions. This was first postulated by chemist Svante Arrhenius (1896) in his proposition of the hypothesis of global warming through an enhanced greenhouse effect, notably from carbon dioxide (CO_2). It was also hypothesised by Hansen et al. (2005), Holland and Bitz (2003), and Manabe and Wetherald (1975).

Data Sources and Methods

The surface temperature measurements used in this study were downloaded as unadjusted winter values (June–July–August) from the Goddard Institute of Space Studies (GISS), which is at Columbia University and part of NASA.

All temperatures were converted to anomalies (in this case from the mean winter temperatures from 1981 to 2010), as is standard practice in climate science in order to determine whether winter temperatures each year are warmer or cooler than average. Winter surface temperature trends were determined through simple linear regression based on mean winter anomalies considering the periods 1959 to 2018, and 1989 to 2018. The period from 1989 represents the last 30 years of measurements.

All the available data was used, except for the five stations on the South Shetland Islands north of the Antarctic Peninsula and the two stations Theresa and Linda, because the temperature series from these locations are too short for the meaningful calculation of anomalies.

Several stations (marked * in Table 7.1) have quantities of missing winter data, making analysis more problematic. For example, data from Dome C is only available from 1996. Possession Island has data for the winters from 1993 to 2009 but none since. Molodeznaja has data from 1963 to 1998, then there is a huge gap with data for only the 2014 and 2018 winters.

Table 7.1 Stations, observation periods, and winter temperature trends from 1959 to 2018

Temperature Trends (°C/100 yrs) at Antarctic Stations from 1959-2018 and from 1989-2018

Station	Start	End	Trend 1959-2018	Trend 1989-2018
Antarctic Peninsula				
Faraday	1950	2018	8.62	1.09
Butler Is	1986	2018	−4.22	−0.63
San Martin	1977	2018	7.50	4.40
Rothera Pt	1977	2018	8.33	3.53
Marambio	1971	2018	2.64	−1.51
Esperanza	1945	2018	3.13	0.00
O'Higgins	1963	2018	4.70	3.29
Antarctic Plateau				
Scott-Amundsen	1957	2018	0.04	3.55
Vostok	1958	2018	1.58	0.97
Dome C*	1996	2018	−11.88	−11.88
Victoria Land (Ross Sea coast)				
McMurdo	1956	2018	2.21	5.29
Marble Pt	1980	2018	−0.38	0.99
Terra Nova*	1989	2018	0.80	0.80
Possession Is*	1993	2009	2.93	2.93

**Temperature Trends (°C/100 yrs) at Antarctic Stations
from 1959–2018 and from 1989–2018**

Station	Start	End	Trend 1959–2018	Trend 1989–2018
Coats Land				
Halley	1957	2018	−0.26	−0.46
Neumeyer	1981	2018	−2.67	−3.42
Belgrano	1980	2018	−0.55	−0.61
Australian Antarctic Territory				
Zhongshan*	1989	2018	−6.83	−6.83
Myrnij	1956	2018	0.89	−3.83
Mawson	1954	2018	0.78	−3.05
Davis	1957	2018	−0.01	−5.35
Casey	1959	2018	−0.39	−7.83
Queen Maud Land				
Syowa	1957	2018	1.60	0.15
Novolazarevsk	1961	2018	2.27	−2.40
Molodeznaja*	1963	2018	2.69	−0.65
All stations			1.91	−0.97
* Limited data				

Source: Schmidt, G., GHCN V3 unadj dataset, viewed 4 March 2019, https://data.giss.nasa.gov/gistemp/stdata/.

The majority of stations are on the coast, with a cluster on the Antarctic Peninsula. Only four stations are in the interior on the Antarctic Plateau, and only three (Dome C, Vostok and Scott–Amundsen) were suitable for use in this study. All data was aggregated for the calculation of continent-wide means. Regional variability was considered by charting the individual time series based on six different regions, as shown in Figure 7.1, which vary in terrain, altitude, latitude, wind direction, cloud cover, and precipitation.

Figure 7.1 Antarctic stations with temperature data used in this study

Source: Schmidt, G., GHCN V3 unadj dataset, viewed 4 March 2019, https://data.giss.nasa.gov/gistemp/stdata/.

Overall winter trend for Antarctica

When surface temperature data from all the stations is aggregated and a mean anomaly calculated for each winter, it is possible to determine the trends for the whole continent. The trend from 1959 to 2018 is warming of +1.91 °C per 100 years. This changes to a cooling trend from 1989 to 2018 of –0.97 °C per 100 years, as shown in Table 7.1.

The pattern of warming and then cooling tracks very closely the values as measured by satellites since 1979, as shown in Figure 7.2. The satellite data is discussed in more detail in Chapter 6 by Roy Spencer.

In summary, considering Antarctica as a whole, temperatures generally increased from the late 1950s to the early 1990s, but there has been no general warming trend over the last 30 years.

Figure 7.2 Antarctic winter anomalies

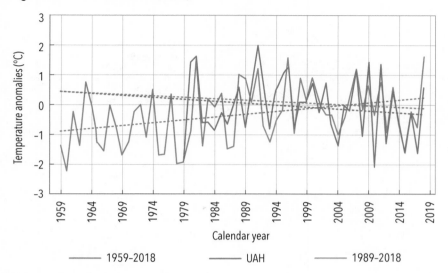

Anomalies calculated from 1981–2010 means.

Sources: Schmidt, G., GHCN V3 unadj dataset, viewed 4 March 2019, https://data.giss.nasa.gov/gistemp/stdata/, accessed 4/3/2019, and National Space Science and Technology Center, viewed 4 March 2019, https://www.nsstc.uah.edu/data/msu/v6.0/tlt/uahncdc_lt_6.0.txt.

Regional variation in winter temperature trends

Table 7.1 summarises trends in the winter temperature data from individual Antarctic stations, loosely grouped by the six geographic regions shown in Figure 7.1. Trends again represent degrees Celsius per 100 years, for the 60 years to 2018, and for the 30 years from 1989 to 2018. Annual values for each station have been charted, and then grouped by each region in Figure 7.3.

1. Antarctic Peninsula

Most stations on the Antarctic Peninsula – including Faraday, San Martin, Marambio, Esperanza, and O'Higgins – show the strongest warming in all of Antarctica. Butler Island alone has a cooling trend. Notably, the longest operating station in Antarctica, Esperanza, showed warming of +3.13 °C per 100 years for the entire 60-year period to 2018, but zero

Figure 7.3 Antarctic winter anomalies grouped by region

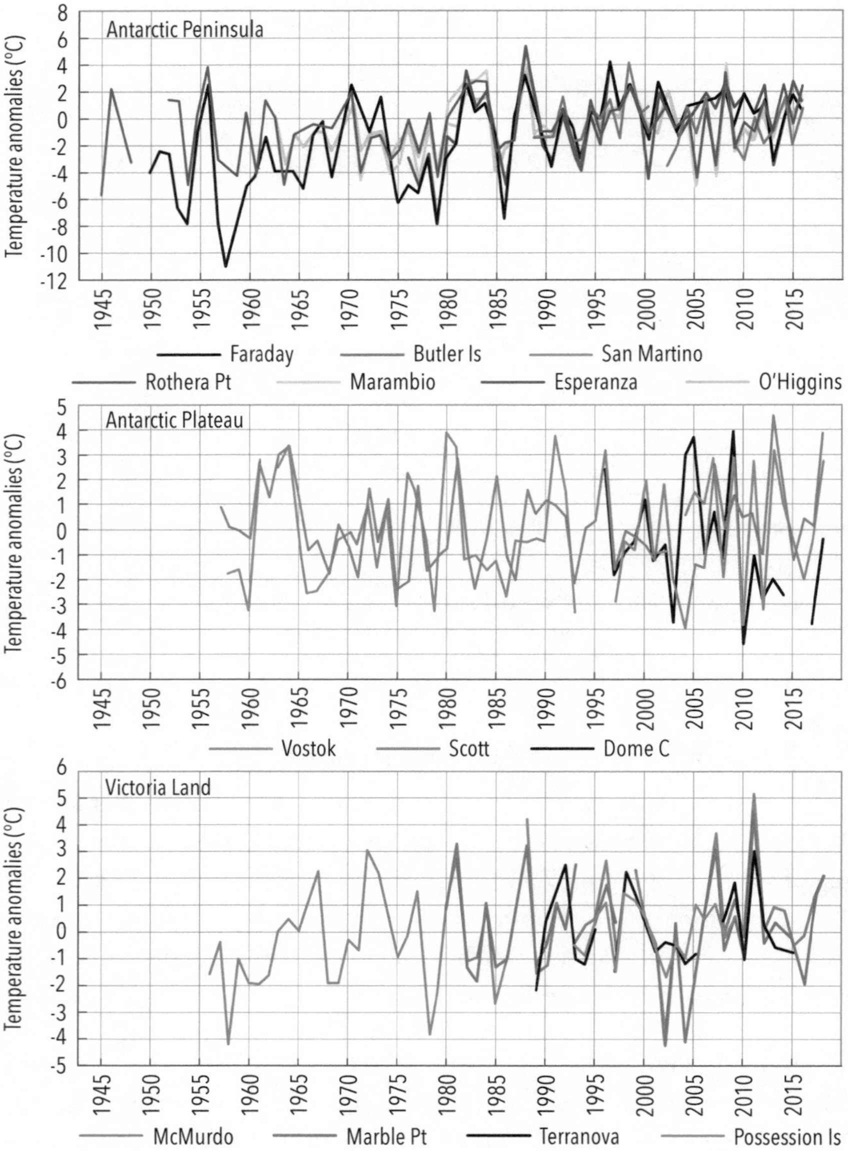

continued

Anomalies calculated from 1981–2010 means.

Figure 7.3 *continued*

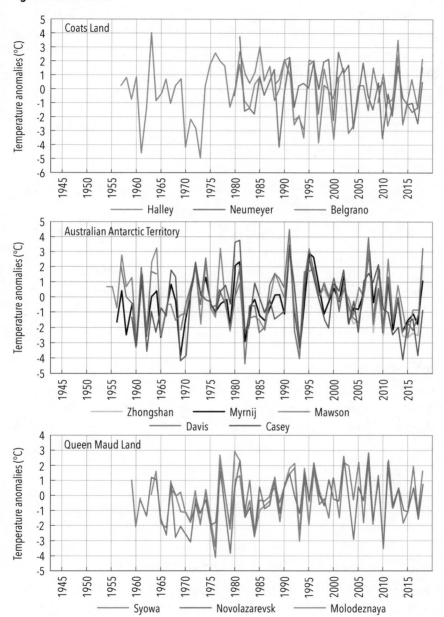

Source: Schmidt, G, GHCN V3 unadj dataset, viewed 4 March 2019, https://data.giss.nasa.gov/gistemp/stdata/.

for the period since 1989. Faraday, with the second-longest data series, has the greatest change in trends, +8.62 °C from 1959 to 2018, reduced to +1.09 °C from 1989 to 2018. This great reduction in trends is worth noting as warming in this region is frequently cited in the media as indicating warming across Antarctica as a whole.

2. Antarctic Plateau

Scott–Amundsen, Vostok, and Dome C, on the high Antarctic Plateau, have an average height of 3000 metres and are furthest from the sea. Winter trends at these three stations vary greatly between winters and across stations. At Vostok there is warming of +1.58 °C per 100 years for the entire period 1958 to 2018, which has reduced slightly to +0.97 °C for the 1989 to 2018 period. Scott–Amundsen, in contrast, shows more warming for the last 30 years of +3.55 °C/ 100 years. Dome C, with a record only beginning in 1996, shows very strong cooling –11.88 °C. These very different apparent trends may simply be an artefact of the start and end year, given the large variability in temperatures between winters, as shown in Figure 7.3.

3. Victoria Land (Ross Sea coast)

Victoria Land, named after Queen Victoria, lies between the Ross Sea, the Ross Ice Shelf and the Antarctic Plateau. The rate of warming at McMurdo has increased since 1989. Terra Nova Bay and Possession Island, which have limited data (and none before 1989), show strong warming since 1989.

4. Coats Land

Coats Land, named after the main supporters of the Scottish Antarctic Expedition of 1902–04, is on the eastern coast of the Weddell Sea, in the eastern region of West Antarctica. All three stations (Halley, Neumayer and Belgrano) show cooling since 1959, which has increased since 1989. This may again be an artefact of the start dates and the substantial inter-winter variability in temperatures, as shown in Figure 7.3.

5. Australian Antarctic Territory

Australia has a vast territorial claim in East Antarctica, and has three stations there: Mawson, Davis and Casey. There are two other stations on the coast within the Territory: Zhongshan, operated by China; and Myrnij, operated by Russia. Since 1959 Myrnij and Mawson show some warming (+0.89 °C and +0.78 °C per 100 years respectively), while the others show no warming at all. All stations show strong cooling since 1989.

6. Queen Maud Land

Queen Maud Land is a territory claimed by Norway. The three stations (Syowa, Novolazarevsk, and Molodeznaja) all show warming trends from 1959 to 2018, which are reduced, or reversed to cooling, since 1989.

Conclusion

Winter temperature trends at Antarctica are a test of anthropogenic global warming theory, because the influences of competing natural causes (including solar radiation, ocean temperatures, and humidity) are minimised.

According to this theory, as the Earth warms, Antarctica *should* continue to show warming, and indeed the warming should be more pronounced here. But the data doesn't show this. Although downplayed by climate scientists wishing to claim otherwise, and in contrast to the Arctic, there is no evidence for polar amplification in Antarctica, and thus no evidence for warming that can be attributable to human emissions of CO_2 and other greenhouse gases.

Overall, Antarctic winter temperatures have increased from 1959 to 2018, and certainly from the late 1950s to the 1990s, but there has been cooling during the last 30 years.

All stations in the Australian Antarctic Territory show strong cooling since 1989. In other regions some stations show strong warming, some show cooling, and some show no trend either way.

A possible reason for the different trends in different regions, especially the warming of the Antarctic Peninsula and Victoria Land, is the circulation of increasingly warm circumpolar deep water moving along the coastline of western Antarctica (Kekesi 2005; Hellmer et al. 2012; IPCC 2013; Schmidke et al. 2014; Spence et al. 2017).

These changes in heat transport by the circulation of circumpolar ocean currents may be a significant influence in regional trends, particularly along the Antarctic Peninsula.

8 No Evidence of Warming at Mawson, Antarctica

Dr Jaco Vlok

The Australian Bureau of Meteorology has measured surface air temperatures at the Mawson weather station in Antarctica since early 1954. Temperatures oscillate within a relatively narrow band, showing no statistically significant long-term warming trend. This is the case whether considering the actual historical measurements, or the temperatures subsequently adjusted by the Bureau before incorporation into other databases. The other weather stations in the Australian Antarctic Territory with long records are at Davis and Casey. The temperature series from these locations move up and down in synchrony with the temperatures recorded at Mawson and also show no warming or cooling trends, whether considering the actual measurements or the homogenised series. The temperature series from Mawson, Davis and Casey represent an amalgamation of measurements from different instruments, which could affect trends. In order to properly assess the equivalence of measurements from the electronic probes versus mercury and alcohol thermometers, there would need to be some assessment of values measured at the same time in the same shelter – known as the parallel data. This data is not publicly available.

Mawson temperature record

Mawson was the first official Australian Bureau of Meteorology weather station in Antarctica, and was named after Sir Douglas Mawson, an Australian geologist and Antarctic explorer (Bureau of Meteorology 2013). This is one of 49 weather stations (measuring air temperature), that have been operated by the Bureau in Antarctica at different times, with data from sixteen of these stations publicly available (Bureau of Meteorology 2019a). The locations of these sixteen weather stations are shown in Figure 8.1 as red crosses. Some red crosses represent more than one weather station.

Figure 8.1 Map of Antarctica with weather station locations

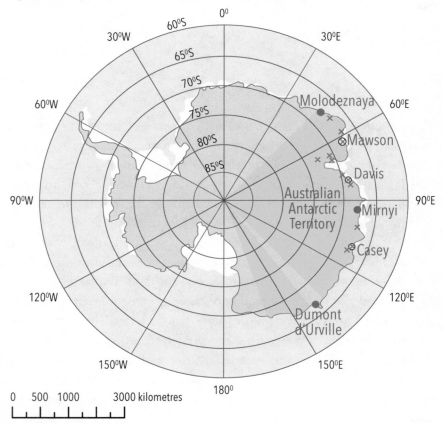

The Australian Antarctic Territory spans nearly 5.9 million km² of Antarctica, south of 60°S and between 45°E and 160°E, excluding the French Terre Adélie (between 136°E and 142°E). The red crosses are all the locations that have surface air temperature data available in the Australian Data Archive for Meteorology (ADAM). Some of the red crosses represent the location of more than one data series and/or weather station. The blue dots are the locations of two Russian and one French weather station that are used as reference stations during homogenisation of Mawson, Davis and Casey data.

Locations sourced from Australian Bureau of Meteorology, March 2020, http://www.bom.gov.au/climate/data/stations/.

The historical temperature record of Mawson is shown in Figure 8.2, including monthly and twelve-month moving averages. The twelve-month moving average is a measure of annual change in surface maximum and minimum temperatures. Maximum temperatures fluctuate between an annual average of –6.7 °C and –10.5 °C. Minimum temperatures fluctuate between an annual average of –12.2 °C and –17.3 °C.

The metadata for Mawson indicates that temperature measurements have been taken at the same location since 1954 (Bureau of Meteorology 2019b) with temperature measuring equipment mounted in a Stevenson screen similar to the shelters shown in Figure 8.3.

Over the period of the temperature record, different equipment was placed in the Stevenson screen at Mawson, and used to measure air temperature. According to the Antarctic station catalogue, daily extremes at Mawson were initially obtained from a thermograph in a Stevenson screen and since early 1973 also from a Fielden (Bureau of Meteorology 2013). This is an instrument with remote temperature measuring equipment and a transmitter that was used especially during heavy snowfall or

Figure 8.2 Temperature data measured at the Mawson weather station since March 1954

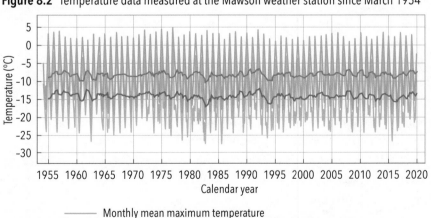

The monthly mean maximum and minimum temperature series were calculated from historical daily observations available in the ADAM database.

Data from Australian Bureau of Meteorology, March 2020, http://www.bom.gov.au/climate/data/.

Figure 8.3 Stevenson screens used to house weather recording equipment in Antarctica

Stevenson screen at Casey (left) in June 2014 and Davis (right) in May 2015. Stevenson screens are used globally to house thermometers and standardise weather measurement.

Source: Australian Antarctic Division, March 2020, http://www.antarctica.gov.au/.

when the Stevenson screen could not be accessed (Jovanovic et al. 2012). Liquid-in-glass thermometers were then used from January 1992, specifically a mercury thermometer to measure maximum temperatures and an alcohol thermometer to measure minimum temperatures (Bureau of Meteorology 2018). According to the online station metadata, however, these liquid-in-glass thermometers were in place since February 1954 (Bureau of Meteorology 2019b). So, there is some inconsistency in the available metadata regarding what equipment was actually used, and when.

The online station metadata furthermore indicates that an electronic probe (also known as a platinum resistance thermometer, PRT) was installed in the Stevenson screen in January 1994. This probe became the primary measuring instrument on 1 November 1996 (Bureau of Meteorology 2012). So, from 1 November 1996 the highest and lowest values from the probe were recorded as the maximum and minimum temperature each day in the ADAM database. These daily values are used to compile temperature statistics for this chapter, as shown in Figures 8.2 and 8.5.

I have summarised the equipment changes as best I can for Mawson as a Gantt chart (Figure 8.4), showing the timelines over which different instruments were used to measure temperatures at Mawson. Clearly

Figure 8.4 Thermometer usage Gantt chart for the Mawson weather station

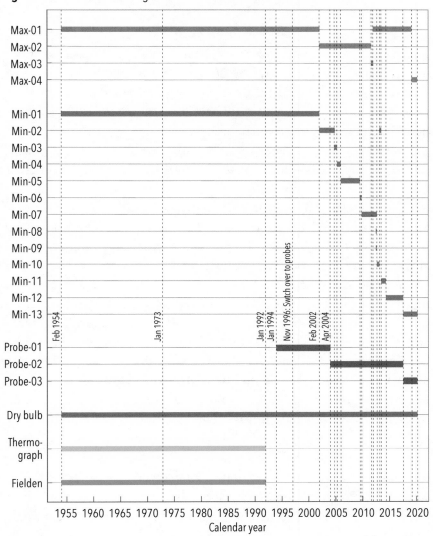

Four mercury thermometers (measuring daily maximum), thirteen alcohol thermometers (measuring daily minimum), three electronic probes (also known as platinum resistance thermometers, with the one instrument measuring both maximum and minimum temperatures) and one dry bulb mercury thermometer (measuring air temperature continuously) are listed as having variously been used to measure temperatures. The Antarctic station catalogue furthermore details the use of a thermograph and the Fielden, consisting of remote temperature sensors and a display.

Until 31 October 1996, measurements from the mercury and alcohol thermometers were entered into the ADAM data base as the maximum and minimum temperatures, respectively. From 1 November 1996, measurements from the electronic probe were recorded in ADAM as the maximum and minimum. While both probes and thermometers were present, the parallel data is not available. Analysis of the parallel data for the period January 1994 to February 2002 would enable some comparison of the values from the first probe with values from the mercury and alcohol thermometers.

Source: Data from the Mawson online climatological station metadata (Australian Bureau of Meteorology 2019b) and Antarctic station catalogue (Bureau of Meteorology 2013) were used to construct this chart.

there have been many changes to the equipment at this one location. The historical temperature record of individual weather stations such as Mawson (Figure 8.2) should therefore typically be viewed as a collection of measurements taken by different instruments over different time periods, and sometimes at different locations.

To make any conclusions regarding long-term patterns or temperature trends it is important to ensure equivalence between the different instruments used over time. The Bureau has taken parallel measurements at several weather stations across Australia, specifically recording temperatures as measured by both liquid-in-glass thermometers and electronic probes (PRTs) at the same location. This is ostensibly done to enable analysis of the difference between these two types of instruments and to ensure equivalence between them (Trewin 2012). This information has not, however, been published by the Bureau. It has therefore not been established that the recordings are equivalent. To be clear, while mercury and alcohol thermometers were always in place, and maintained in parallel with an electronic probe at Mawson since January 1994, this parallel data for Mawson has not been made publicly available. Neither has there been any report from the Bureau providing an indication of the equivalence between the measurements taken from the liquid-in-glass thermometers (mercury and alcohol) versus the electronic probes (PRTs) at Mawson, or at any of the other Antarctic weather stations.

Adjustments are made to individual temperature series following equipment changes, but this is not done based on an assessment of the parallel data. Rather it is made through an assessment of the series relative to neighbouring sites through a process known as homogenisation.

Homogenised temperature series

The Bureau's official temperature reconstruction, known as the Australian Climate Observations Reference Network – Surface Air Temperature (ACORN-SAT) dataset includes the three longest temperature series from the Australian Antarctica Territory: Mawson, Davis and Casey after they have been homogenised. This ACORN-SAT dataset is used by the Bureau, and also CSIRO, to report annually to the government on climate change.

The annual mean maximum and minimum temperature series for Mawson for both the ACORN-SAT and actual measurements from the ADAM database are shown in Figure 8.5.

Homogenisation is ostensibly done to remove non-climatic effects that may be present in the temperature record (Peterson et al. 1998), for example, changes in measuring methods and instrumentation, such as replacing a thermometer with an electronic probe (PRT), as has occurred at the Mawson weather station.

Through the process of homogenisation, actual measurements are adjusted after the identification of breakpoints in individual time series. According to published papers detailing the technique, adjustments

Figure 8.5 Historical observations versus homogenised temperature series for Mawson

——— Actual historical observations (max)	·········· 0.287°C / 100y
– – – Homogenised (max)	·········· 0.613°C / 100y
——— Actual historical observations (mean)	·········· -0.284°C / 100y
– – – Homogenised (mean)	·········· 0.396°C / 100y
——— Actual historical observations (min)	·········· -0.854°C / 100y
– – – Homogenised (min)	·········· 0.179°C / 100y

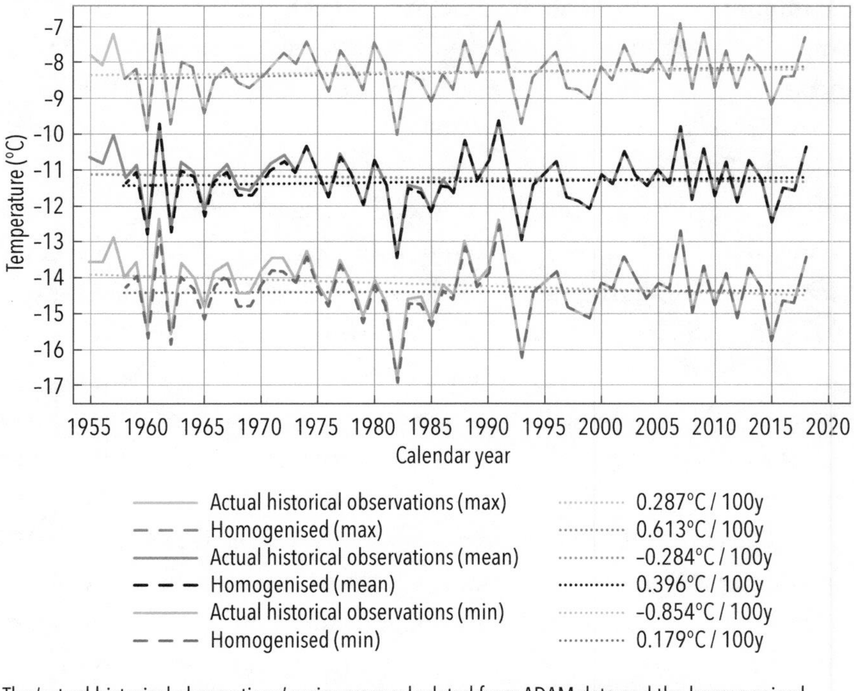

The 'actual historical observations' series were calculated from ADAM data and the homogenised series from ACORN-SAT data. Each mean series in green is the average of the corresponding maximum and minimum series.

Data as daily values was downloaded from Australian Bureau of Meteorology, March 2020, http://www.bom.gov.au/climate/data/ and http://www.bom.gov.au/climate/data/acorn-sat/.

could be made based on statistical analyses of individual temperature records in isolation, or in combination with measurements at neighbouring locations (reference stations), or both together, along with an analysis of metadata or documents describing the history of each weather station (Ribeiro et al. 2016). This theoretically depends on the availability of suitable reference stations and reliable metadata, or any other evidence supporting a potential breakpoint. In the case of the data from Mawson and other Antarctic stations, the adjustments are made with reference to neighbouring stations and the metadata.

As can be seen in Figure 8.5, the Bureau created the official or homogenised Mawson series from the actual historical observations by discarding the first three years of observational data, and dropping the minimum series down over the period 1958–1991. Specifically, the annual mean minimum temperature values were reduced by 0.35 °C and 0.15 °C for the two periods: 1958–1972 and 1973–1991, respectively. These changes are the result of unique monthly adjustments shown in Table 8.1 below.

The Bureau homogenised the Antarctic records ostensibly by identifying statistical breakpoints through comparison of meteorological data from neighbouring sites and analyses of metadata (Jovanovic et al. 2012). The 'homogeneity' of the Mawson data was apparently tested through comparison with neighbouring reference sites, specifically Davis and Russia's Molodeznaya, shown in Figure 8.1. The first three years of Mawson were excluded, as reference data for this period does not exist.

The effect of this homogenisation process, as applied to the Mawson record, is an increase in the warming trend in the maximum temperature series. Homogenisation (the adjustment of the values) also changes the long-term cooling to a warming trend in the minimum series, as shown in Figure 8.5. Overall, the mean temperature trend at Mawson in the historical observations is changed from cooling at a rate of 0.284 °C per century to warming of 0.396 °C per century in the official ACORN-SAT record.

Table 8.1 Adjustments (°C) made to Australian Antarctic records during homogenisation

Period	Jan	Feb	Mar	Apr	May	Jun	Jul	Aug	Sep	Oct	Nov	Dec	Avg
Mawson minimum temperature													
Jan 1958-Dec 1972	-0.4	-0.2	-0.2	-0.9	-1.7	1.0	-0.1	-0.4	-0.7	-0.1	0.0	-0.5	-0.35
Jan 1973-Dec 1991	-0.9	0.1	0.4	-0.6	-0.6	0.8	0.6	0.0	-0.2	-0.3	-0.2	-0.9	-0.15
Davis maximum temperature													
Jan 1958-Jan 1970	2.0	0.2	-0.1	0.3	0.6	0.1	0.2	0.5	1.1	0.3	0.8	0.9	0.575

Each calendar month within a given period is uniquely adjusted by the Bureau to create the official ACORN-SAT record. For example, in the Mawson minimum record, all January values from 1958 to 1972 are adjusted downwards by 0.4 °C, and all July values over 1973 to 1991 are adjusted upwards by 0.6 °C.

Data, as daily values, was downloaded from the Australian Bureau of Meteorology, March 2020, http://www.bom.gov.au/climate/data/ and http://www.bom.gov.au/climate/data/acorn-sat/.

Other long temperature records

While nearly 50 weather stations have been operated at one time or another by the Bureau at Antarctica, temperature data is only publicly available for sixteen of these, and only five have records spanning several years. The five longest records shown in Figure 8.6 include the three locations included in ACORN-SAT and two stations that are no longer active, including Wilkes and Casey (The Tunnel) that were operated in close proximity to the current Casey station (Australian Antarctic Division 2018).

Considering the twelve-month moving average for these five series (unadjusted), including Mawson (unadjusted), there is significant inter-annual variability, while overall there is no clear warming or cooling trend, as shown in Figure 8.6. The synchronisation between the temperature series based on the historical observations from the different locations is remarkable. For example, maximum and minimum temperatures rise in the early 1980s, then drop suddenly before rising again in unison, before dipping once more in the mid-1990s.

The temperature measurement history at Davis and Casey is similar to Mawson; initially a thermograph and the Fielden were used (Bureau of Meteorology 2013). Liquid-in-glass thermometers were also used, and automatic weather stations were deployed at all three sites in early 1994, with the installation of electronic probes (PRTs).

The actual historical temperature data available in the online ADAM database (Bureau of Meteorology 2019a) for Davis and Casey, as for Mawson, were therefore taken using different equipment over time. There were also site relocations, particularly at Casey. Initially measurements were taken at 'The Tunnel' from February 1969 to December 1988, after which a new weather station was opened nearly a kilometre away (Bureau of Meteorology 2013). Measurements were taken at both locations for a period of thirteen months, after which the old station was closed down.

Every change, including equipment changes and weather station relocations, can potentially undermine the reliability of long-term temperature trends. The standard practice in climate science to mitigate

Figure 8.6 Temperature trends at the Bureau's five longest-running Antarctic stations

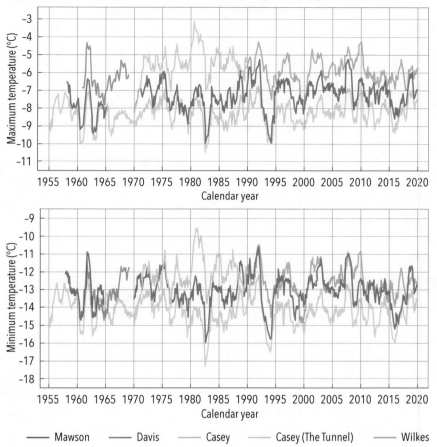

— Mawson — Davis — Casey — Casey (The Tunnel) — Wilkes

The twelve-month moving average series, calculated from the BoM's longest Antarctic records, show similar patterns and they all track together over time. None of these shows a clear warming pattern.

Data sourced from the Australian Bureau of Meteorology, March 2020, http://www.bom.gov.au/climate/data/.

the effect of these changes is to remodel the temperature measurements through homogenisation, as mentioned in the previous section (Homogenised temperature series).

The Bureau has homogenised the Davis record by adjusting the maximum measurements upwards from 1958 to 1970 using the values given in Table 8.1, and leaving the minimum record unchanged.

The effect of these adjustments is a reduction of the long-term warming rate in the Davis maximum record of 1.281 °C to 0.609 °C per century. The remodelling of the Davis temperature series was performed using Mawson and Russia's Mirnyi as reference stations, with a breakpoint identified in February 1970 aligning with the installation of the Fielden (Jovanovic et al. 2012).

Homogenisation of the Casey record was performed somewhat differently by the Bureau. No adjustments were made to the actual measurements as the record was found to be homogenous using Mirnyi and France's Dumont d'Urville as reference stations. The ACORN-SAT Casey record consists of the monthly means observed at 'The Tunnel' from January 1970 to January 1989, and at the current Casey weather station thereafter – without any adjustment, despite multiple instrument changes and a site relocation. There are thirteen months of overlapping measurements corresponding to the period when the two weather stations were operating. My comparison of maximum temperatures for this period indicates the new site is on average nearly 0.5 °C cooler. To be clear, the Bureau made no adjustment to the data to account for this difference.

No evaluation of measurements from different instruments

Temperature measurements from different instruments (for example, mercury thermometers versus electronic probes) and different site locations are variously combined by the Bureau and then evaluated for potential breakpoints through comparison with neighbouring weather stations. Specifically, if the difference between the data series being homogenised and neighbouring series changes over time, a potential breakpoint is identified. Adjustments are then made according to these breakpoints and differences, if supported by metadata (Jovanovic et al. 2012).

The equivalence of measurements from the different instruments used at each weather station is not evaluated. This could be done through an assessment of the parallel data. For example, both an electronic probe and mercury and alcohol thermometers were in place at Mawson for the period January 1994 to February 2002, as shown in Figure 8.4. Parallel data for this period should therefore exist, assuming the Bureau took daily measurements from both instruments.

Rather, measurements from the electronic probes (PRTs) from 1 November 1996 onwards were simply added to the measurements from mercury and alcohol thermometers before 1 November 1996. This change affected all three ACORN-SAT sites in Antarctica at the same time. Using Mawson and Davis to homogenise each other will, therefore, likely result in no detection of a potential breakpoint at this switch-over date. To know the true underlying long-term trend, it might be necessary to adjust individual records based on the equivalence between different instruments used at a given weather station at the same time. Such an analysis could only be undertaken if the parallel measurements were made available, which they have not been. Only one series is ever archived in the ADAM database.

Statistical analysis

Despite the potential limitations of the temperature series resulting from them being an amalgamation of measurements from different instruments, I nevertheless undertook a statistical analysis of each of the longer temperature series to test for a warming trend.

Long-term temperature trends are generally reported using the best linear fit (IPCC 2013), which is obtained through linear least-squares regression. The slope b of the regression line expresses the rate of temperature change over time, which can be viewed as an estimate of the true temperature change or slope β, an unknown value. Given a series of temperature observations over time, hypothesis testing concerning the slope can be used to infer whether there is significant warming or cooling over time (James 2006). A two-tailed hypothesis test can thus be formulated in terms of the true slope β as:

H_0: $\beta = 0$ (There isn't sufficient evidence of a trend)
H_1: $\beta \neq 0$ (There is a statistically significant trend)

The standard approach to perform hypothesis testing is to calculate a test statistic t and the associated p-value, expressing the probability that the statistic could have been obtained under the null hypothesis H_0. A smaller p-value therefore indicates stronger evidence that H_0 should be rejected. Statistics of the slope and the p-values are shown in Table 8.2.

Table 8.2 Temperature trends statistical significance test

Station	Temperature series	Period	Data points N	Slope b	Standard error of b (x10⁻³)	95% CI of b	p-value
Mawson	Max (ADAM)	1955–2018	64	0.287	4.643	−0.641, 1.215	0.538
(300001)	Max (ACORN)	1958–2018	61	0.613	4.946	−0.377, 1.603	0.220
	Min (ADAM)	1955–2018	64	−0.854	5.448	−1.943, 0.235	0.122
	Min (ACORN)	1958–2018	61	0.179	5.813	−0.984, 1.342	0.759
Davis	Max (ADAM)	1958–2018	55	1.281	6.860	−0.094, 2.657	0.067
(300000)	Max (ACORN)	1958–2018	55	0.609	6.768	−0.748, 1.966	0.372
	Min (ADAM)	1958–2018	55	−0.176	7.941	−1.768, 1.417	0.826
	Min (ACORN)	1958–2018	55	−0.176	7.941	−1.768, 1.417	0.826
Casey	Max (ADAM)	1970–2018	49	−1.489	7.580	−3.013, 0.036	0.055
(300006 and 300017)	Max (ACORN)	1970–2018	49	−1.478	7.577	−3.003, 0.046	0.057
	Min (ADAM)	1970–2018	49	−1.388	9.354	−3.270, 0.494	0.145
	Min (ACORN)	1970–2018	49	−1.387	9.352	−3.268, 0.495	0.145

Statistical values were calculated using linear regression (SciPy). The unit of the slope b is in °C per century. The statistical significance level was chosen as $\alpha = 0.05$, with associated 95% confidence interval.

Data was sourced from the Australian Bureau of Meteorology, March 2020, http://www.bom.gov.au/climate/data/ and http://www.bom.gov.au/climate/data/acorn-sat/.

The p-values for the actual measurements as archived in the ADAM dataset, and the homogenised ACORN-SAT temperature series for Mawson, Davis and Casey, are all greater than the chosen threshold level $\alpha = 0.05$. Therefore, the null hypothesis H_0 should not be rejected. In other words, there is no statistically significant warming or cooling trend in the three longest temperature series as measured by the Bureau in the Australian Antarctic Territory, whether using actual historical observations from the ADAM dataset or the homogenised ACORN-SAT values.

9 Deconstructing Two Thousand Years of Temperature Change at Antarctica

Dr John Abbot & Dr Jennifer Marohasy

During the 12th century, the English philosopher, diplomat and bishop, John of Salisbury (1120–1180), wrote that we see more and farther than our predecessors, not because we have keener vision or greater height, but because we are lifted-up and borne aloft on their knowledge.

In this chapter, we are not claiming to be experts on Antarctica, nor on historical temperature reconstructions, or even artificial intelligence (AI). Rather we have taken two published reconstructions of Antarctic temperatures that extend back 2000 years – one for East Antarctica and one for West Antarctica, both undertaken by recognised mainstream climate scientists – and then used an established technique to break them down into their component parts. After this, we input the components into off-the-shelf AI software and forecast what the temperatures of East and West Antarctica might have been for the period 1880 to 2000 AD if there had been no industrial revolution. The difference between the published temperature reconstruction and our forecast temperatures potentially gives an indication of the human-caused global warming component from 1880 to 2000 AD.

These temperature reconstructions of East and West Antarctica were undertaken by Barbara Stenni and seventeen of her colleagues. The scientists are from the Australian Antarctic Division, British Antarctic Survey in Cambridge, Arctic and Antarctic Research Institute in St Petersburg, and so the list goes on. Together, they conclude that there has been a general cooling trend for 1900 years, until 1900 AD, as shown in Figure 9.1.

Figure 9.1 Temperature reconstructions for East and West Antarctica

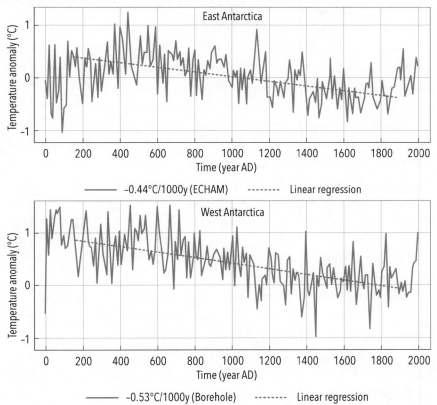

Anomalies calculated from 1981–2010 means. Composite temperature reconstructions (Temperature anomalies in °C) for East and West Antarctica using ten-year averages and the different temperature scaling approaches as detailed in Stenni et al. 2017. Linear trends are calculated between 165 and 1900 AD only.

Source: Stenni et al. 2017.

Stenni et al. (2017) also explain that the proxy temperature reconstruction for East Antarctica from 1900 shows warming to about 1920, then cooling to about 1960, and warming again since then. The pattern of temperature change over the 20th century is somewhat different for West Antarctica, which shows significant warming only since about 1960.

In order to determine what percentage of the recent warming in the East and West Antarctica reconstructions has likely been caused by human activity, we have attempted to separate out any warming that could be a consequence of natural climate cycles.

We have previously used this technique to evaluate the likely human contribution to 20th century warming for the Canadian Rockies, Switzerland, Tasmania, New Zealand and South America – but not Antarctica (Abbot & Marohasy 2017a). In this earlier study, published in the international climate science journal *GeoResJ*, we detailed how temperature series could be decomposed into sets of sine waves and then used as input into a statistical model to generate a temperature forecast based on natural climate cycles. We hypothesised that the difference between the forecast, and what was observed, would be the human-caused component of recent warming. The largest deviation was 0.2 °C, and from this we calculated an Equilibrium Climate Sensitivity of approximately 0.6 °C. This is the amount we calculated temperatures would rise given a doubling of levels of atmospheric carbon dioxide (CO_2) based on our AI-based analysis of the temperature record in the reconstructed proxies for the different regions. Our 2007 study in *GeoResJ* built on our experience using AI, specifically artificial neural networks (ANNs), to forecast rainfall (Abbot & Marohasy 2012; 2013; 2014; 2015a, b, c, d; 2016a, b; 2017b, c). Crossing over to temperature forecasting did not require a major change in our techniques, but it did prove to be far more controversial (Marohasy 2018).

The history of science and technology is replete with examples showing extraordinary political pressure applied to prevent competition from new technologies that could displace established power structures (Kuhn 1962; Acemoglu & Robinson 2012). It is interesting how quickly the theory of human-caused global warming has come to dominate climate science, despite its evident lack of practical utility, and in particular the absence of any measurable improvement in the skill of rainfall, and other climate-related, forecasts. This is despite an extraordinary investment in the tool used – general circulation models (GCMs). AI represents a different method, a different tool, that is currently being actively resisted.

AI represents a radically different approach to forecasting both rainfall and temperatures. AI is much more dependent on the integrity and quality of the statistics and/or data that is used, rather than theory. This method makes no assumptions about physical processes, for example the role of CO_2, and does not attempt to simulate atmospheric and oceanic processes. Rather our method is about mining historical data and/or reconstructions – in this case study the values used by Stenni et al. (2017) in their reconstruction of Antarctica's climate – to find patterns, and then build statistical models based on these patterns. In this study we were specifically interested in making a forecast for temperatures at East and West Antarctica had there never been an industrial revolution.

Proxy data and temperature reconstructions

Instrumental temperature series, for example using thermometers, do not extend back even a century for Antartica. Proxy temperature records for the Antarctic are much longer because they can be derived from ice cores (Petit et al. 1999; Masson et al. 2000; Stenni et al. 2017). Specifically, water samples from within the ice samples can be characterised by stable isotope ratios of oxygen and hydrogen. Both of these parameters within ice cores provide information on past atmospheric temperatures and can reveal oscillatory behaviour that may continue over millennial, centennial and decadal time scales (Münch & Laepple 2018; Yiou et al. 1997; EPICA Members 2004). The methodology assumes that water containing different isotopes will evaporate at different rates, depending on the temperature.

Stenni et al. (2017) began with water stable isotope data from Antarctic ice cores from 112 isotopic records and reported temperature reconstructions for the last 2000 years at five- and ten-year resolutions, for seven climatically distinct regions at Antarctica. They also produced composites and reconstructions for broader regions including East Antarctica and West Antarctica at ten-year intervals, as illustrated by the temperature series shown in Figure 9.1.

It is important to understand that these reconstructions did not simply involve the adding together of the 112 isotopic records (the proxy data) from different places for the ten-year intervals. The reconstruction had to be much more elaborate than this because only nine of the 112 isotopic records actually extended back to the beginning, back all

2000 years. So, weighting systems were applied after a complex system of calibration, glaciological issues were resolved through screening, and after all of this the isotopic records (the proxy data) was run through a general circulation model, specifically ECHAM5-wiso. The subsequent simulation was used as the reconstruction for East Antarctica, as shown in Figure 9.1. In the case of West Antarctica, where there is also borehole temperature data, output from the simulation was further scaled using this temperature measurement. Borehole data are direct measurements of temperature from holes drilled into the Earth's crust.

Periodic oscillations and their component parts

Measurements of variables associated with weather and climate, when arranged chronologically, tend to show patterns of recurring oscillations. This is also the case with the Stenni et al. (2017) reconstructions for East and West Antarctica, as shown in Figure 9.1. The oscillations may not be symmetrical, but they generally channel between an upper and lower boundary. In the case of East Antarctica, the decadal-scale oscillations are generally less than 1 °C. The channel moves up, peaking just after 400 AD, and then trends down to 1900 (see Figure 9.1). Such oscillations – at the decadal, centennial and millennial scales – can be decomposed into sets of sine waves of varying phase, amplitude and periodicity.

The oscillations may, or may not, represent real-world phenomena, for example changes in the Sun's magnetism, or the varying eccentricity of the Earth's orbit around the Sun – to mention just two. At one of the shortest intervals, there are cycles of daily temperature change because the night follows day. This is a consequence of the Earth being a sphere that rotates once relative to the Sun each 24-hour period. This in turn creates diurnal changes such that temperatures are generally coolest just before dawn and warmest in the early afternoon of each day. Considering the sine wave in Figure 9.2, dawn would represent a trough in temperatures, while early afternoon, the crest. This daily cycle repeats indefinitely, but with variations.

Considering just amplitude (measured as the peak deviation of the function from zero/the anomaly/how hot it gets each day), this will change through the year – especially at higher latitudes, because the amount of solar radiation reaching higher latitudes changes with the season because

Figure 9.2 Components of a sine wave

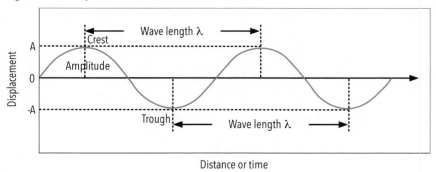

Distance or time

the Earth is tilted relative to its axis, creating a changing declination of the Sun's rays. (A common misconception is that the Earth is closer to the Sun in summer. In reality, the Earth is closest to the Sun in January which is winter in the Northern Hemisphere. Earth is closest to the Sun in January because of the trajectory of the elliptical revolution.)

When particular phases of different cycles align (for example, daily, annual and orbital) more extremes of temperature will be experienced. For example, all other things being equal (for example, latitude, altitude), it is likely to be hotter at midday in summer in the Southern Hemisphere, than at midday in winter in the Northern Hemisphere.

The general concept is illustrated in Figure 9.3, showing two sine waves with identical wavelength and amplitude, but with the two waves having a phase displacement. But it is when there is alignment, not displacement, that there is more likely to be an extreme climate event.

It is possible to combine two or more sine waves together, taking account of amplitude, wavelength and phase, to produce a single resultant oscillatory wave. This is illustrated in Figure 9.4 showing the resultant wave after the addition of Wave 1, and Wave 2. The resultant wave exhibits greater total displacement in regions where peaks or troughs of the individual sine waves overlap, and lower total displacement from the axis in regions where the directions of displacement of the individual sine waves are opposite.

It is possible to generate complex oscillatory patterns by the addition of a set of sine waves with different values of amplitude, wavelength

Figure 9.3 Phase displacement between two sine waves

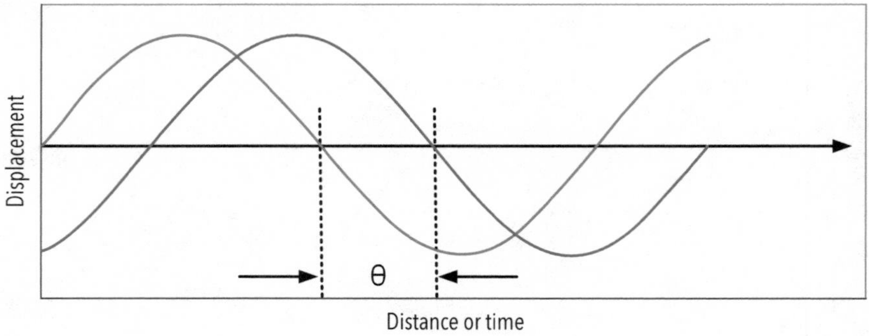

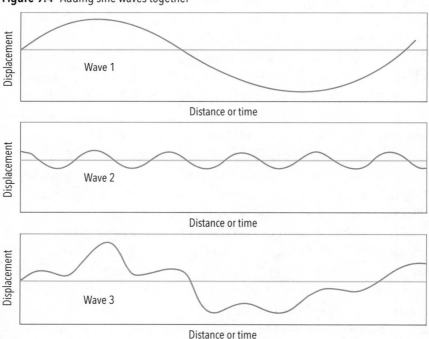

The two waves have a phase displacement of θ.

Figure 9.4 Adding sine waves together

and phase. This is illustrated with the set of six sine waves shown in Figure 9.5. The addition of this set of sine waves generates the resultant complex oscillatory pattern shown in Figure 9.6.

It is also possible to start with a complex oscillatory pattern that may be derived from the observations of natural phenomena – such as atmospheric temperature – and decompose the overall profile into a set of component sine waves. Recombining the set of component sine waves will then produce an approximate representation of the original pattern. This is the reverse of the process illustrated below in Figures 9.5 and 9.6.

The oscillations shown in Figure 9.1, representing Stenni et al.'s (2017) East and West Antarctica proxy temperature reconstructions, can be decomposed into a set of sine waves. Using signal analysis software (AutoSignal), we decomposed the original complex oscillatory patterns

Figure 9.5 Set of six sine waves with different amplitude, wavelength and phase

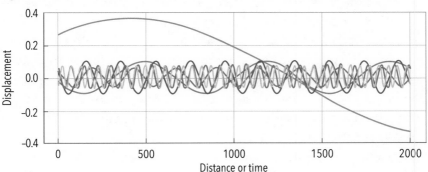

Figure 9.6 Complex oscillatory pattern generated by addition of six sine waves

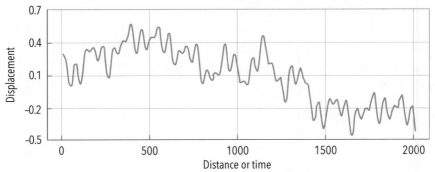

in Figure 9.1 to generate datasets theoretically based on their component parts that could be subsequently used as input data for our ANN.

Finding cycles in Antarctic proxy data

In our Antarctic study, the two temperature-proxy reconstructions from Stenni et al. (2017) were first digitised using UN-SCAN-IT software. The digitised time-series were then examined by spectral analysis using AutoSignal software. The results of this spectral analysis are shown in Table 9.1. In each case, the number of sine curves applied to reconstruct the total signal was increased until the improvement in fitting, estimated by correlation coefficient, showed only marginal improvement. This explains why there are four sine waves reported for East Antarctica and seven for West Antarctica.

In each case, the optimisations were undertaken from the proxy record start date through to 1830 AD. This can be regarded as the pre-industrial period, when there were only natural influences on climate (without anthropogenic/human-caused contributions). In each case, this data based on the sinusoidal analysis was used as the input data for subsequent machine learning using ANNs. In particular, the data was provided as input to Neurosolutions Infinity software. Application of the ANNs generated a superior fitting to the original temperature profile, than simply using a recombination of the component sine waves (see Abbot & Marohasy 2017a).

It is important to understand that although these complex oscillatory patterns can mathematically be approximately represented as the combination of a set of sine waves, this does not mean that the underlying natural phenomena are themselves individual sine waves. The use of combinations of sine waves is a mathematical technique that allows complex oscillatory

Table 9.1 Significant periodicities found from spectral analysis of temperature reconstructions

Region	Significant Periodicities (years)
East Antarctica	2367, 232, 195, 93
West Antarctica	137, 451, 291, 155, 99, 73, 41

patterns to be represented, and potentially enables projections to be made into the future by assuming the underlying cyclical processes continue, without knowing the exact underlying natural cycles.

Considering the peer-reviewed scientific literature we noted that Zhao and Feng (2014) examined the relationship between solar sunspot number and the local temperature in Vostok in East Antarctica. They found that the variations of solar sunspot number and temperature have some common periodicities, such as the 208-, 521-, and 1000-year cycles. An investigation of periodicity in Antarctic ice-core records (Mason et al. 2000) showed oscillations over a millennial time scale during the Holocene (the past 12,000 years). These records show several common intervals of significant periodicities in the multi-decadal to centennial mode, as discussed by Yiou et al. (1997) between 70 and 240 years. Lüdecke et al. (2013) examined Antarctic ice-core data reported by Petit et al. (1999) and found significant periodic cycles at 188, 499 and 948 years.

Davis et al. (2018) reported on a study of East Antarctic temperature-proxy records from ice cores extracted from four drill stations on the East Antarctic Peninsula and described a previously unexplored natural temperature cycle recorded in ice cores that has oscillated for at least the last 226 millennia. They named this natural cycle the Antarctic Centennial Oscillation (ACO). Focusing on the Holocene, Davis et al. (2018) found that centennial-scale spectral peaks from temperature-proxy records at Vostok over the last 10,000 years occur at similar frequencies in three other paleoclimate records from the East Antarctic Plateau, with six periodicities of the Vostok ice-core records in the range between 193 and 1096 years.

Davis et al. (2018) concluded that centennial-scale paleoclimate cycles comprise a 'natural' source of temperature forcing that is free from anthropogenic influences. Furthermore, characterising past cycles of temperature fluctuation can elucidate the distinction between natural (non-anthropogenic) and anthropogenic forcing of climate in the present. Properties of the ACO are potentially capable of explaining the current global warming signal and that a 'natural global warming' (NGW) hypothesis may be a viable alternative to human-caused global warming.

Forecasting using cycles and neural networks

So, the existence of periodic oscillations in temperature records provides a method for generating temperature forecasts. To reiterate, this assumes that a combination of sufficient numbers of identifiable periodic cycles can provide an adequate representation of the overall temperature profile, and that the underlying cycles generated from the analysis of past behaviour continue into the future.

Using the Antarctic proxy temperatures reconstructions, we could have constructed a composite signal from the sets of component sine waves through simple addition of the sinusoidal components as described in Abbot and Marohasy (2017a). This composite signal could have been used to make the temperature forecasts. However, better forecasts are obtained when there is more skilful fitting of the historical temperature reconstruction based on the proxies to the sine wave components, and these in turn used as input data, as detailed in Abbot and Marohasy (2017a). We thus used this method to generate temperature forecasts for East and West Antarctica.

The pink lines in Figures 9.7 and 9.8 show the outputs generated from our Artificial Neural Networks (ANNs). For this analysis, the input training data corresponded to the periods between 0 and 1830 AD, matching the pre-industrial era when it can be assumed that anthropogenic influence on climate would be minimal. For the period between

Figure 9.7 Output based on spectral analysis for East Antarctica

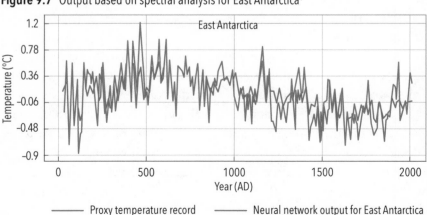

Figure 9.8 Output based on spectral analysis for West Antarctica

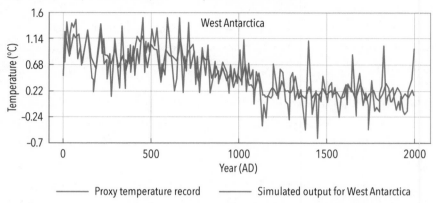

Proxy temperature record —— Simulated output for West Antarctica

1880 and 2000 AD, the pink lines were generated as forecast simulations based on projecting the sine waves as input beyond 1880, and again generating an output from the ANN based on the relationships established for the prior training period between 0 and 1830 AD. In this way, the pink lines generated between 1880 and 2000 AD theoretically represent the temperature profiles that could be attributed to cyclical natural causes.

Results

Our neural network simulations were able to follow the general patterns of the proxy temperature profiles. The results are shown in Figures 9.7 and 9.8. The simulation was more skilful for East Antarctica. The peaks and troughs in the proxy temperature reconstructions for West Antarctica were not well replicated by our simulation (our neural network output) as shown by the pink line in Figure 9.8. That the peaks and troughs throughout the proxy temperature reconstruction (turquoise line), but especially after 1000 AD, were not replicated by our West Antarctic simulation (pink line) suggests that there are intermittent natural events that are non-cyclical and consequently not replicated by the analytical methods used, or there could be a problem with the actual input statistics (with the Stenni et al. proxy reconstruction).

Intermittent non-cyclical effects could include, for example, volcanic activity (Antoniades et al. 2018) and complex cycles that link phenomena in other parts of the globe as far away as the tropical Pacific Ocean. The mismatch is much more pronounced than anything found for other regions (South America, Tasmania, New Zealand, etc.) that were examined using the same analytical methods in our previous study (Abbot & Marohasy 2017a).

Distinguishing human-caused versus natural warming

For both East and West Antarctica the projected neural network outputs are relatively flat after 1880, as shown in Figures 9.9 and 9.10 (pink lines). The corresponding proxy temperature reconstruction (turquoise lines) show temperature increase. This increase is not continuous through the industrial era as might be expected if there were a simple explanation associated with a steady increase in the concentration of atmospheric greenhouse gases, resulting in a proportional global warming effect. It is difficult to explain the warming in East Antarctica during the period from 1910 to 1930, see Figure 9.9. (Most of the volcanic activity is in West Antarctica.) Also, it is unlikely that an anthropogenic effect associated with greenhouse gases would produce an effect that is double the impact on West Antarctica compared to East Antarctica, which is apparent from the Stenni et al. (2017) results.

The displacements of proxy temperatures (turquoise line) in the period 1950–2000 relative to the projected simulation (pink line), as shown in Figures 9.9 and 9.10, are significantly higher than previously

Figure 9.9 Projection in absence of an industrial revolution, East Antarctica

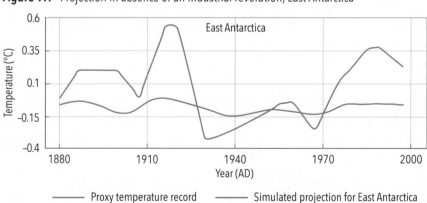

Figure 9.10 Projection in absence of an industrial revolution, West Antarctica

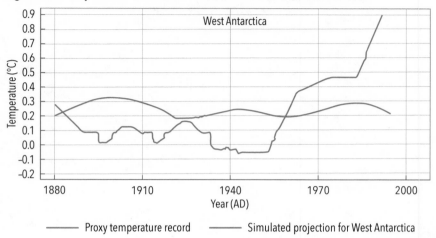

— Proxy temperature record   — Simulated projection for West Antarctica

found for other regions (Abbot & Marohasy 2017a). In other parts of the world, we found displacements in the range 0.1 °C to 0.2 °C. Comparison of the turquoise lines in Figures 9.9 and 9.10, shows that atmospheric warming in East Antarctica since 1950 has been approximately 0.5 °C. It is even larger in West Antarctica at approximately 1.0 °C. This would suggest a very big effect from greenhouse gases from anthropogenic global warming.

It is the case, however, that actual temperatures as measured at West and East Antarctica using satellites, and also thermometers, do not show any such dramatic increase in temperatures – at least over recent decades, as shown and discussed in Chapters 6, 7 and 8.

Interestingly our neural network output has generated a flat line consistent with the observational data discussed in Chapters 6, 7 and 8, but is not consistent with popular theory or the Stenni et al. proxy temperature reconstruction. Perhaps there is an issue with the proxy temperature values used by Stenni et al. (2007) that only begins after 1830? These more recent statistics (1830 to 2000) were never inputted into our neural network model.

SECTION II

TOWARDS A NEW
THEORY OF CLIMATE

10 Cosmoclimatology
Dr Henrik Svensmark

Climate changes of the past were so large that they affected the conditions for life on Earth. It may seem crazy, but who would imagine that these changes are influenced by energetic atomic particles that rain down on us from distant reaches of the Milky Way? After all, the total energy represented by the particle rain is as feeble as the starlight on a clear night. British science writer Nigel Calder originally coined the name for this new emerging field – *cosmoclimatology*.

The story began in 1996 with the unexpected discovery that variations of global cloud cover correlates with the intensity of galactic cosmic rays penetrating the Earth's atmosphere to cloud altitudes (Svensmark & Friis-Christensen 1997). Since clouds are very important for the radiative energy balance of the Earth, any systematic change in cloud cover will affect the temperature of the atmosphere and, ultimately, the climate. The scientific challenge of the discovery was to consolidate the result and to understand how particles from space are able to affect clouds. Now, after more than twenty years of research, we have proposed a consistent theory, with implications we are only beginning to comprehend.

The cosmic ray – cloud hypothesis
Cosmic rays originate from stars that have exploded far away in the galaxy, and consist mainly of protons and a smaller fraction of heavier atomic nuclei. Before colliding with the Earth's atmosphere, the particles

have to penetrate the huge magnetic bubble known as the heliosphere, which is created by the Sun and carried beyond the most distant planets by the solar wind. Cosmic rays are electrically charged and are affected by magnetic fields, which tend to screen them out. When solar activity increases, so does the magnetic field of the heliosphere, which decreases the intensity of cosmic rays reaching the Earth. The discovery of the relations between cloud cover and cosmic rays then suggests a correlation between the variations in solar activity and changes in the climate.

In fact, there is plenty of evidence that links solar activity to Earth's climate. In the 1970s, John Eddy noticed a relation between solar activity and the European climate over the previous millennium (Eddy 1976). For example, the Little Ice Age (1300–1850 AD) was a cold period that took place while the Sun was particularly inactive. The Medieval Warm Period (1000–1250 AD), on the other hand, occurred while the Sun was active. Another example is the beautiful multi-millennial agreement between the temperature of the Indian Ocean as mirrored in the ratio between different oxygen isotopes of stalagmites in a cave in Oman, and solar activity, estimated from the concentration of the cosmogonic carbon-14 isotope produced by cosmic ray variations (Neff et al. 2001). Based on these and many additional studies, it is safe to conclude that over the Holocene period (the last 10,000 years) solar activity has had a significant impact on the climate.

One study discussed the strength of the link between climate and solar activity changes by using the oceans as a calorimeter over the (approximately) eleven-year solar cycle. They estimated the forcing induced by solar variations to be 1.0–1.5 W/m^2, which is nearly ten times larger than the forcing expected from direct solar irradiance variations. This forcing cannot be explained without an amplification mechanism, of which cloud cover modulations is a possible candidate (Shaviv, 2008).

The ideas soon focused on the formation of aerosols and on the growth of cloud condensation nuclei, somehow affected by the ionisation from cosmic rays. Aerosols are minute particles of liquid or solid suspended in the atmosphere (1–1000 nm). Aerosol concentrations typically vary from approximately 100 cm^{-3} in maritime air to approximately 1000 cm^{-3} in unpolluted air over landmasses. When aerosols are larger than about 50 nm they can serve as cloud condensation nuclei, which provide the surface that is needed for water vapour to condense. In pristine areas of the globe, a large fraction of aerosols are typically formed from sulphuric acid by photochemistry in the atmosphere. A change in the number of cloud condensation nuclei will change both the radiative properties and the lifespan of clouds, in turn influencing the radiative energy budget of the Earth.

The cosmic ray – cloud hypothesis is summarised in Figure 10.1. First, solar variability manifests itself as changes in the magnetised solar wind. Then the solar wind modulates the cosmic-ray flux, which ionises the atmosphere producing positive and negative ions. The ions help to form and stabilise new seed aerosols from trace gases in the atmosphere, and these grow into cloud condensation nuclei that assist the formation of clouds.

Figure 10.1 The cosmic ray – cloud hypothesis

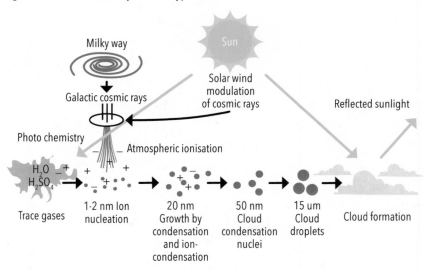

nm = nanometre. A nanometre is one millionth of a millimetre.
The physical mechanism linking solar activity variations to climate change. In summary, the link is:
a) a more active Sun, b) stronger solar wind, c) fewer cosmic rays, d) less atmospheric ionisation, e) less ion-nucleation of aerosols and slower growth, f) fewer clouds condensation nuclei, g) clouds with fewer droplets and shorter lifespans, h) less reflectivity, i) less reflection of sunlight and a warmer Earth.

Source: Adapted from Svensmark 2019.

Evidence for the cosmic ray – cloud link

In 2004 it became possible to experimentally test these ideas by using a relatively large 8 m^3 reaction chamber. The chamber was filled with air resembling the air over the tropics, with fixed trace gas concentrations of water vapour, ozone and sulphur dioxide. In addition, it was possible to mimic the effect of solar radiation by illuminating the trace gases with ultraviolet radiation, resulting in a photo-chemical reaction leading to the production of sulphuric acid in the gas phase. Sulphuric molecules can be seen as a form of 'super glue', which helps molecules stick together to form small aerosols. Finally, and most importantly, the reaction chamber could be radiated with gamma rays of varying intensity by producing positive and negative ions in the chamber, which have the same effect as cosmic-ray particles.

Initially it was tested whether the presence of ions would help the formation and stabilisation of small molecular aerosol clusters (greater than 3 nanometers). Elaborate experiments unambiguously showed that ions help promote the formation of small aerosols. Therefore, it was clear that the first part of the hypothesis was correct (Svensmark et al. 2007). However, the immediate reactions to these results from a number of scientists was that although we see small molecular aerosol particles formed in the experiment, they would be unimportant in the real atmosphere as there are sufficient numbers of cloud condensation nuclei already. The opinion that ions are unimportant in cloud condensation nuclei formation found further support when the output from computer models showed a failure of growth of the small ion-nucleated aerosols to cloud condensation nuclei (e.g. Pierce & Adams 2009).

The argument for the lack of response to ions is this: by increasing the ionisation, additional small aerosols are formed, but then there are more aerosols competing for the same material from which to grow and they, therefore, grow more slowly. This means that the probability of a small aerosol being lost to a larger aerosol particle increases and fewer of the small aerosols survive to reach cloud condensation nuclei sizes. If this is true it would mean the end of the cosmic ray – cloud idea.

We had good reasons to believe that this pessimistic view was wrong. First, the prediction could be tested experimentally. The idea was to let small aerosols grow to the sizes required for cloud condensation nuclei in our experimental chamber and see what the effect would be by changing the ionisation in the chamber. A number of experiments simulated what would happen in the atmosphere without the presence of cosmic rays. Small neutral aerosols of 6–7 nm were injected continuously into the experimental volume and their growth towards cloud condensation nuclei sizes was observed. In Figure 10.2, the top panel shows how the small molecular clusters fail to grow sufficiently to change the number

Figure 10.2 Experimental tests of aerosol growth into cloud condensation nuclei

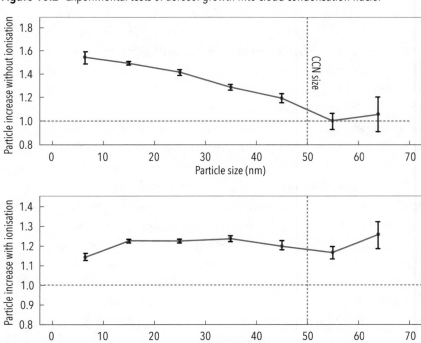

Top panel: without ionisation. The added small aerosols gets lost as the aerosol particles grow to cloud condensation nuclei sizes; Bottom panel: with ionisation. In this case all the small aerosols grow to cloud condensation nuclei sizes.

Source: Adapted from Svensmark 2012.

of cloud condensation nuclei (larger than 50 nm). The small particles were lost to larger particles and to the walls of the experimental chamber before they reached the cloud condensation nuclei size of 50 nm. Therefore, these experiments confirmed the negative results of the numerical simulations. But in a second series of experiments, when the air in the chamber is exposed to ionising radiation simulating the effect of cosmic rays, all additional formed small aerosols (3 nm) grew to cloud condensation nuclei sizes, contradicting the numerical simulations (see Figure 10.2, bottom panel). This is direct evidence that the prevailing theory is insufficient to explain the growth, and an unknown mechanism operates by which ions help the growth of aerosols.

Even more remarkable is the fact that one can observe the whole chain – from solar activity to cosmic rays to aerosols to clouds – in the real atmosphere. It starts on the surface of the Sun where on rare occasions 'explosions', called coronal mass ejections, result in a plasma cloud passing the Earth, with the effect that within a day the flux of cosmic rays decreases suddenly and stays low for a week or two. Such events, with a significant reduction in the flux of cosmic rays, are called Forbush decreases, and are ideal to test the link between cosmic rays and clouds. By finding the strongest Forbush decreases and applying three independent cloud satellite datasets and one dataset for aerosols, a clear response to Forbush decreases can be seen (as shown in Figure 10.3). There is a delay of four to seven days between cosmic rays and the aerosol-cloud parameters. Such a delay is expected and is due to the time it takes for small aerosols to grow to cloud condensation nuclei size in the atmosphere (Svensmark et al. 2009; Svensmark et al. 2016).

So the evidence is clear; ions assist in the formation of new aerosols and they also assist with their growth to cloud condensation nuclei sizes. But five years ago, the answer of 'why' the small aerosols could grow to cloud condensation nuclei sizes was still unsolved. Our initial working hypothesis was that ions produced extra sulphuric acid through ion chemistry, which increased the aerosol growth. But after two years of convincing theoretical results, but endless failed experimental verifications, we had to conclude that nature did not like this solution, and we had to start again. Only recently did we find the

Figure 10.3 Forbush decreases effect cloud formation

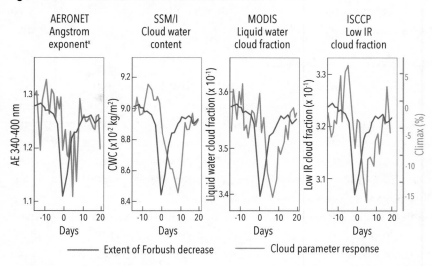

The daily averages of aerosol and cloud parameters preceding and during Forbush decreases in the flux of cosmic rays from the Climax neutron monitor station (purple curve), averaged over the five strongest events between 1987 and 2005. From left to right: Angstrom exponent from the Aerosols Robotic Network (AERONET) ground observational data (indicating the density of aerosols in the atmosphere); cloud water content over the oceans (data from the Special Sensor Microwave/Imager (SSM/I)); liquid water cloud fraction (data from Moderate Resolution Imaging Spectroradiometer (MODIS)); and low IR-detected clouds (data from International Satellite Cloud Climatology Project (ISCCP)). The cloud data (blue curves) are from satellite observations.

Source: Svensmark 2009.

correct solution. The explanation is a so-far ignored contribution to the growth of aerosols that comes from the mass of the ions. Although ions are relatively scarce in the atmosphere, they exert electromagnetic interactions with the aerosols that compensate for their scarcity and make fusion between ions and aerosols much more likely. This means that at typical concentrations of sulphuric acid in the atmosphere, about 5% of the growth rate of aerosols is due to ions. In the case of a near supernova, ionisation can be much greater, and the ion effect can be responsible for more than 50% of the growth rate. This can have a profound impact on clouds, and the Earth's temperature and climate (Svensmark et al. 2017).

In summary, ions assist the formation of new small aerosols, but ions are also important in the growth process and help aerosols to survive to cloud condensation nuclei sizes, at which size they can influence clouds.

A link to the Milky Way

What has been presented so far involves cosmic rays modulated by solar activity. But over many millions of years there can be much larger variations in the cosmic-ray flux entering the Earth's atmosphere. In the course of time the solar system travels, as do all the other disk stars, around the Galactic Centre of our Milky Way on a journey that takes about 240 million years. This journey brings the solar system in and out of regions with high and low star formation. Regions of high star formation are also regions where stars of masses more than eight times the mass of our Sun can form. Such massive stars end their lives after a relatively short time in a large explosion as supernovae. Since the supernova shock fronts are the main accelerators of cosmic rays, temporal variation in the number of supernovae in the solar neighbourhood translates into variations in the cosmic-ray flux on million-year time scales. Astrophysicist Shaviv (2003) linked icy episodes on the Earth during the Phanerozoic Eon (the last 542 million years) to the solar system's encounters with the spiral arms of the Milky Way as it orbited around the Galactic Centre. Since spiral arms are regions of increased star formation and, therefore, also the heavy stars that become supernovae, they are regions of elevated Galactic cosmic rays and provide an independent test of the cosmic ray – cloud link. So far, this connection looks quite promising with evidence accumulating over the last two decades.

A recent study used open star clusters in the solar neighbourhood to construct supernova history during the last 500 million years (Svensmark 2012). Open stellar clusters are a loose, irregular grouping of stars held together by gravity that originated from a single molecular cloud. Open star clusters typically contain from 100 to several thousand stars and are confined to the disk of the galaxy. The Pleiades is the best known open cluster. Open star clusters in our stellar neighbourhood, born many millions of years ago, clump together at times of high star formation rates, corresponding with our planet's encounters with the bright star forming regions of the Milky Way. These are the scenes of intense star formation, including massive stars that explode as super-novae and shower our planet with energetic Galactic cosmic rays. Based on the cosmic ray – cloud link one would expect that at such times the Earth would be cold, because cosmic rays promote cloud forma-tion that reduces sunshine at the surface. In darker regions of the Milky Way, between the star-forming regions, the Earth would be warm – far hotter than today. Figure 10.4 shows the supernova rate reconstructed from an open star cluster in the solar neighbourhood during the last 500 million years (red curve). The variations in supernovae rate over this period correlates with the variations in climate. The coloured band on top indicates climatic periods: warm periods (red), cold periods (blue), and glacial periods (grey). Note the correspondence between high super-novae activity and a cold or glacial climate (for more information, see Svensmark 2012).

Figure 10.4 Marine biodiversity correlates with supernova history

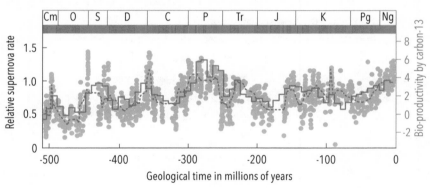

The red curve is the variation in the local supernova rate and, therefore, also the variation in cosmic-ray flux during the last 500 million years. The coloured band indicates climatic periods: warm periods (red), cold periods (blue), glacial periods (grey bars). The proportions of carbon-13 in sediments ($d^{13}C$ in parts per million) over the past 500 millon years, shown in the scattered points (blue dots), reflect changes in the carbon cycle. $d^{13}C$ carries information on the burial of organic material in sediments, and is therefore also a record of bio-productivity. The green-dashed curve is the smoothed $d^{13}C$. Abbreviations for geological periods are Cm – Cambrian; O – Ordovician; S – Silurian; D – Devonian; C – Carboniferous; P – Permian; Tr – Triassic; J – Jurassic; K – Cretaceous; Pg – Palaeogene; Ng – Neogene.

Source: Adapted from Svensmark 2012.

Now to the most remarkable implication of cosmic rays and climate: it is not just climate that is affected. Life on Earth responds to climatic variations following cosmic ray variations. Figure 10.4 compares the cosmic-ray history (red curve) with a carbon-13 isotope history obtained from marine sediments (grey symbols, blue-dashed curve is the smoothed carbon-13). Carbon-13 follows changes in the carbon cycle and carries information on the burial of organic material in sediments, and of bio-productivity, and so it also reflects the condition of life on Earth, overall. Despite the large variability in carbon-13 in each time interval, the influence of the supernova rate plainly overrides all the complex practical and theoretical reasons why such coherence with carbon-13 might not be expected. The effects are counter-intuitive, with the most generous bio-productivity occurring in the cold intervals associated with the star-forming regions. So the high bio-productivity and subsequent burial of organic material is likely connected to the more vigorous mixing and recycling of nutrients during cold times. A cold climate means that the temperature difference between the equator and the poles is large, which strengthens wind and ocean currents.

Life's connection to the supernova rate does not stop with the bio-productivity. The long-term diversity of life in the sea is also found to depend on the sea level set by plate tectonics and the local supernova rate set by astrophysics, and on virtually nothing else. This is a surprise, since countless factors have been suggested to control long-term diversity. Sea level is important, because a higher sea level will result in flooding of low inland areas, increasing the total length of coastline and the area of the shallow seas. This results in more disconnected habitats in which species can evolve, leading to an increase in diversity.

Figure 10.5 (top panel) shows the variation, at genera level, of marine invertebrate diversity during the last 440 million years (Alroy et al. 2008). The middle panel shows the corresponding sea level over that time. The rough correspondence between diversity and sea level seen in Figure 10.5 is plainly not the whole story. Here the novel assumption is that variations in the supernovae (SN) rates will also change conditions for the evolution of life by changing the climate and thereby altering, for example, the circulation of nutrients and the variety of habitats between the equator

Figure 10.5 Supernovae history and marine biodiversity

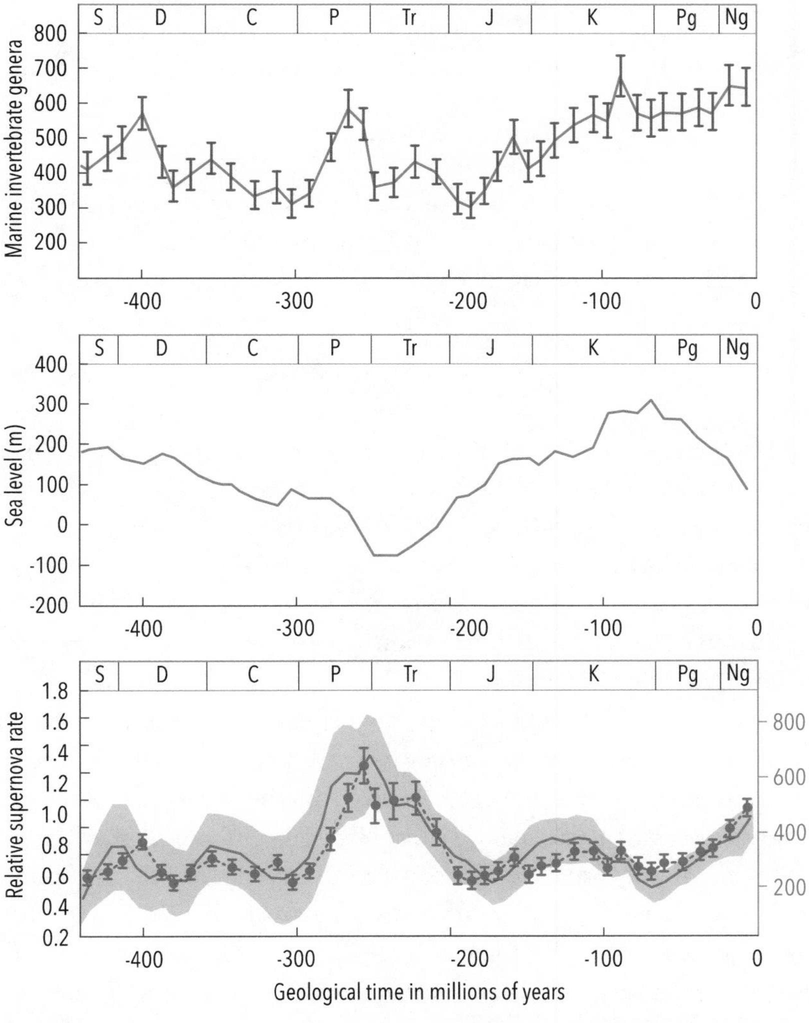

Supernovae history and marine invertebrate genera. Upper panel: genus-level diversity of extant and extinct marine invertebrates as an index of variable biodiversity (data from, Alroy et al. 2008, based on a sampling-standardised analysis of the paleobiology database). Middle panel: first-order sea-level variations attributable to tectonics (adapted from Haq & Shutter 2008). Bottom panel show supernovae rates smoothed by 20 million years (red curve). The grey area is the one standard deviation (1σ) variance of the supernovae rates calculated from a Monte Carlo simulation. The blue curve shows marine genera normalised with the sea level. The error bars on the genera (blue curve) show a minimum of one standard deviation of uncertainty since an error estimate is not available for the sea-level data. Evidently, marine biodiversity is largely explained by a combination of sea level and astrophysical activity.

Source: Adapted from Svensmark 2012.

and the poles. The idea is that if both sea level and the changes in climate caused by supernovae are important for the temporal evolution of marine invertebrate diversity, one may assume a simple relation between the three:

$$N(t) = \Gamma\ (SN(t))\ \Lambda(h_{Sea}(t)),$$

$\Gamma\ (SN(t))$ is a linear function of the supernova rate, where the $SN(t)$ shown in Figure 10.4 (red curve) $\Lambda(h_{Sea}(t))$ is a linear function of the sea level, and $h_{Sea}(t)$ is shown in Figure 10.5 (middle panel). Dividing the above equation $(\Lambda(h_{Sea}(t))$ we obtain:

$$N(t)/\Lambda(h_{Sea}(t)) = \Gamma\ (SN(t)).$$

The interesting point of this equation is that on the left-hand side are only quantities related to the Earth, whereas on the right-hand side are only astrophysical quantities. $\Gamma(SN(t))$ is therefore taken as a 30 million year average of the supernova rates to account for average recovery time in diversity after mass extinctions (for details see, Svensmark 2012). Performing the above normalisation of the genus-level diversity, the normalised diversity shows a remarkable agreement to variable super-novae rates, as implicated by the above equation. Figure 10.5 (bottom panel) as an index of variable biodiversity shows how, during the past 440 million years, the changing rates of supernova explosions relatively close to the Earth (red curve) have strongly influenced the biodiver-sity of marine invertebrate animals (blue curve), from the trilobites of ancient times to the lobsters of today. The astonishing thing to note is that after 'correcting' the genus-level diversity for the sea-level varia-tion, the remaining variation closely resembles the supernovae variation (Svensmark 2012).

Some geoscientists want to blame the drastic alternations between hot and icy conditions during the past 500 million years on increases and decreases in the carbon dioxide (CO_2) concentration in the atmosphere. But, as shown in Svensmark (2012), the changes in the concentration of CO_2 during the Phanerozoic varies in accordance with the supernova rates. This suggest that climate and life are controlling past CO_2 concentrations. When the climate is cold, nutrients are provided through river runoff and circulated in the oceans, making life plentiful, and a larger fraction of CO_2 gets drawn into the oceans and is stored in life. In contrast, during warm climates there is less circulation both in the atmosphere and oceans, so nutrients are not circulated as readily and less CO_2 is stored in the oceans and in life. In short, climate and life control CO_2, not the other way around, and the varying supernova rate controls the climate.

Conclusion

Progress has been made in understanding how the Sun influences Earth's climate. Galactic cosmic rays – atomic particles coming from supernova remnants left by exploded stars – play a major part. By ionising the air, cosmic rays help to form aerosols that may grow into the cloud condensation nuclei that are required for water droplets to condense and create low-altitude clouds. As these exert a strong cooling effect, increases or decreases in the cosmic-ray flux, and, thereby, in cloudiness can significantly lower or raise the Earth's temperature. Details of the mechanism have been identified, both theoretically and experimentally, and show that ions help the formation of new small aerosols, but also accelerate the growth to cloud condensation nuclei. The potential importance of the mechanism can be seen in, for example, the frequent changes between states of low and high solar activity during the last 10,000 years, clearly seen in empirical climate records. Of these climate changes, the best known are the Medieval Warm Period (950–1250 AD) and the Little Ice Age (1300–1850 AD), which are associated with a high and low state of solar activity, respectively. The temperature change between the two periods is of the order of 1.0–1.5 °C. The mechanism also manifests itself on time scales of days during Forbush decreases. These are sudden decreases in cosmic-ray flux caused by solar eruptions of plasma clouds, which shields temporally against cosmic rays. This makes it possible to observe the response in each stage in the chain of the theory – from solar activity, to changes in cosmic-ray flux, to ionisation changes, to aerosols, and finally to cloud changes. The impact of cosmic rays on the radiative budget over the approximate eleven-year solar cycle is found to be an order of magnitude larger than the total solar irradiance changes.

Additional support for a cosmic ray – climate connection is lent by the remarkable agreement that is seen on timescales of millions and even billions of years, during which the cosmic-ray flux is governed by changes in the stellar environment of the solar system; in other words, it is independent of solar activity. As for palaeo-biology, remarkable connections to the long-term history of life and the carbon cycle have shown up, unbidden. Biodiversity and bio-productivity ($d^{13}C$) all appear so highly sensitive to supernova in our galactic neighbourhood that the biosphere seems to contain a reflection of the Milky Way.

This leads to the conclusion that a microphysical mechanism involving cosmic rays and clouds is operating in the Earth's atmosphere, and that this mechanism has the potential to explain a significant part of the observed climate variability over the history of the Earth.

11 The Important Role of Water and Water Vapour

Dr Geoffrey Duffy

Water is unique. It is the only fluid in the world that floats on itself when it freezes. It is the only fluid in the atmosphere that changes phase. Water evaporates and then humidifies the entire atmosphere. As the warm moist air rises it can condense to form mists, fogs, or clouds. Clouds cover 65% to 70% of the planet at any one time. They can act as umbrella-shades or thermal shields in daylight hours. They can reflect some incoming solar radiation back out into space, but also can re-radiate some longwave electromagnetic energy emitted from Earth, back to Earth, as shown in Figure 11.1. Clouds have different forms, sizes, shapes and structures, and the higher clouds are formed when micro-crystals of ice are formed at sub-cooled temperature.

Clouds can grow, disperse or amalgamate depending on the local atmospheric moisture conditions and temperature. They then can precipitate as rain, sleet, snow, or hail, and cool the atmosphere and then the planet's surface. In addition, the precipitation can scrub the air of dust, and even dissolve some gases like carbon dioxide (CO_2) to form weak carbonic acid (H_2CO_3).

In addition, large water vapour concentration changes can occur locally with small temperature variations. There is a unique simultaneous transfer of heat energy and mass (water quantity) while other atmospheric gases retain an almost constant concentration within that particular domain. These concentration changes of water vapour (humidity gradients) occur even in small volumes, making it almost impossible to model

Figure 11.1 The greenhouse effect

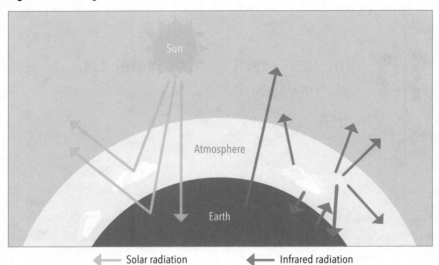

⟵ Solar radiation   ⟵ Infrared radiation

Some of the infrared radiation emitted from the Earth's surface passes through the Earth's atmosphere, while some is absorbed and re-emitted by greenhouse gases.

Source: Sandra Anastasi.

the atmosphere accurately on a larger scale. For example, a 1 °C increase in temperature over minutes can cause the water vapour concentration to change by an amount equal to the total increase in CO_2 over an entire century. As almost 71% of the world is ocean, it is no wonder why water and water vapour are dominant globally in affecting our weather. Because all other atmospheric gases are non-condensable, they perform none of these actions.

If all the above were not enough, water vapour has the highest concentration of all greenhouse gases. We will see it is by far the strongest absorber and re-radiator of both incoming solar energy as well as electromagnetic energy coming back up from Earth.

Varying humidity

Atmospheric humidity is defined as the mass of water vapour (gas) per mass or volume of dry air. The amount of water vapour depends both on temperature and pressure. At saturation (100 RH%), the pressure exerted by the water vapour is termed vapour pressure, and the temperature at that condition is termed the 'dew point'. When temperatures are below the dew point, water vapour concentrations can vary and change quickly and a partial pressure of water vapour contribution is made. Hence the relative humidity (RH%) is often used in percentage terms rather than partial pressures.

It is the large atmospheric volumetric domain between the ocean surface and the cloud droplets that is highly significant because it can vary markedly and quickly, and have the largest effect on electromagnetic radiation that other greenhouse gases cannot possibly have. Electromagnetic radiation is a form of energy that occurs at different frequencies and wavelengths, as shown in Figure 11.2.

Figure 11.2 The electromagnetic spectrum

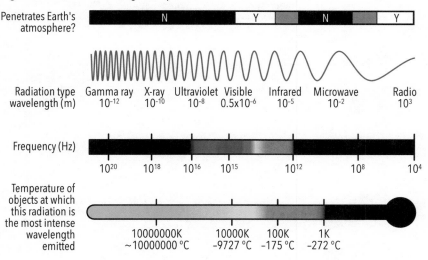

Infrared radiation is part of the electromagnetic spectrum. To clarify the spectroscopic terminology, the wavelength is the distance between the peaks of a wave, the spatial frequency is the number of wavelengths within a given unit distance and the frequency (temporal) is the number of wave peaks per second observed at a point. Thus, in the infrared spectrum the wavelengths are of the order of 1 μm to 200 μm, the spatial frequencies of order 50 cm⁻¹ to 10,000 cm⁻¹ while the frequencies range from 1.5×10^{12} Hz to 3×10^{14} Hz. Hertz (Hz) refers to one cycle per second.

Source: NASA self-made, Wikimedia Commons, viewed 30 May 2017, https://en.wikipedia.org/wiki/Electromagnetic_spectrum. Reprinted under Creative Commons CC-Share Alike 3.0 Unported license.

An RH% value of zero means bone-dry air, which never really occurs in the lower atmosphere. At 0 °C, and say 75 RH%, there are 3.15 grams of water per kilogram of dry air (g/kg), which is 3150 parts per million by mass (ppmm) (0.31%). This is much higher than atmospheric CO_2 at 0.04% and man-made CO_2 at 0.002%. At 20 °C and 75 RH% there is now 10.96 g of water vapour per kg of air (see Table 11.1). This is 10,960 ppmm or 1.1% of the atmosphere – again very much higher (270 times) than atmospheric CO_2.

For a typical hot summer day, with a tropical temperature of, say, 35 °C and 75 RH%, there is 27.93 g/kg, which is 27,930 ppmm or 2.79% of the

Table 11.1 Examples of humidity change with and without temperature change

Temperature °C	Relative humidity %	Atmospheric water content ppmm	Mass %
-35	100	470	0.047
-10	100	2250	0.23
0	100	4100	0.41
10	75	5880	0.59
14	60	6030	0.6
14	70	7040 [16.7% increase]	0.7
20	75	10,960	1.06
25	75	14,970	1.5
25	85	16,970 [13.3% increase]	1.7
30	75	20,450	2.05
30	100	27,270 [33.3% increase]	2.73
35	80	29,790	2.98
40	90	45,780	4.58

Humidities expressed in parts per million by mass (ppmm) and by mass percentage. The temperature of -35 °C is equivalent to the temperature at about 7 km to 8 km elevation. At any particular temperature there is a maximum amount of water vapour that air can contain, but this percentage is variable.

Source: Lenntech Relative Humidity Calculator
https://www.lenntech.com/calculators/humidity/relative-humidity.htm.

atmosphere (about 70 times greater than CO_2, or about 700 times greater than man-made CO_2) (see Table 11.1). In the tropics, at, say, 40 °C and 80 RH%, the water vapour concentration is more than 100 times greater than CO_2 at almost 4.1% (40,690 ppmm) of the atmosphere.

At all temperatures above about –35 °C (at polar regions, or equivalent to about 7 km to 8 km in elevation), the concentration of water vapour is larger than the concentration of CO_2. CO_2 probably exceeds water vapour in concentration above about 7 km elevation, but more than 50% of the total mass of the atmosphere is below about 6 km height anyway, where water vapour is far stronger in radiation absorption over a wider range of wavelengths. Even at –60 °C the water vapour content is still 100 ppmm. CO_2 with two narrow active absorption bands can only possibly absorb about 8% of the total incoming solar radiant energy anyway at these at high altitudes.

Small temperature differences occur throughout all bulk atmospheric mixing actions (breezes, winds, storms) as well as within and around clouds. These smaller and larger-scale temperature gradients can not only result in phase-change, but can also affect the local radiation competitiveness. Weather changes are, therefore, made far more complex by phase-changing water vapour. This also makes accurate mathematical modelling virtually impossible, and hence an overall parameter (or factor) is used that makes the theoretical mathematical models more empirical-type models.

Varying concentration

Most textbooks define the atmosphere on a 'dry-basis'. However, the atmosphere is never dry and this omission has led many to discount water vapour's influence. In fact, water vapour is usually number three in concentration after nitrogen (N_2) and oxygen (O_2) as shown in Table 11.2. Argon (Ar) usurps this ranking at low temperatures.

N_2, O_2, water vapour, and Ar are the top four gases making up 99.8% of the entire atmosphere on a moist-air basis. Condensable water vapour is the only greenhouse gas in the top four. This means that water vapour is the highest concentration greenhouse gas by far, being five to over 100 times greater in concentration than its nearest radiation rival, CO_2.

Table 11.2 Composition of the moist atmosphere

GAS		Volume ppmv	Volume %	Comment
Nitrogen	N₂	780,840	77.17	
Oxygen	O₂	209,460	20.70	
Water Vapour	H₂O	10,000	0.99	Basis: 1% at 20 °C 75 RH%. Varies from about 0.001% to 4.5% (tropics) (10 ppm to 45,000 ppm)
Argon	Ar	9340	0.923	Largest inert gas: 23 times greater than CO₂
Carbon Dioxide	CO₂	405	0.0399	Far less than 6% is man-made. Plankton and photosynthesis moderate the CO₂ levels all the time
Neon	Ne	1828	0.0018	Inert gas
Helium	He	5.24	0.00052	Inert gas
Methane	CH₄	1.79	0.00018	About 40% natural: <60% man-made
Hydrogen	H₂	0.5	0.000049	Trace reactive gas
Nitrous Oxide	N₂O	0.3	0.00003	About 60% natural: <40% man-made
Ozone	O₃	0.04	0.000004	Very reactive gas

Composition of the moist atmosphere including water vapour taken at about 1% (basis:1% at 20 °C; 75 RH%). Volumetric concentration in parts per million by volume (ppmv). Trace pollutants, such as man-made volatile chemicals, and chemicals, such as hydrofluorocarbons, and dust, are excluded.

Source: Google Wikipedia, 'Atmosphere of Earth' https://en.wikipedia.org/wiki/Atmosphere of_Earth; and the Lenntech Relative Humidity Calculator https://www.lenntech.com/calculators/humidity/relative-humidity.htm.

It is more than 5000 times greater in concentration than the next greenhouse gas, methane (CH_4), as shown in Table 11.3.

The sources of water vapour are numerous. The oceans cover 70.9% of the Earth's surface and there are many lakes, rivers and other bodies of water as well. Evaporation also occurs from land; water vapour can come from plants and trees (termed the transpiration mechanism). Hence, the humidity (water vapour in air) varies widely with location,

Table 11.3 All radiation-absorbing and/or emitting greenhouse gases

GAS		Volume ppmv	Greenhouse Gas Volume %	Comment
Water Vapour	H_2O	10,000	96.1	Almost 25 times greater than CO_2 at 20 °C (250 times greater than man-made CO_2). Can be over 100 times greater than CO_2 in the tropics, which is over 1000 times higher than man-made CO_2
Carbon Dioxide	CO_2	405	3.88	About 225 times greater than methane. Total CO_2 is about 405 ppm; naturally produced CO_2 is about twenty times more than man-made CO_2 (man-made CO_2 is less than 25 ppm)
Methane	CH_4	1.79	0.017	About 40% naturally produced: wetlands, soil, ground, natural gas, sediments, wildfires, ocean floor, volcanoes. Less than 60% man-made: Industry, agriculture, waste processing, landfills
Nitrous Oxide	N_2O	0.3	0.0029	60% naturally occurring from nitrogen cycle, fossil fuels, industry. Less than 40% from human activities: agriculture, wastewater treatment, combustion
Ozone	O_3	0.04	0.00038	Vital in protecting UV-B from reaching the Earth

All greenhouse gases make up 1.028% of the total atmosphere, of which water vapour is 1% or 96% of the greenhouse gases (20 °C and 75 RH%).

Source: Google 'Greenhouse gas' https://en.wikipedia.org/wiki/Greenhouse_gas.

but changes markedly with elevation because the temperature decreases 6.5 °C per kilometre rise (termed the thermal lapse rate, TLR).

Air and water temperatures are higher in the tropics and atmospheric water vapour can reach well over 40,000 ppm (more than 4% of the atmosphere). This is more than 100 times greater than CO_2. Water vapour in colder climates (or at high elevations) can still be high; more than 1000 ppm at −20 °C and 100% relative humidity. The lateral and vertical temperature differences induce significant changes of phase only with condensable water vapour and not with any other atmospheric gases.

Absorption of radiation

How much electromagnetic energy does each greenhouse gas actually absorb and re-radiate? Figure 11.3 shows the radiation transmitted by the atmosphere as ultraviolet (UV), visible light and infrared radiation. Clearly, as shown in this graph, our atmospheric gases only absorb in part of the overall range between about 0.6 and 3.0 micrometres. The height of each peak denotes the quantity of solar radiant energy that each gas can possibly absorb over that specific wavelength peak range. The width of each graphical peak clearly shows the wavelength range over which the particular greenhouse gas is solar radiant energy active.

For CH_4 and nitrous oxide (N_2O), the absorption peaks are both extremely small (less than 10% absorption peaks) and are very narrow. This shows CH_4 and N_2O are almost immune or unreactive to all incoming solar radiation. CO_2 only has two absorption peaks and only one potentially can absorb 100% across a narrow wavelength band. This peak also coincides with the wide and strong water vapour peak, and hence it would have to compete with the far larger amount of water vapour in the atmosphere. In contrast, water vapour has seven absorption peaks, three operating at 100% level and wider than all others, and two more above the 50% absorption level.

It can be concluded that for incoming solar energy, water vapour is the dominant absorbing greenhouse gas. Based on the total energy absorbed (relative total area under the curve), water vapour is five times greater in absorptive ability than CO_2. As shown, the absorption potentials of both CH_4 and N_2O are miniscule, or virtually non-existent.

Less than 60% of all solar radiation ever reaches the planet's surface. When it does, the solar radiation 'impacts' the Earth's surface by 'exciting' surface molecules (land and sea) to various extents. These molecules vibrate, rotate, twist, or oscillate more and emit long-wavelength radiation. It is this re-radiation that greenhouse gas molecules can more readily interact with. The energised atmospheric molecules themselves can now also move faster and collide more frequently with all other atmospheric molecules.

Figure 11.3 Absorption of incoming solar radiation

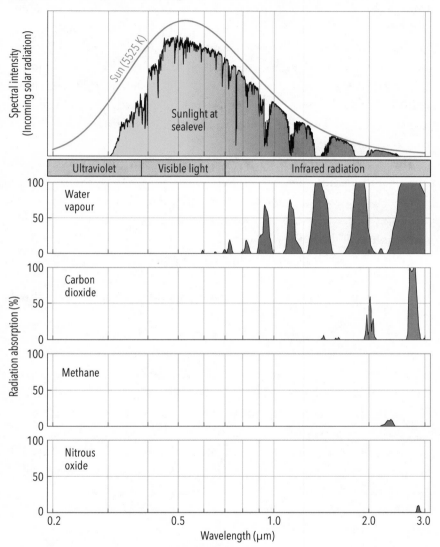

Emissive spectral density of the sun (5525 red curve), and the spectral density (solid yellow-red envelope) for radiant energy arriving in the atmosphere. The electromagnetic wavelength range for solar energy covers the range from 0.2 to 3.0 micrometres. Each greenhouse gas is labelled alongside its respective absorption spectrum in the bottom four panels.

Source: Adapted from Robert A. Rohde, Wikimedia commons.

Greenhouse gases within the first few hundred metres elevation absorb some of this radiant energy in various ways depending on the structure of the molecule. None absorb over the total span of wavelengths, now 3 to 70 micrometres. Each gas is sensitive to a specific frequency band as shown in Figure 11.4. As mentioned, the height of each peak indicates the magnitude of energy absorbed at each specific wavelength. The peak width indicates the sensitivity span of each greenhouse gas and thus the potential portion available to absorb the emitted longwave re-radiation from the planet's surface.

The low-concentration gas CH_4 (1.8 ppm) has two narrow bands, each absorbing less than 50% of radiant energy. They compete with water vapour anyway (water vapour is more than 5500 times higher in concentration). N_2O (0.3 ppm) has three very narrow bands, with only one peaking at about 90% at just one wavelength point (water vapour concentration >33,000 higher). CO_2 (405 ppm) only has a further two bands that are slightly wider than CH_4 and N_2O, but operate over a very small proportion of the total re-radiation span from Earth (<8% of the total wavelength span).

In stark contrast, water vapour is fully active across 85% of the total span (note log scale). Water vapour is at a far higher concentration (10,000 ppm at 20 °C and 75 RH%). If radiation is regarded as the key issue in changes of weather, then the focus must simply be on water vapour in the atmosphere and the absorption of longwave radiation. CO_2 makes a very small contribution, and CH_4 and N_2O and ozone (O_3) are really almost inconsequential. In fact, by determining the total energy absorbed (area under the curve), water vapour is twelve times more effective than CO_2 in longwave radiation absorption.

While it is most valuable to assess the impact of radiant energy on weather change, it would be totally wrong to just isolate 'radiation-water vapour' and 'radiation-CO_2' in weather change as the main competitive issues as we have done so far. Water vapour actually forms clouds that reflect a lot of incoming solar energy. It is the water vapour plus fine water droplets in the clouds that reflect and scatter a lot of solar radiation. Some is transmitted back to outer space. Some is re-radiated back to Earth. It is the subsequent cloud-forming, sun-shielding, and

Figure 11.4 Emission of re-radiated thermal radiation

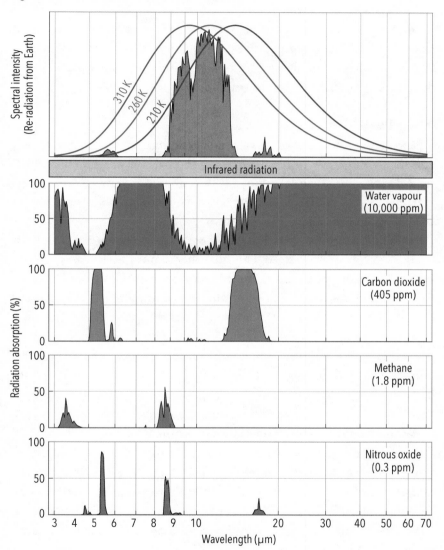

Electromagnetic radiation re-emitted from Earth in the 3 to 70 micrometre wavelength range. The top panel shows longwave infrared emissions not absorbed by greenhouse gases. The bottom four panels show absorption spectra for each respective greenhouse gas, as labelled.

Source: Google File: Atmospheric Transmission.png Radiation Transmitted by the Atmosphere
https://commons.wikimedia.org/wiki/File:Atmospheric_Transmission.png.

precipitation processes that are also highly significant. Snow, rain and hail reflect a lot of solar radiation also.

Although we have focused so far on water vapour and radiation, and its dominance, the main takeaway is that water and water vapour play additional and extensive roles in global atmospheric energy transfer and, therefore, in all the changes in weather.

An hypothetical

One very stark and clear way of understanding just how water vapour powerfully impacts weather on Earth is to envisage water vapour theoretically as a non-condensable gas like all other atmospheric gases. Untrue, of course, but examining this contrast amplifies the overall effectiveness of this unique contributor.

The only condensable atmospheric gas is water vapour. If it did not condense and acted instead like all other atmospheric gases, clearly weather and climate would be far different, perhaps catastrophic!

We could assume that while the concentration of N_2, O_2 and Ar would remain fixed, other gases like water vapour, CO_2, CH_4 and N_2O would continue to vary, because of their on-going supply sources. In particular, the oceans would go on delivering water vapour (evaporation) and CO_2 (degassing) as they do now (with smaller amounts of both from the land). The other non-condensable gases CH_4 and N_2O both have natural and man-made origins, and could also vary.

Without condensation no clouds would form. Cloud umbrella-thermal shielding would not exist to protect humans from the direct impact of sunlight during the day. There would be no rain, hail, or snow by precipitation. Any ice formed on land could only result because of low temperature freezing conditions. As there would be no precipitation, there would be no atmospheric liquid water droplets, no air cooling, no atmospheric water-dust scrubbing, and no cooling of the planet's surface by rain. There would be no storms, cyclones or tornadoes with water in them as we have now.

Far more direct sunlight would impact the planet unimpeded all day, every day. Daily surface temperatures would increase markedly, even in the sub-tropics. We would have Sahara Desert-like conditions: hotter days only without cloud protection, and very cold cloudless nights. Breezes, winds and cyclones would consist only of non-condensable gases and be usually dry and perhaps often quite dusty. Dust would accumulate in the atmosphere, and some dry storms would be unbearable and dangerous for health. Atmospheric pollution and toxic effluent from natural and human generation would accumulate without rainwater scrubbing. If water were present in the air at all, it would be as entrained droplets swept up from the oceans and lakes.

Air density and gas concentrations would continue to decrease with increasing elevation alongside the subsequent lowering of total pressure with elevation in the troposphere. The lowering of temperature with increasing elevation (the TLR) would still occur as it does now, as the gas molecules and atoms become further apart (the mean-free path length between collisions increases causing the subsequent lowering of density, pressure, and temperature). Bulk mixing in winds and the selective radiation absorption by the low-concentration greenhouse gases (CO_2, CH_4 and N_2O) would prevail and play a big part in the weather as no water vapour would exist. CO_2 would become the dominant greenhouse gas even though its concentration would still be low and its range quite limited as it is now. Vital processes like transpiration may either be different or not occur.

To make matters worse, the Earth's surface temperature would increase, and more energy would be stored energy in the land and sea. Thermal re-radiation back from Earth would then be far stronger. The atmosphere would be heated more by a sizeable increase in rising hot air, as well as the large increase in radiant electromagnetic energy. Unless the greenhouse gas concentrations increase (CO_2, CH_4 and N_2O), they would not absorb much more radiant energy than they do now as they are very limited already to specific wavelengths. However, the amount of the now non-condensable water vapour would increase as the Earth's surface temperature increased. As it is already active over 80% of the total radiation span, water vapour would now absorb more because of its higher atmospheric concentrations. Of course, much more radiant energy would be transmitted directly to outer space as well.

Living in tropical zones would become almost unbearable without the advantage of condensable water vapour and thus clouds and rain.

The diurnal temperature differences and changes would be far greater than presently experienced due to the current advantage of water vapour, clouds, storms, and rain mechanisms. Thermal radiation would govern the weather and thus climate.

One may consider this hypothetical consideration of water as a non-condensable gas initially to be irrational or extreme. But it dramatically and strongly highlights that water vapour has a far more powerful role in weather change and climate patterns than just thermal radiation. Water vapour at 1% (can vary from 0.2% to more than 4% at the tropics) coexists with the major gases (N_2, O_2 and Ar) at more than 99.8% of the entire atmosphere; whereas the sum of all the other non-condensable greenhouse gases is only 0.0041%. Indeed, the vast benefits of water (liquid), water vapour, condensation (clouds), and precipitation (rain) are in sharp contrast with the other far weaker contributions of low-concentration non-condensable greenhouse gases.

The importance of water vapour

Several additional and powerful mechanisms come into play as soon as water vapour is added to the atmosphere as the one-and-only phase-change gas.

Water evaporation from land and sea, followed by condensation as clouds, and precipitation as rain, dominate weather dynamics. Clouds reflect some incoming solar radiation away from Earth, and provide 'umbrella-type' protection on sunny days. Storms, cyclones and torna-does behave far differently with water vapour and water dominant. Atmospheric cooling from rain and snow lowers the planet's surface temperature as well. But there is also the powerful effect of humidifi-cation of the atmosphere. The other greenhouse gases simply cannot contribute to any of these.

Small and large temperature changes and thermal gradients induce additional water vapour transfer within the atmosphere. In addition to temperature changes, there are strong thermal and kinetic energy transfer mechanisms causing the simultaneous mass transfer from the liquid to vapour phase (evaporation), and then from the vapour to the liquid state (condensation to clouds or fogs). When precipitation occurs as rain, snow or hail, there is the simultaneous transfer of energy and vapour from the rain droplets or ice particles that can have a sizeable effect on the atmosphere, and thus weather. If all the above were not enough, water vapour is the dominant atmospheric greenhouse gas in the absorption and emission of solar energy and also the electromagnetic radiation from Earth.

It might be argued that breezes, winds and thermals would affect all atmospheric gases equally. But that is not the case. If there are the smallest temperature differences within or around these bulk mixing processes, then differences in humidity (water vapour concentration) would occur affecting potential phase-change with concomitant thermal energy transfer, as well as changing the radiation absorption potential. In the atmosphere, storms, cyclones and hurricanes – as well as jet streams and trade winds – convectively transfer large amounts of energy. But, in addition, the changes of phase (liquid water-water vapour-liquid water) must be added to the assessment. This makes accurate modelling almost impossible, certainly imprecise. In addition, the atmosphere does not exist by itself without interacting with the oceans and land. The interactive effects of the oceans (conveyor belts, waves, upwellings, tides and more) must be factored in.

12 Tropical Convection: Cooling the Atmosphere

Dr Peter Ridd & Dr Marchant van der Walt

For those of us who live in tropical Australia, deep convection in the form of thunderstorms and cyclones can be entertaining, awe inspiring, and even fear inducing. Storms make a spectacle with towering clouds, intense rainfall, thunder and lightning.

It is rarely appreciated that such deep tropical convection is also the main driver of the general circulation of the atmosphere. It is the giant 'heat engine' driving the wind and weather. The engine's fuel is water vapour that has evaporated from the tropical and subtropical oceans.

Deep convection drives energy from near the Earth–ocean surface, where it has accumulated due to the absorption of solar energy, to the upper atmosphere where it can be radiated to space as infra-red (IR) emissions. This is one of the energy transfer pathways that cools the surface. Deep convection ultimately reduces the surface temperature of the Earth by about 45 °C, from what it would be if there were no air motion in the atmosphere; if the only mechanism to drive thermal energy upwards was by IR radiative transport (Thomas & Stamnes 1999; Manabe & Strickler 1964). Without the air circulation caused by deep convection the Earth would be uninhabitably hot.

If the Earth's temperature increases from whatever cause (natural or human influences), increased evaporation over the ocean provides more fuel to the convection heat engine, and the surface cooling effect it drives. We make this analogy with an engine and calculate that 'the engine' becomes more powerful by about 10% per degree rise in temperature.

Figure 12.1 Deep convection in the form of a thunderstorm

The sprawling anvil cloud in this picture has formed from a main cumulonimbus tower. Sometimes anvil clouds are formed from multiple cumulonimbus towers. In this picture there are other 'fluffy' clouds at lower altitudes caused by rising air in minor convection cells.

Source: NASA/Tsado, image ID: KFCWA2 https://www.alamy.com/stock-image-anvil-thunderstorm-cloud-formation-africa-cumulonimbus-cloud-over-164396458.html.

Therefore, in the tropics the heat engine is a very strong negative feedback mechanism, acting to counter the effects of increasing concentrations of greenhouse gases.

Tropical deep convection

Thunderstorms are highly non-linear processes that can be set off by a small trigger. It is possible to tell when and where they are likely to form by measuring the vertical variations of air temperature and humidity. When the temperature and humidity near the surface get above a critical point, all that is needed is a trigger to set the storm off – the air needs only to be lifted.

When air is lifted, its pressure drops, and it cools. This is because as the air rises it gains gravitational potential energy, but this gain is at the expense of its thermal energy and so it cools. At first it cools at a rate of 9.8 °C for every kilometre it rises (the dry adiabatic lapse rate). However, because the air is very humid, as it cools the water vapour condenses forming cloud and releasing latent heat into the air. This is the opposite process to sweat evaporating from our skin. Evaporation cools the air, but condensation of water heats it. The latent heat added to the air means that as the air rises it now no longer cools at 9.8 °C/km but at only about 4–5 °C/km (the moist adiabatic lapse rate).

The air surrounding the storm cell changes temperature at roughly 6.5 °C/km, so the point is rapidly reached when the centre of the storm is significantly hotter than the surrounding air. This hotter air in the centre of the rising column of air consequently keeps rising, condensing even more water, and further heating the air column, thereby driving the formation of the storm cell. After this process has been triggered,

Figure 12.2 Convection cells in the equatorial trough and convergence zone

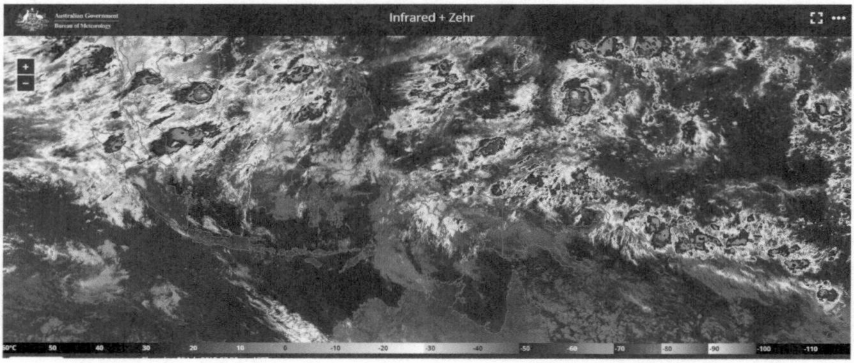

Video from satellites shows deep convection (the intense updrafts) as resembling vigorously boiling water in a saucepan. These storms usually have a life of a few hours as they move across 'hot' land or 'hot' water sucking in vast amounts of moisture to keep the system going.

The tops of the storms of the deep convection are red in satellite imagery. These are the giant 'pistons' of the atmospheric 'heat engine'. At any given time, there are in the order of 1000 of these storms.

The white cloud around the convection cells is high-level cirrus streaming out from the tops of cumulonimbus and/or anvil clouds.

Source: Source data from Himawari-8 satellite 23 July 2019. Australian Bureau of Meteorology website at http://satview.bom.gov.au/.

the air will keep rising until it reaches a height where its gain in gravitational potential energy will equal the total amount of energy it had at the ground, in the form of heat, and latent heat. This is the top of the storm cloud, which is usually well over 10 km in height. Temperatures at the top of the storm can drop to considerably less than −50 °C.

Most deep convection occurs in the equatorial trough, which is the area of relatively low pressure above the hot, mainly oceanic, tropics within 10 degrees latitude each side of the Equator.

Energy is also imported into the equatorial trough via the trade winds that bring in moisture-laden air. The water vapour carried by the trade winds can be thought of as more 'fuel' to drive the atmospheric heat engine.

Greenhouse gases

The atmosphere is heated from the Earth's surface by the absorption of sunlight. Most of this energy gets into the lower atmosphere by the evaporation of water over the oceans (latent heat), direct conduction, and also IR radiation from the surface that gets absorbed by greenhouse gases in the lower atmosphere.[1] These concepts are shown schematically in Figure 12.3.

So, most of the Sun's energy is initially 'dumped' in the lower atmosphere. Some of this energy is radiated directly back to space, but most will be transported high into the atmosphere to a level (the emission level) where it can be radiated to space.[2]

The role of greenhouse gases in the air beneath the emission level is often misunderstood. It is often stated that greenhouse gases 'trap the heat', but this is neither accurate nor particularly useful. The greenhouse gases at all levels emit IR, and do so in all directions including upwards and downwards (Stefan Boltzmann Law).

If we had eyes that could see IR, we would see the air glowing. If we looked downwards from an airplane with our 'IR eyeballs', we would barely be able to see the ground because of the glowing air. If we looked

---

1 Excluding that energy radiated to space directly through the IR spectral windows.
2 The emission level is a crude measure of the level above the ground from which IR radiation is emitted to space. In reality, emission of IR to space occurs from all levels but, on average, it is dominantly by the mid and upper troposphere.

Figure 12.3 Simplified schematic of the vertical fluxes of energy in the tropics

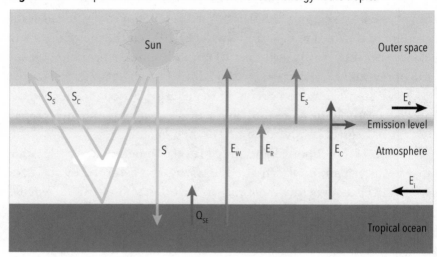

Much of the Sun's energy (S) is absorbed by the sea. Some is reflected from clouds (S_c) and the surface (S_s). Most of the Sun's energy is returned to the air via evaporation over the ocean (Q_{se}) but some (E_w) is radiated directly back to outer space through the spectral 'windows' (parts of the IR spectrum that are not absorbed by greenhouse gases). Energy reaches the emission level either by IR radiation (E_R = 4.6 PW) or through the vertical updrafts in the deep convection (E_c = 8.7 PW). Once the energy has reached the upper atmosphere around the emissions level, 7.9 PW is radiated to outer space (E_s); 6.6 PW is exported towards the poles as potential energy (E_e). Energy is also imported in the form of latent heat from higher latitudes (E_i = 5.1 PW). PW = petawatt 1 PW = 10^{15} watts (or a million billion watts).

Source: The flux data comes largely from Riehl and Simpson 1979. Uncertainty is in the order of 10%.

up, the air would glow less than the air beneath us because the temperature of the air above us is less than the air below. So, at any given level in the atmosphere, there is more IR going up than down. (The air below us radiates upwards, more than the air above us radiates downwards.) Net IR radiation transports energy upwards, from high to low temperature, via the greenhouse gases, in a similar way that thermal energy is transported from hotter to colder regions of a metal, through the mechanism of conduction.[3] This is the radiation pathway for cooling the surface of the Earth.

---

3   This is ultimately a manifestation of the second law of thermodynamics.

The Rosseland approximation

Higher greenhouse gas concentrations would result in a proportionate increase in the temperature change that is required to transport a given amount of energy vertically upwards, at least for what is called an 'optically thick' atmosphere. This is sometimes called the Rosseland approximation and is reasonably valid in the middle atmosphere, especially in the dominant spectral absorption bands of the greenhouse gases.[4]

We can think of greenhouse gases as changing the 'resistance' to the radiation transport.[5] So rather than greenhouse gases trapping heat, it is perhaps more useful to think of them as inhibiting the flow of energy vertically up to the emission level from the Earth's surface. The more greenhouse gases, the larger the vertical temperature change required. Given that the emission level high in the atmosphere remains at a constant temperature, a higher greenhouse gas concentration requires a higher temperature at the Earth's surface to drive energy upwards to this level via IR radiation.

In other words, the radiation of energy upwards at any given level depends on the temperature gradient and the greenhouse gas concentrations. This in turn raises two questions: What are the vertical temperature changes in the troposphere? And are they closely related to greenhouse gas concentrations?

If the Earth's temperature profile was totally dominated by the effect of the greenhouse gas concentration, we would expect to see greater vertical temperature changes where greenhouse gas concentrations are high.[6] In

---

4 Well away from the ground or the emission level.
5 See Appendix 12.2. Formally, the Rosseland approximation was written in the same form as Ohm's law for electricity, i.e. current density is proportional to the electric potential gradient. http://jullio.pe.kr/fluent6.1/help/html/ug/node486.htm. In Ohm's law, the resistance of a circuit determines the current in the circuit for a given voltage applied. High resistance means low current. The Rosseland approximation is also equivalent to Fick's Law of Diffusion. The Rosseland approximation is rarely invoked for Earth's atmosphere calculations as far more accurate formulations are available, but it is a useful concept when it comes to understanding the vertical radiative transport due to greenhouse gases in terms of the vertical temperature gradient.
6 From the Rosseland approximation, higher greenhouse gas concentrations increase the 'resistance' to energy flow thus requiring a larger temperature gradient to transport the same amount of energy.

particular, the most important greenhouse gas is water vapour, which varies in concentration dramatically with altitude and latitude, being about ten to 100 times more concentrated in the tropics close to the ocean than it is in the colder air above, or in regions away from, the tropics.

It is one of the most remarkable and surprising features of the troposphere that, despite the enormous variations in greenhouse gas concentrations (including water vapour), the vertical temperature changes are almost constant, at about 6.5 °C/km.[7] In other words, as one ascends in the atmosphere by 1 km, the temperature reduces by 6.5 °C. This temperature gradient hardly changes with altitude despite the order of magnitude changes in greenhouse gases (mostly water vapour) concentration. It is almost the same in the tropics as it is in the mid-latitudes and the North and South Poles, again despite very large differences in water vapour content. The temperature gradient changes over shorter periods, such as the day–night cycle near the ground, but for averages over longer periods, the vertical temperature gradient hardly changes with time.

The reason for the remarkable sameness of the vertical temperature change is that the rough average figure of 6.5 °C/km is determined by the dynamics of the atmosphere, which is driven by deep convection.

The atmospheric heat engine

There is a flow of energy from the Earth's surface to the emission level, where IR radiation is emitted to space. The heat engine in the form of the deep convective cells converts some of this energy to kinetic energy: wind, which is the movement of air (Peixoto & Oort 1992; Renno & Ingersoll 1996).

The flow rates of the air in the deep convection, and also the trade winds, are greatly influenced by the balance between the power of the heat engine, and the rate of frictional energy loss of the air movements. A more powerful engine will drive faster winds.

Importantly, if the atmospheric heat engine becomes more powerful, it drives more energy from the tropical surface to the emission level, thereby cooling the lower atmosphere.

---

7   See US Standard Atmosphere, 1976, https://ntrs.nasa.gov/search.jsp?R=19770009539.

If there was no flow of air in the atmosphere, the temperature of the surface of the Earth would be around 60 °C. It is the motion of the air due to the heat engine that keeps the surface temperatures low and makes the planet liveable.

Two essential features of any heat engine are that there is:

1. a flow of energy into and out of the air through the cycle, and
2. that energy is added when the air is hot and lost when the air is relatively cool.

For the atmosphere, energy input is via water condensation in hot tropical air. It is lost to space via radiation when it is high in the atmosphere and very cool. It is critical that, for the atmospheric heat engine to run, the air in the deep convection cells must be raised to about the emission level or higher so that heat can be lost at the cool part of the cycle.[8] This situation is well satisfied in the tropical atmosphere where the energy of the moist air near the Earth–ocean surface can ensure that it can rise to well above the emission level.

Why is it useful to consider the atmosphere as a heat engine? Because this relatively simple physics gives us insights that are not possible with the ultra-complicated black boxes of the simulation models used by most climate scientists.

Nowadays, car manufacturers have very sophisticated models that simulate every facet of their engines – the heat transfers, and gas flow, the ignition, and the mechanical movements. These models are like the general circulation models used by climate scientists, except that the engineers know all the physical process inside their engines very accurately. The analogy is explained in more detail in Table 12.1.

It is possible to make some very useful predictions about car engine performance based on very simple calculations. For example, increasing the displacement (size) of an engine will increase the power because it can extract energy from more fuel. Increasing the compression ratio will also increase the efficiency and power by well-known amounts.

---

8 The classic and equivalent explanation for the heat engine to run is the generation of convective instability: the cooling of the high troposphere by net radiation loss and the warming of the surface by solar heating.

Table 12.1 The analogy between the petrol engine cycle and the atmospheric heat engine cycle

Petrol engine cycle	Atmospheric convection heat engine cycle
Air is compressed making it hot. Fuel is added.	Air near the surface of the tropics is at a higher pressure and relatively warmer than the thin air at the top of the atmosphere. It also is loaded with fuel by way of water vapour.
Ignition: a fuel–air mixture is ignited by spark plugs heating the gas.	Air is lifted, causing condensation of water. This is the ignition of the water vapour fuel. Latent heat release will heat the air.
Power stroke: hot, very high-pressure gas pushes piston and drives engine and the car.	Power stroke: the relatively hot air is buoyant and moves upward, then poleward, driving much of the circulation of the atmosphere.
Remaining heat is discharged to the exhaust.	Energy lost to outer space by IR radiation emitted from greenhouse gas. This offsets the solar radiation entering the atmosphere.
Fresh air is now compressed.	Air now sinks back to the surface and is warmed by compression as the pressure rises.
Fuel is added ready for ignition and the next power stroke.	Air moves back towards the tropics and more water vapour fuel is added for the next deep convection power stroke. The water vapour fuel is replenished by solar radiation heating of the surface and causing evaporation, mostly over the ocean.

For the atmosphere, we will simply consider what will happen if we run the engine with a little more fuel in the form of hotter, more humid tropical air which may be caused by some perturbation in the climate. We can easily estimate how much more air motion it will produce and how much energy it will transport from the Earth's surface to the emission level – thereby providing extra cooling of the lower atmosphere.

A simple calculation

There is a very large transport of energy from near the surface vertically upwards to the emission level by two mechanisms. The first mechanism is the flow of 4.6 PW of IR radiation due to the vertical temperature change (E_R).[9] The second pathway is the 8.7 PW carried by the vertically rising air in storms – deep convection (E_c).

It is notable that the deep convection pathway transports more energy to the emissions level than the radiation pathway – convection is more important than radiation in cooling the tropical lower atmosphere.

If the greenhouse gas concentration increases due to humans burning fossil fuels, the 'resistance' to the radiation flow increases, and the surface temperature rises because less energy is moving vertically up to the emission level. However, this temperature increase will cause more deep convection therefore increasing the flow of energy from the surface to the emissions layer via the convective pathway. This is because an increased temperature will cause more evaporation and increase the humidity of the air, providing more fuel for the heat engine to run more powerfully. We can show from simple energy considerations that this convective flow, E_c will vary according to the formula below, where q^* is the saturation vapour pressure of water vapour for the air near the surface and K is a constant:[10]

$$E_c = Kq^{*\ 32}$$

It turns out that for tropical air q^* varies by about 7% per °C.[11] Thus, from this formula, the energy transported by the convective pathway will increase by about 10% for a 1 °C increase in the tropical surface temperature, i.e. about 0.9 PW for the equatorial trough. This change in the convective energy pathway due to a temperature increase of just 1 °C is very large: about 20% of the entire radiative pathway (4.6 PW) in the tropics. This demonstrates the importance of deep convection relative to IR radiation for cooling in the tropics.

It follows that a modest rise in the surface temperature can increase the convective pathway to counteract a major reduction in the energy flow in the radiation pathway.

---

9 PW Petawatt = 10^{15} W. Also, values quoted are for half of the equatorial trough. See Appendices 12.1 and 12.2.

10 See Appendix 12.2. The mechanical energy generated by the heat engine to move air around the globe is ultimately dissipated as friction at a rate that varies with the cube of the characteristic wind speed of the atmospheric circulation.

11 https://en.wikipedia.org/wiki/Clausius%E2%80%93Clapeyron_relation.

We use the Rosseland approximation to estimate by how much the tropical radiative pathway will reduce for a doubling of greenhouse gases. Water vapour is the dominant greenhouse gas and assuming that carbon dioxide (CO_2) represents about 30% of the greenhouse effect (an upper estimate), then the radiative pathway will drop by about 0.7 PW for a doubling of CO_2.[12,13] The increase in the vertical heat flux in the convective pathway from a 1 °C temperature increase (0.9 PW) calculated previously can therefore more than offset this throttling-back of the radiative pathway due to a doubling of CO_2.

The above analysis ignores other changes that might occur with increasing CO_2, specifically a temperature rise due to CO_2 will also increase the water vapour content (humidity) of the air by around 7% for every 1 °C increase. Water vapour is a strong greenhouse gas and thus the radiative pathway will be further restricted. This is a positive feedback effect.

If we again use the Rosseland approximation, the water vapour increase as a result of a 1 °C increase in temperature will throttle back the radiative pathway by a further 0.3 PW. Importantly, as the humidity increases, and the radiative pathway is reduced, the convective pathway inevitably throttles up and becomes more active with more deep convection.

A 1 °C increase in temperature will increase the convective pathway by 10%.[14] Under such circumstances we would expect more clouds, which will reflect more sunlight.[15] Sunlight that is reflected before it reaches the Earth's surface means that less energy needs to be transported upwards. A simple balance of energy in the equatorial zone shows that a modest change in the clouds can easily compensate for the positive feedback effect of increasing humidity.

---

12 https://www.giss.nasa.gov/research/briefs/schmidt_05/

13 The radiation 'resistance' increases by 15% of 4.6 PW. Using a more elaborate radiative transfer model gives a similar result with mid-tropospheric net-upwards radiation dropping from 35W/m² to 19 W/m² (or 0.7 PW) as CO_2 rises from 300 ppm to 600 ppm (e.g. http://clivebest.com/blog/?p=4345). Note the area of half of the equatorial trough is 440 million km².

14 From Appendix 12.2, equations 1 and 2, this would cause a 3% increase in wind speed, which is too small to see in meteorological records.

15 The albedo increases, which means the amount of light reflected to outer space increases.

Conclusions

Rising towers of thunderstorms can be thought of as the giant pistons of an engine, driving air upwards and, ultimately, away from the tropics at levels high in the atmosphere. At any given time, there are in the order of 1000 of these pistons, or storms. Some are much bigger than others (Riehl & Simpson 1979).

This movement of air takes energy with it, with potential energy exported to the temperate regions of the globe ultimately warming these regions.[16] The deep convection also sucks in air driving the trade wind systems that flow towards the tropics. These trade winds, in turn, drive much of the ocean currents that also transport large quantities of heat from the tropics to the poles. The atmospheric heat engine cools the lower atmosphere by around 45 °C. This atmospheric engine has ample capacity to run a little faster and more powerfully acting as a strong negative feedback mechanism to rising temperature from any cause, including rising greenhouse gas concentrations. Its power increases by 10% for every 1 °C increase in the tropical temperature.

Convection, especially deep convection, is also responsible for cloud generation. There is still considerable debate about how clouds respond to increasing CO_2 concentrations.

Complicated general circulation models used by many climate scientists attempt to simulate the role of convection and clouds, but the problem is that most of the convection cells are far smaller than the grid-scale of the models and therefore they cannot be resolved in the models. Instead, deep convection is 'parameterised'. The question is whether this parameterisation accurately simulates reality. If the models understate the possible increase in deep convection that might occur due to higher greenhouse gases closing down the radiative upwards energy pathway, they will likely over-estimate the temperature increase.

We have considered the effect of energy exchange as being driven by moisture and latent heat, which is appropriate for tropical regions and

---

16 It is exported out of the tropics mostly as potential energy high in the atmosphere. When the air ultimately falls, the potential energy is converted to thermal energy and the air warms. Without this export, the poles and subtropics would be considerably colder.

Figure 12.4 Hurricane Florence over the Atlantic Ocean

Source: https://www.istockphoto.com/au/photo/hurricane-florence-over-the-atlantics-close-to-the-us-coast-gaping-eye-of-a-category-gm1049802580-280745908.

which we suggest also dominates effects at higher latitudes. Furthermore, although the use of the heat engine analogy and the Rosseland approximation is a considerable simplification of the physics, it is a useful first order approximation for relatively small perturbations from the present conditions. We hope that it serves to encourage further work using the analogy of tropical convection as a heat engine for the atmosphere.

13 Reflections on the Iris Effect

Professor Richard S. Lindzen

This chapter describes the discovery and subsequent history of a process, the 'iris effect', which provides a strong negative feedback that seems capable of greatly diminishing the importance of carbon dioxide-induced warming. This effect deals with the response of upper-level cirrus clouds to temperature. Although rarely noted, these clouds are extremely important greenhouse substances. According to the iris effect, these clouds diminish in response to warming, thus cancelling much of the warming.

Carbon dioxide (CO_2) has long been looked at as a possible source of anthropogenic influence on climate. However, until the early 1970s, it was generally dismissed as being a minor factor. The reason was simply that the impact of a doubling of CO_2 was found to be small compared to other factors. Doubling CO_2 only perturbed the natural climate forcing by about 2% and this was expected to change global mean temperature by only about 1 °C. The equilibrated response to a doubling of CO_2 is referred to as the climate sensitivity. Moreover, because of saturation, each doubling has the impact of the previous doubling.

However, a paper by Manabe and Wetherald (1975) offered the possibility that the influence of CO_2 might be much greater. By assuming relative humidity remained constant, while ignoring clouds, they obtained a positive feedback that doubled the sensitivity. The idea was that the small warming due to CO_2 would, with constant relative humidity, lead to an increase of the most important greenhouse gas, water vapour. Moreover, the addition of any other, smaller, positive feedback would lead to a dramatic increase in the climate sensitivity, leading to the famous claim of climate sensitivities ranging from 1.5 °C to 4.5 °C (see Appendix 13.1 for an explanation of feedbacks). Yet, the iris effect acts to cancel and even exceed the water-vapour feedback.

This is consistent with the general expectation that long-surviving systems are dominated by negative feedbacks. The iris effect, moreover, provides a good explanation for the remarkable stability of tropical temperatures in the presence of truly major climate changes – those associated with major glaciations, the warmth of the Eocene 50 million years ago, and the reduction in solar output by about 20% 2.5 billion years ago. The changes in global mean temperature among these climates was almost entirely due to changes in the temperature differ-ence between the tropics and the poles while tropical temperatures changed little.

The iris effect

In 2001, I wrote a paper with Ming-Dah Chou and Arthur Hou that continues to be referenced, although references are usually accompanied by the claim that our results are controversial. Our paper (Lindzen et al. 2001) introduced what we referred to as the 'iris effect'. By this, we simply meant that regions with high cirrus and high humidity could change their areal coverage. This feature is of fundamental importance. In particular, it shows that it is no longer possible to disentangle the water-vapour feedback from the cloud feedback – especially since, in regions with high cirrus, the high cirrus rather than the water vapour determines the infrared emission level (viz. Figure 13.1). Instead, it is necessary to consider a long-wave (or infrared) feedback that incorpo-rates the combination of cloud, water vapour and areal responses to temperature. (This was subsequently the approach of Lindzen & Choi 2009; 2011.)

These were the fundamental points of our paper, and they were essentially rigorous and even obvious. The paper nonetheless proved controversial mainly, I suspect, because it found that upper-level cirrus area decreased as temperature increased and that this would provide a

Figure 13.1 Effect of upper-level cirrus clouds on the characteristic emission level of infrared radiation.

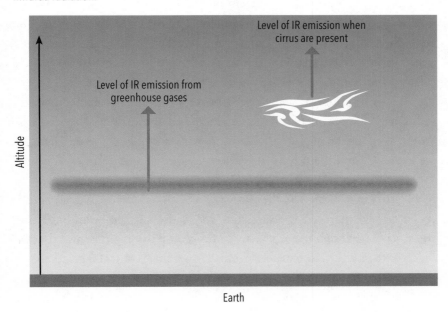

Earth

Note that the greenhouse effect of these clouds is particularly strong because temperature decreases with altitude. As a result these clouds more strongly reduce the ability of the atmosphere to cool.

Source: Modified from Lindzen et al. 2001.

powerful negative feedback that would cancel the water-vapour feedback that is essential to climate alarm.

This chapter will review the historical background of how this paper came to be; it will briefly deal with the comments claimed to have refuted our results and rendered them controversial; it will discuss where things appear to stand at the moment; and, finally, it will briefly describe some areas in particular need of more work and the relevance of the iris effect to the more fundamental questions concerning major climate changes in the Earth's history (which, incidentally, seem to bear little relation to CO_2 as a causal factor). Some of these things have been dealt with earlier (Lindzen 2004; 2012).

Historical background

Like most people working on climate, I had pretty much started by assuming that relative humidity remained fixed as climate changed, and that, therefore, following Manabe and Wetherald (1975), there was a water-vapour feedback (Lindzen et al. 1982). However, by 1990, I had begun wondering about the water-vapour feedback, and in Lindzen (1990), in response to a review, I suggested a possible mechanism for reducing it. Subsequent calculations showed the proposed mechanism would be ineffective. In collaboration with a student, De-Zheng Sun, we began a careful study of the water-vapour budget (Sun & Lindzen, 1993). Microwave observations of water vapour by Spencer and Braswell (1997) made it clear that there was an additional factor that was going to be crucial: namely, the fact that above the surface boundary layer, there was a clear delineation between moist and dry regions, and that the behaviour of any water-vapour feedback would have to take account of the relative areas of these regions. My own familiarity with satellite data was inadequate for me to undertake such a study, but a kind invitation by Franco Einaudi permitted me to spend part of a sabbatical at the Goddard Space Flight Center where I could collaborate with Ming-Dah Chou and Arthur Hou who both had the requisite expertise.

As the resulting paper shows, our original emphasis was, indeed, on the behaviour of the water vapour, but an important practical consideration forced us to focus on cloud cover. Namely, in order to observe the detailed response of area to surface temperature, we had to use geostationary data; the available satellite (Japanese GMS5) measured brightness temperature, which could be used to estimate cloud cover as a function of height, but could not directly measure water vapour. It was also unable to detect thin cirrus. However, we used measurements of thicker cirrus as surrogates for total cirrus. Our first instinct was to use the relation found by Udelhofen and Hartmann (1995) to assume a relation between high cloud and humidity, but, in the course of our calculations, we realised that the clouds, themselves, were a major factor in the long-wave budget, and allowed for water vapour and clouds to be unrelated. This, of course, was a conservative choice that minimised negative feedbacks. The feedbacks were negative because we found that high thin cloud cover, when normalised by a measure of cumulonimbus mass flux, diminished with increases in temperature. It is this behaviour that was referred to as the 'iris effect'.

The normalisation was crucial; without normalisation the results would simply reflect the concentration of cumulus activity over warm regions. Note that total cumulus convection is determined by the overall energy budget, and cirrus outflow per unit convection is what is relevant to climate (viz. Figure 13.2).

Much of the rest of the paper was devoted to the description and implementation of simple calculations to infer the impact of the iris effect on the climate sensitivity. The limitations of the study were carefully described, and, for the most part, assumptions were conservative. However, the results suggested a very strong negative feedback that would cancel the positive feedbacks in current models. The paper also showed that the few models we examined differed sufficiently dramatically from the observations to overwhelm the uncertainties in the observations.

Our final suggestion was that modelers should incorporate the iris effect to see how it would impact results. It took about fourteen years for Mauritsen and Stevens (2015) to take up this modest suggestion.

In the next section, I will describe the rather strange reception to our paper.

Initial criticism of the iris effect

The response to our paper came as something of an unpleasant surprise. The editor of the *Bulletin of the American Meteorological Society* was immediately replaced. The next issue of the *Bulletin* contained an attack, not in the form of a letter (to which we could have immediately replied) but as a separate article (Hartmann & Michelsen 2002). The title of the article was 'No Evidence for Iris', with the new editor appending a subtitle 'Careful analysis of data reveals *no* shrinkage of tropical cloud anvil area with increasing sea surface temperature (SST)'.

The basic assertion of Hartmann and Michelsen was that the stratiform clouds we were looking at did not originate in cumulonimbus convection. Rather, they were clouds associated with extratropical systems that intruded into the tropics. Such clouds, they argued, were associated with

Figure 13.2 Cirrus cloud cover, when normalised by a measure of cumulonimbus mass flux, diminishes with increases in temperature

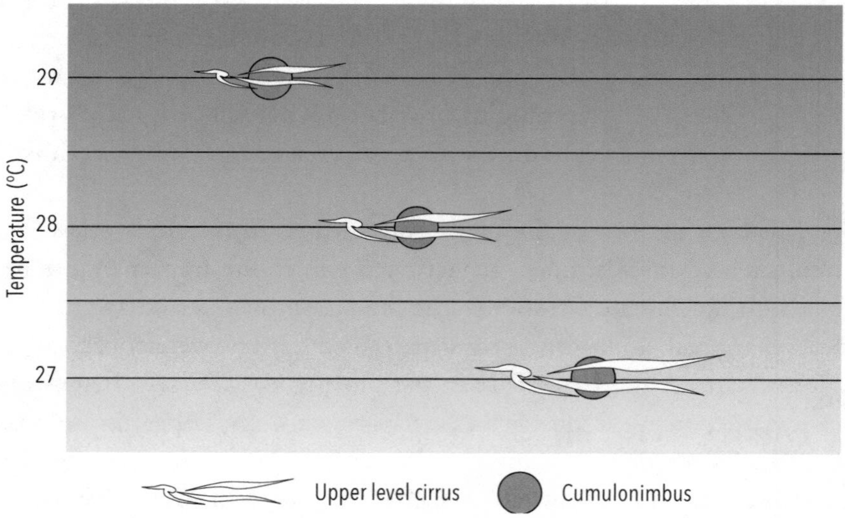

Schematic illustrating change in cloud-weighted SST due to cloud systems moving from the central position to colder and warmer regions. Horizontal lines correspond to isotherms. Units are nominally °C.

Source: Adapted from Linzen et al. 2001.

colder sea surface temperature, and hence would produce an increase in cloud cover associated with a decrease in cloud-weighted sea surface temperature. We had, in fact, explicitly tested for this in our paper.

It is easy to dismiss the Hartmann–Michelsen assertion on several counts:

1. If Hartmann and Michelsen were correct, the iris effect would diminish as we restricted ourselves to regions closer to the equator. If anything, the effect increased.
2. Close examination of GMS5 data permitted us to identify cumulonimbus cores associated with cirrus in all parts of the domain.

Our response to Hartmann and Michelsen appeared six months later and is rarely referenced (Lindzen et al. 2002).

Del Genio and Kovari (2002) noted that when they looked at a warmer region and a cooler region in the tropics, they found more upper-level cloud cover in the warmer region in contrast to what the iris effect would suggest. Unfortunately, they failed to distinguish cloud cover from cloud cover per unit cumulus. The iris effect refers to the latter. As Rondanelli and Lindzen (2008) noted, when the normalisation was applied, they actually found a stronger iris effect than we had estimated.

Two other critical papers (Fu et al. 2002; Lin et al. 2002) (and again published as separate papers so as to avoid our response) argued against our treatment of the radiative effects of the upper-level cirrus. As noted in our responses (Chou et al. 2002a; 2002b), both these criticisms argued that we – Lindzen, Chou and Hou – had misrepresented the radiative properties of upper-level cirrus by over-emphasising the infrared effect and understating the visible effect. Lin et al. (2002) went so far as to claim that the main impact of upper-level clouds was in the visible spectrum.

They essentially arrived at these conclusions by ignoring the fact that we were taking Ac(260) (i.e. the area with brightness temperature 260 K) as a surrogate for all tropical upper-level cirrus, and instead asserting that the clouds defined by Ac(260) were essentially the only clouds we were considering. (N.B. Virtually all documents dealing with the impact of clouds on climate agree that for upper-level cirrus, the primary radiative impact is in the infrared, viz. Choi & Ho 2006). Thus, they focused on the clouds in the immediate vicinity of the cumulus cores, which are atypically thick and do have a high albedo. We have already noted that separating thick clouds from thinner clouds did not change the magnitude of the effect. Neither did changing the reference brightness temperature.

In brief, these criticisms managed to change the results obtained by us (Lindzen, Chou and Hou) by claiming that we had followed an irrational procedure, different from the one actually employed.

Perhaps the most peculiar criticism was that of Lin et al. (2004). As noted by Chou and myself (2005), Lin et al. assumed differences compared with Lindzen, Chou and Hou implied the absence of negative feedbacks, but the negative feedbacks implied by their results are larger than those in our paper. Similarly, they assumed that the discrepancies between their specification of cloud radiative properties and the observations is due to their use of the iris result that cloud cover diminishes 22% per 1 °C. What they failed to note was that to bring their values into line with the observations, they would need much more than 22%.

In brief, Lin et al. failed to recognise how conservative Lindzen, Chou and Hou had been. They assumed (without apparently checking) that we had attempted to maximise negative feedbacks, and that any deviation would eliminate negative feedbacks. This is clearly not the case.

Subsequent studies and remaining questions

The collaboration with Chou and Hou ended when Chou returned to Taiwan, and Hou assumed responsibility for NASA's precipitation measurement program. However, together with students and postdoctoral researchers, work continued at Massachusetts Institute of Technology (MIT), and other work contributed to slow but reassuring progress.

Studying feedback effects on clouds inevitably involves at least two intrinsic difficulties: measurements are always to some extent indirect (that is to say one does not in general measure cumulus mass flux or stratiform cloud area directly), and, perhaps more importantly, clouds depend on many factors other than surface temperature so that attempts to correlate cloud cover with temperature have large scatters and relatively low correlations. As Rondanelli and Lindzen (2008) noted: 'small values of the regression coefficients such as the ones we observe do not imply a small effect but rather the presence of noise.' Nevertheless, as shown by Choi et al. (2014), sufficient noise can render signal detection difficult if not impossible. These matters will be discussed further later in this section.

Rondanelli and I used the ground-based radar at Kwajalein as well as the Tropical Rainfall Measuring Mission (TRMM) satellite data to examine aspects of the iris effect (Rondanelli & Lindzen 2008). In particular, the iris effect implied that precipitation efficiency in cumulonimbus towers would increase with increasing sea surface temperature, and that the area of stratiform detrainment from these towers would decrease. The Kwajalein ground radar offered good resolution and coverage. As usual, one needed to be careful about adequate time intervals. At the least, one had to avoid sampling only one part of the time evolution of the cloud system. However, we found that at any given time the radar was sensing a large number of systems at different stages of evolution, and this minimised the issue.

We found that convective precipitation increased by approximately 10%/K while stratiform area decreased by approximately 22%/K – consistent with the earlier results from Lindzen, Chou and Hou (2001). TRMM data provided coverage over the whole tropics, but with poor time resolution and other problems, perhaps most notably the inability to detect low levels of precipitation. Still, the TRMM data also showed increases of convective precipitation (approximately 7%/K) and decreases of stratiform area (approximately 5%/K) albeit at a reduced level (though TRMM showed great regional variability with some regions showing much larger values). Thus, a variety of independent approaches (Lindzen et al. 2001; Rondanelli & Lindzen 2008; Horvath & Soden 2008; and Del Genio & Kovari 2002) have all confirmed the iris effect, although there is uncertainty as to the exact magnitude of the effect. As already noted, Del Genio and Kovari (2002) required normalisation to show the effect. The reason for increased precipitation efficiency with increasing sea surface temperature is, of course, not resolved by these studies.

The issue of normalisation warrants some discussion. The normalisation used by Rondanelli and Lindzen (2008) was total precipitation. Lindzen, Chou and Hou (2001) recommended cumulus mass flux. Whatever the choice, as noted by Rondanelli and Lindzen:

> The normalization is a necessary condition to obtain a meaningful result since to the first order the SST distribution organizes the spatial variation of convection in the tropics. Cloud properties such as the area of detrainment, total rainfall, and cloud radiative effects can be considered extensive with respect to the amount of convection. The normalization simply takes into account the fact that the amount of convection in a region of a given temperature is not indicative of the amount of convection in a climate with the same temperature, since the global amount of convection is determined by global energy balance considerations. It remains an open question whether total precipitation, mass convective flux, or some other measure of convection is the most adequate normalization factor with regard to climate effects.

Lindzen, Chou and Hou noted that if relative humidity in the tropical boundary layer remained constant as climate changed, the cumulus mass flux was the appropriate normalisation. However, small changes in relative humidity would lead to different choices, and until we have a good basis for predicting such changes, the exact choice of normalisation remains uncertain.

As an application of the iris effect, Rondanelli and Lindzen (2010) considered its possible role in resolving the 'Early Faint Sun Paradox' (Sagan & Mullen 1972). Briefly, Sagan and Mullen noted that according to the standard model of the sun, solar output between 2 and 3.5 billion years ago should have been 20–30% less than today's solar output. (Remember that a doubling of CO_2 represents only a 2% change in radiative balance.) This should have led to an ice-covered planet, but the geological evidence indicated flowing water and possibly the complete absence of ice. The most commonly proposed solutions involved large quantities of greenhouse gases, but each of these had profound difficulties. For example, the amount of CO_2 required was at least ten times greater than permitted by geological evidence from paleosols and other proxies. We examined the potential of upper-level cirrus in the tropics to resolve the problem and found that this was a possible solution. Invoking the iris effect with a magnitude of $-22\%/K$ allowed resolution without any additional greenhouse gas, while with $-5\%/K$, the mechanism allowed for an ice-free tropics while allowing complete resolution with the addition of CO_2 well within what was allowed by geological data.

As already mentioned, one important implication of the highly variable, tropical, upper-level cirrus (regardless of the iris effect) is that it is no longer possible to isolate the water-vapour feedback from the infrared cloud feedback associated with upper-level cirrus. It is, rather, more appropriate to simply look for the infrared (long-wave) feedback. Similarly, it is reasonable to look for a short-wave feedback as well. Lindzen and Choi (2009; 2011), using satellite data from the ERBE (Earth Radiation Budget Experiment) and from both ERBE and CERES (Clouds and the Earth's Radiant Energy System) respectively, investigated the dependence of variations in both long-wave and short-wave radiation, as seen from space, on sea surface temperature. As is common in feedback studies in other fields, it was necessary to restrict oneself to fluctuations over time intervals that were long compared to the time scale for the feedback processes, while being short compared to the time for climate equilibration. Simple regressing over the whole time record can, and often does, lead to totally misleading results. What we found was a fairly clear negative long-wave feedback, an ambiguous signal for the short-wave feedback, and large differences between model responses to specified sea surface temperatures compared to what was observed.

A subsequent study by Choi et al. (2014) showed that the noise (i.e. processes other than sea surface temperature that caused changes in clouds) dominated the attempted measurements of short-wave feedback leaving the feedback essentially undetectable. Noting that part of the problem was that cloud changes could produce rapid changes in the skin temperature of the ocean (which is what satellites are essentially measuring), Cho et al. (2012) found that using sea surface temperature – weighted according to the absence of clouds – could reduce noise but not enough to get a good handle on the short-wave feedback.

The negative long-wave feedback is compatible with the results of Lindzen, Chou and Hou (2001). Trenberth and Fasullo (2009) independently ascertained that something (as they note, possibly the iris effect) basically eliminated the positive long-wave feedback that has long been essential to high positive feedbacks in models. Mauritsen and Stevens (2015), in introducing the iris effect into their model (by making precipitation efficiency increase due to clustering of convection in mesoscale systems), also found that the long-wave feedback was essentially eliminated, but found that, in their model, short-wave feedback still provided a net positive feedback that brought climate sensitivity to the low end of the IPCC range. Several recent studies (including some from major data centres) confirm the implausibility of higher climate sensitivities (Lindzen 2014; Lewis & Curry 2014; Fyfe et al. 2012; Stott et al. 2013). Currently, there appears to be a search for a short-wave positive feedback to replace the original infrared feedback of Manabe and Wetherald that originally permitted climate alarm.

Concluding remarks

So, where do things stand? I think it is fair to say that the iris effect remains a viable mechanism. To be sure, there remain such questions as the exact mechanism for increased precipitation efficiency, and the proper normalisation for application to climate change. However, neither of these appears to have the capacity to change the major conclusions. More important, perhaps, would be to have instrument platforms better suited to the measurements we have described. Microwave sensors on a geostationary satellite would be a significant help, but this does not seem likely in the immediate future. As for the need to replace such features as the water-vapour feedback with a more general long-wave feedback when calculating climate sensitivity – that remains obviously correct. Moreover, the iris effect not only profoundly reduces sensitivity, it also greatly narrows the range of possible sensitivities (see Appendix 13.1) and it provides an explanation for the stability of tropical temperatures over the history of the Earth. For almost a generation, the obsession with CO_2 in climate has precluded the essential development of an understanding of the truly major climate changes that have characterised the Earth's history. These have been characterised by large changes in the temperature difference between the tropics and the poles with the tropics remaining remarkably stable. The iris effect provides an explanation for the latter. Understanding the former is actually the major problem calling for an explanation. As noted by Lindzen and Farrell (1980) and Lindzen (2018), this almost certainly is related to the influence of ice on the vertical profile of temperature.

SECTION III

REAL SCIENCE AND OUR FAILING INSTITUTIONS

14 Rainfall-Powered Aviation
Dr John Abbot

Not all ideas for a clean energy future are stupid; using hydroelectricity to produce hydrogen as a fuel for aviation, rather than using kerosene, is one that has merit. But for this approach to achieve financial sustainability in Australia, key government institutions would first have to admit their current methods have singularly failed to predict both climate and rainfall patterns. They would also have to be open to newer and more effective forecasting techniques such as Artificial Neural Networks (ANNs). How hydrogen-fuelled aviation and forecasting rainfall with Artificial Intelligence (AI) are connected becomes clear as this chapter progresses. Stick with us!

It has been predicted that the COVID-19 pandemic, coupled with pressures to reduce carbon dioxide (CO_2) emissions, may accelerate the development of a revolution in the aviation industry, possibly leading to development of hydrogen-powered planes (Whitley 2020; Donovan 2020). Hydrogen gas can be generated on an industrial scale by electrolysis of water, that is the passage of an electric current through clean water. When this is from hydroelectricity – from the force of water flowing through turbines – no CO_2 emissions are generated when the hydrogen is burned to provide thrust for planes (or turbines in power stations).

ZeroAvia, a company based in Hollister, California, aims to develop aircraft that run on hydrogen-powered electricity using fuel cells. It claims its planes will eventually be cheaper to manufacture and fly than standard planes powered by jet fuel, while also producing zero carbon emissions (Hawkins 2019). Another advantage of hydrogen-powered aircraft compared to conventional aircraft is noise reduction (Royal Aeronautical Society 2020). ZeroAvia believe they can get a 10 to 15 decibel reduction in the noise outputs compared to a turboprop of a similar size. A $5 million grant from the UK government announced in September 2019 will fund a test program based in Cranfield, England (Moore 2020). ZeroAvia is sharing the cost equally with the UK government, and hopes to conclude the program in September 2020 with a 320 km demonstration flight using hydrogen power.

Tasmania, in the far south-east of Australia, is ideally placed for industrial-scale renewable hydrogen generation (COAG 2019; Department of Industry, Science, Energy and Resources 2020; Tasmania Department of State Growth 2019; Hydro Tasmania 2019). The state's existing electricity generation supply is overwhelmingly hydropower, with 30 power stations and more than 50 dams with a combined capacity of 2283 MW, operated by state-owned enterprise Hydro Tasmania (Hydro Tasmania 2018). Tasmanian Premier Peter Gutwein has stated that his Liberal Party government has a vision to establish a renewable hydrogen-generation facility by 2024 and be commercially exporting hydrogen by 2030 (Fowler 2020). The government will provide $50 million as part of its Hydrogen Action Plan (Tasmania Department of State Growth 2019) aimed at exploring the potential for a 1000-megawat renewable hydrogen facility in Bell Bay or Burnie.

However, bringing this vision to fruition requires a reliable supply of fresh water to the system.

Until 2005, Tasmania was self-reliant with regards to its electricity supply. That year the Basslink interconnector to the mainland state of Victoria was completed. This link was intended to provide both energy security, in case of drought in Tasmania, and also to provide renewable energy to Victoria, which relies heavily on coal for power generation. Following two years of exporting high volumes of energy to Victoria – in order to take advantage of high mainland power prices – Tasmania experienced an energy crisis. Hydro Tasmania's water storage levels decreased from being at more than 60% capacity in 2013 to below 15% in early 2016, followed by a period of low rainfall. The crisis was exacerbated by a fault in the link to the mainland.

Successful management of dams can also be thwarted by causing flooding. In Queensland, in January 2011, there were delays in releasing water from the Wivenhoe Dam until concerns for its structural integrity led to sudden large releases of water, which contributed to the flooding of South East Queensland, including the capital city Brisbane (Queensland Flood Commission 2012; van den Honert & McAneney 2011). More than 2.5 million people were affected, with approximately 29,000 homes and businesses experiencing some form of inundation. The cost of the flooding is estimated to be more than A$5 billion. A class action lawsuit case found the Queensland government and dam operators negligent and liable for damages that may exceed A$1 billion.

With Jennifer Marohasy, I have developed a technique for skilful monthly rainfall forecasting using artificial intelligence (AI). Through a series of peer-reviewed papers we have demonstrated that our method consistently generates significantly more accurate forecasts than are produced by the Australian Bureau of Meteorology using general circulation models (GCMs) (Abbot 2019; Abbot & Marohasy 2012; 2013; 2014; 2015a, b, c, d; 2016a, b; 2017a, b). It was GCMs that did *not* predict either the flooding that caused the dam overflow in Queensland, or the drought in Tasmania. The forecasts from GCMs for Tasmania have been shown to be consistently less skilful than climatology, which is worse than just calculating the long-term average rainfall and using this month by month.

In this chapter, I show how applying the latest techniques in AI, specifically through the application of ANNs, it is possible to generate skilful (much better than climatology) monthly rainfall forecasts for catchment areas in Tasmania. Such skilful rainfall forecasts would enable dam managers to anticipate floods and droughts. This is critical to the long-term sustainability of hydrogen gas production from hydroelectricity, and, in turn, rainfall-powered aviation.

Artificial intelligence and rainfall forecasting

The mainstream approach to rainfall forecasting involves attempting to simulate the actual physical processes using GCMs. The Predictive Ocean Atmosphere Model for Australia (POAMA) is a GCM developed

209

by the Australian Bureau of Meteorology (Hudson et al. 2011; Schepen et al. 2014). It was run operationally from 2002 until September 2018.

This GCM approach assumes ocean and atmospheric processes are well enough understood that they can be represented mathematically in a physical model to provide skilful weather forecasts. However, GCMs do not generally perform well at forecasting medium-term rainfall and they are very expensive to run. The GCMs produce output corresponding to very large grid areas and, therefore, the results require downscaling before useful localised rainfall (and other) forecast information can be generated. Of more concern, the forecasts are just not very good. The forecast skill for medium-term monthly forecasts generated by POAMA is generally only about equivalent to climatology, that is the long-term monthly average typically measured over a period of 30 years (Hawthorne et al. 2013). This could be simply calculated with a pen and paper, yet the Bureau, with a budget in the hundreds of millions of dollars, is achieving a worse result using an expensive supercomputer.

Over the last two decades, the application of AI methods – particularly ANNs, as an alternative to GCMs – has been researched for rainfall forecasting in many parts of the world (Wu & Chau 2013; Nayak et al. 2013). ANNs have been applied to rainfall forecasting over a wide range of forecast periods from hourly (Wei 2013), to monthly (Abbot & Marohasy 2017a), through to annual (Philip & Joseph 2003).

It is curious that ANNs have not been used to produce operational forecasts by the Australian Bureau of Meteorology – not even on a trial basis to act as a comparison with output from GCMs. Similarly, when Marohasy and our mutual colleague at the Institute of Public Affairs, Scott Hargreaves, visited Tasmania in 2017 their offer to brief staff of Hydro Tasmania – the state-owned enterprise that operates the hydroelectric system – was declined. They did meet with the Minster responsible for energy (then and now), Guy Barnett, but no reply to their follow-up correspondence, which included copies of published (and peer-reviewed) papers on using ANNs for rainfall forecasting was ever received (Hargreaves 2017). This indifference, if not hostility, contrasts with institutions elsewhere in the world. In addition to examples that can be found in the papers cited above, Marohasy and I have, through a collaborative program run by the Queensland Institute of Technology (QUT), trained employees of the Indonesian Bureau of Meteorology (BMKG) in ANN techniques through 2018 and 2019.

ANNs potentially enable complex relationships to be established between a set of input variables and a corresponding desired output variable, as shown in Figure 14.1. The technique is valuable for rainfall forecasting because many different physical factors may contribute to the output. Input data, which Marohasy and I have found most useful to input into ANNs when forecasting monthly rainfall for the Australian east coast, include the Southern Oscillation Index (SOI), which measures

Figure 14.1 Schematic of the ANN process

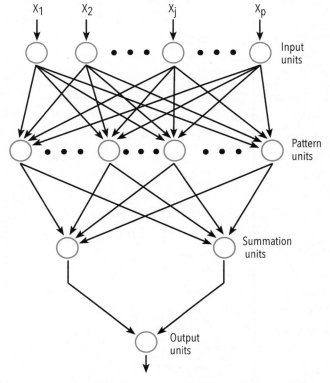

For the datasets used in this investigation, the optimal ANN model automatically selected by *Neurosolutions Infinity* was a general regression neural network (GRNN). The topology is a feedforward network that can be used to estimate an output value Y from a set of input values X1…..Xp.

pressure changes in the western Pacific; the four Niño indices (Niño 1.2, Niño 3, Niño 3.4, Niño 4) associated with surface sea temperature patterns across the Pacific Ocean; and the Dipole Mode Index (DMI), which is a measure of sea surface temperature gradients across the Indian Ocean. In addition to these climate indices, local rainfall and maximum and minimum temperatures, are also used as input variables.

The advanced ANN software that we use automatically investigates, selects and optimises those input variables useful in deriving relationships. Relationships and patterns are subsequently used to train the network so that forecasts can be made, assuming that these relationships and patterns will persist. Our forecasting method is thus not primarily dependent on an understanding of atmospheric circulation – rather it is about data mining. This is a completely different approach to that adopted by mainstream climate scientists, which falsely assumes they can simulate atmospheric and oceanic processes.

A land of droughts and flooding rains

Although Tasmania represents less than 1% of Australia's land area, it has 12% of Australia's fresh water supply and 27% of Australia's freshwater dam storage capacity (Bureau of Meteorology 2012). There are many long rainfall records for Tasmania, with some dating as far back as the mid-1800s well distributed over the state. The Hobart Botanical Gardens has rainfall records as far back as 1841. This makes Tasmania particularly suitable for ANN rainfall forecasting. For this study, 50 sites in Tasmania were selected from the Australian Bureau of Meteorology database with rainfall records generally extending back approximately 100 years. The locations of these sites are shown in Figure 14.2.

Like continental Australia, there is a long history of droughts and flooding rains, extending into recent years. It is common to hear opinions that present climatic conditions are 'unprecedented', whereas this turns out to be untrue when records of long enough duration are examined. Droughts are a recurring natural feature of the Australian and Tasmanian climate, with evidence of droughts dating back thousands of years (Barr et al. 2014; Gergis & Ashcroft 2013; Vance et al. 2015). Three periods of prolonged droughts have occurred in southe-astern Australia during the

Figure 14.2 Sites with long rainfall records used to forecast rainfall

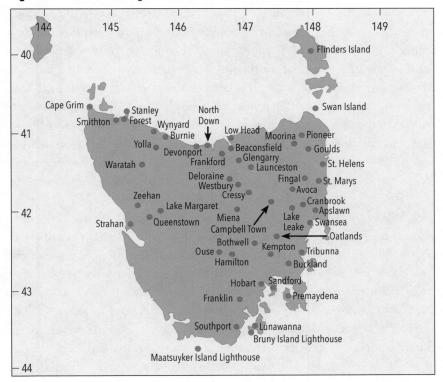

period of the Bureau's rainfall records. These are known as the 'Federation drought' (1895–1902), 'World War II drought' (1937–1945) and the 'Big Dry' (1997–2010).

Particular phases of climate indices are associated with droughts and floods. The SOI is a measure of the intensity of the Walker circulation, which relates to the strength of El Niño and La Niña events – with the former associated with drought conditions and the later with flooding (Tularam 2010; McBride & Nichols 1983). For most Australian regions, individual climate drivers associated with particular climate indices generally account for less than 20% of the monthly rainfall's variability (Risbey et al. 2009). Different multiple combinations of climatic processes represented by the climate indices need to be inputted into ANNs (Nicholls 2010; Kiem & Verdon-Kidd 2009) in order to forecast periods of extended drought, or to forecast flooding.

Although linkages between climate change and recent Australian droughts have been claimed with a high degree of certainty (Steffen et al. 2018), an examination of the scientific literature would suggest a more cautious approach is warranted (Cai et al. 2014). It is very important to take into consideration natural fluctuations in the climate before concluding that periods of prolonged drought are unprecedented and, therefore, must be attributable to climate change. For example, studies of the Millennium drought during 2003–2009, and the record-breaking rainfall and flooding in the summer of 2010–2011 in eastern Australia, by Cook et al (2016), found limited evidence for an impact from climate change. They found that rainfall variability during the Millennium drought fell within the range of natural variability over the last 500 years. Van Dijk et al. (2013) studied the Millennium drought in south-east Australia (2001–2009) and concluded that prevailing El Niño conditions explained about two-thirds of rainfall deficit in eastern Australia, but results for the state of South Australia were inconclusive with a contribution from global climate change plausible, but unproven.

Cycles of droughts followed by flooding rains tend to repeat, so not surprisingly more skilful rainfall forecasts are generally obtained using long duration data series (Abbot & Marohasy 2017b). The longer rainfall records generate more skilful forecasts when inputted into ANNs because there are likely to be examples of diverse behaviour in the training dataset of conditions encountered in the past – conditions that the ANN can potentially learn from, and subsequently apply to generating a forecast. Shorter rainfall records of only a few decades may not represent the full range of conditions previously encountered, including prolonged periods of both drought and flooding rains that are a common feature of the Australian climate.

The high variability of Tasmanian rainfall, both geographically and temporally, can be illustrated by reference to records for the towns of Queenstown, Ouse and Swansea, as shown in Figures 14.3, 14.4 and 14.5. Queenstown is located on the western side of the island in a region of Tasmania with high annual rainfall. Figure 14.3 shows the variation in annual rainfall for Queenstown between 1907 and 2019, with an average of 2534 mm, a maximum of 3295 mm and a minimum of 1752 mm (by way of comparison, this average is nearly two-and-a-half times the

Figure 14.3 Annual rainfall

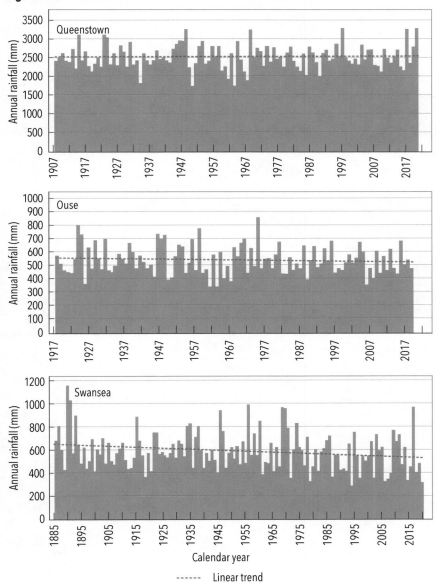

Source: Data from the Australian Bureau of Meteorology.

Figure 14.4 February monthly rainfall

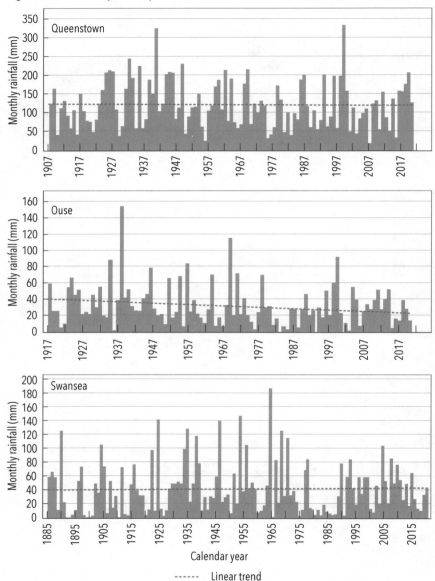

Source: Data from the Australian Bureau of Meteorology.

Figure 14.5 August monthly rainfall

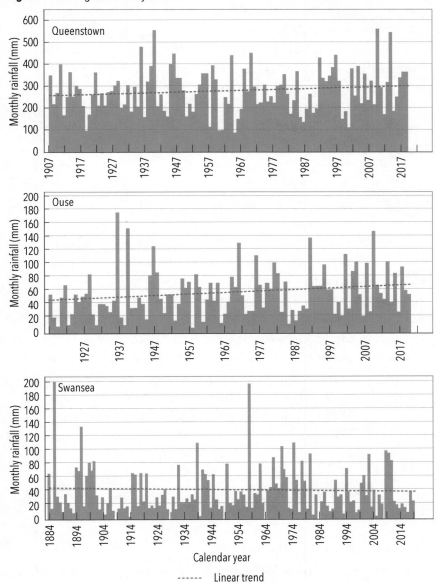

Source: Data from the Australian Bureau of Meteorology.

average rainfall of Galway, on the Atlantic coast of Ireland, and slightly more than Singapore, which is in the tropics). The trend line shows a slight increase in rainfall over the period.

Lower annual rainfall is generally observed on the east coast of Tasmania and in the central regions of the island. Ouse, located in the central region of Tasmania, has an average annual rainfall of 549 mm, with a maximum of 697 mm and a minimum of 348 mm, with annual rainfall between 1917 and 2019, shown in Figure 14.3. Swansea, located on the east coast of Tasmania, has an average annual rainfall of 590 mm, with a maximum of 1148 mm and a minimum of 247 mm, with annual rainfall between 1884 and 2019, as shown in Figure 14.3 (bottom). The trend lines in Figures 14.3 (middle) and (bottom) show a decrease in rainfall over the period.

The high variability monthly rainfall in summer and winter in these three locations in Tasmania is illustrated for the months of February and August respectively, in Figures 14.4 and 14.5 respectively. Queenstown has a minimum rainfall in February of 15 mm, a maximum of 330 mm, and an average of 132 mm. Ouse has a minimum rainfall of 1 mm, a maximum of 153 mm, and an average of 32 mm. Swansea has a minimum monthly rainfall of 0 mm, a maximum of 186 mm and an average of 42 mm. Trend lines show decreasing rainfall for Queenstown and Ouse, with increasing rainfall for Swansea during February.

For August, Queenstown has a minimum monthly rainfall of 91 mm, maximum of 542 mm and an average of 273 mm. Ouse has a minimum of 6 mm, maximum of 173 mm and an average of 56 mm. Swansea has a minimum of 4 mm, maximum of 197 mm and an average of 43 mm. Trend lines show increasing rainfall for Queenstown and Ouse in August, and decreasing rainfall for Swansea.

Forecasting rainfall for Tasmania

Climatology refers to the long-term average of temperature and rainfall, and is typically calculated over a 30-year period. It is easy enough to calculate this statistic for a particular location if measurements have been taken. Forecast accuracies, using different rainfall forecasting methods, can be compared after calculating percentage skill scores relative to climatology. A skill score can be calculated by application of the equation (1), using values of the root mean square error (RMSE) derived from values of the observed and forecast rainfall.

$$\text{Skill score} = \frac{\text{RMSE (climatology)} - \text{RMSE (model)}}{\text{RMSE (climatology)}} \times 100\% \quad (1)$$

The RMSE value is a measure of the difference between the forecast value and that calculated according to a particular modelling technique. Applying this equation, it follows that if the calculated values of RMSE from climatology and for a particular model are equal, the forecast skill score will be zero. For a perfect forecast, the RMSE for the model will be zero, and the calculated skill score 100%. Negative values calculated from equation (1) indicate a forecast skill score worse than climatology.

It is difficult getting the Bureau to provide its rainfall forecasts in a way that enables the calculation of a skill score, though there is a published paper by Hawthorne et al. (2013) that provides this information for different geographic areas of Australia, including Tasmania. In Hawthorne et al. (2013), the skill scores include forecasts made eight months in advance, which refers to the lead time. At this eight-month lead time, the forecast skill for rainfall across Tasmania using POAMA was below 0% for all months except June, typically in the range 0% to –4% (Hawthorne et al., 2013). This means the skill of POAMA is less than climatology.

For the month of June, the skill level using POAMA was in the range 0% to 4% for Tasmania. Hawthorne et al. (2013) described the skill of their monthly rainfall forecasts as 'low' and concede that monthly rainfall forecasting with POAMA remains 'a challenge'. Yet the questions are never asked as to why mainstream climate scientists persist with this method, particularly when there are better methods for forecasting, for example ANNs.

In order to be sure we can have a sustainable hydrogen industry, it is important to have some idea of the likely longer-term rainfall pattern within a catchment area. So far, GCMs are rarely used to forecast more than three months in advance, and at most eight months. Yet Marohasy and I have demonstrated the capacity of ANNs to forecast up to eighteen

Table 14.1 Average forecast skill scores for monthly rainfall forecasts for 50 locations in Tasmania with twelve months lead time.

Month	Average (%)
February	62.4
May	58.6
August	61.2
November	56.2

months in advance (Abbot & Marohasy 2016b). Specifically, we have shown that it is possible to achieve a skill score of between 40% and 70% with ANNs for various regions of Australia for lead times between one and twelve months (Abbot & Marohasy 2012; 2013; 2014; 2015a, b, c, d; 2016a, b; 2017a, b). In all cases, the skill scores were above 20%.

Table 14.1 shows average skill scores calculated using ANNs for all 50 sites in Tasmania for the months of February, May, August and November, with twelve months lead time (i.e. twelve months in advance). The skill scores lie in the range 56.2% to 62.4%, indicating that the forecasts are significantly better that using climatology.

Considering just three of these 50 sites in Tasmania, let us visually compare the actual monthly rainfall with the ANN forecast. This is illustrated by simply charting the actual and forecast amounts of rain that fell for the months of February at Campbell Town, May at Franklin and August at Zeehan, over a twelve-year test period as shown in Figure 14.6.

Each figure shows a set of blue bars representing the observed rainfall amounts at the location for the month specified. There is a corresponding red bar adjacent to each blue bar that shows the ANN forecast amount of rainfall for that site with a twelve months lead time, so that the ANN is predicting how much rainfall to expect one year ahead of the observation. Visual inspection shows that, in general, the ANN forecasts are able to predict instances of higher and lower rainfall in accordance with the amounts actually observed. For example, Figure 14.6 shows that for Campbell Town in the month of February, significantly less rain fell in 2007 compared to 2010, and this was predicted by the ANN using only input data available one year prior to the

Figure 14.6 Observed and forecast monthly rainfall

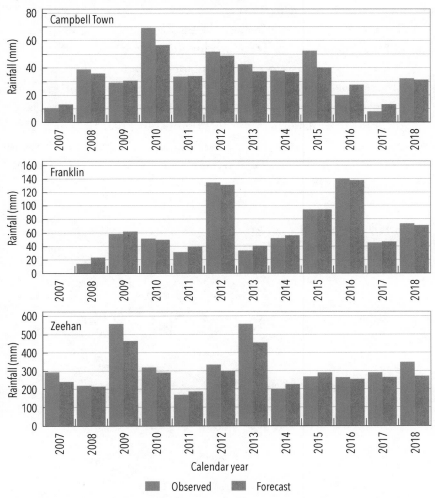

Top: Campbell Town during February; Middle: Franklin during May; Bottom: Zeehan during August. Twelve-month lead time over test period 2007–2018.

observations. Figure 14.6 (middle) shows that for Franklin during the month of May, low rainfall (less than 20 mm) was observed in 2008, while in 2012 rainfall was much higher than average (above 120 mm). The ANN provided forecasts enabling the low and high rainfall events to be differentiated as shown in Figure 14.6 (middle), in each case providing the ANN with data available only up to twelve months before the required prediction.

Because there are many regions in Tasmania with long records, it is possible to generate regional forecasts for entire catchments and the entire state. The ANN forecasts were generated for each of the 50 individual sites distributed over the state (see Figure 14.2). The results from these ANN runs were then entered into TeraPlot software – as previously detailed for South East Queensland in Abbot & Marohasy (2017a,b) – in order to generate a regional contour map for the entire state of Tasmania, as shown in Figure 14.7. This is a rainfall forecast for August 2017 at twelve months lead time derived from 50 individual forecasts.

Figure 14.7 ANN rainfall forecast map for Tasmania for August 2017

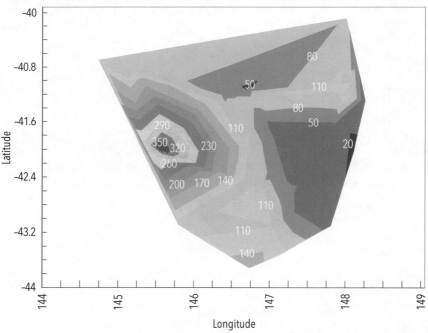

The isohytes show the predicted rainfall for that part of Tasmania for August 2017 as derived from 50 ANN runs. Much more rain was predicted for regions in the west, than east of the state of Tasmania.

Source: Image created using output from ANN with TeraPlot software.

Conclusion

Regardless of the potential of a hydropower–hydrogen nexus, it should be of some concern to policymakers and taxpayers that the Bureau's very expensive GCM, which was developed over a period of two decades, provides a worse rainfall forecast than would be provided by relying simply on the long-term average rainfall that could be generated with a handheld calculator from historical data.

That a real opportunity could go begging – and a $50 million invest-ment go to waste – because of an institutional unwillingness to innovate among both the Bureau and organs of the Tasmanian government should also be of interest and concern to those who genuinely want to work towards a clean energy future.

15 Wildfires in Australia: 1851 to 2020

Dr Jennifer Marohasy

The word 'unprecedented' is often applied to destructive climate-related events, as though they could not have been predicted, and have never happened before. This creates anxiety because it follows that there really is a climate emergency and that we have pushed nature out of balance. For example, the forest fires in the south-east of Australia in the summer of 2019–2020 have become a symbol of catastrophic human-caused global warming. The words 'unprecedented' and 'catastrophic' were used by the President of the Australian Academy of Science, Professor John Shine, who said, 'the scale of these bushfires is unprecedented anywhere in the world'.[1]

Severe bushfires are weather-related in that they are usually preceded by high temperatures, low relative humidity, strong winds and a rainfall deficit; but there must also be critically high 'fuel loads'. Fuel is live and dead vegetation that accumulates over time. When fuel loads are kept low through regular hazard reduction burning, fires are unlikely to kill all the understorey plant species or the tree layer. The trunks and roots of eucalyptus trees will only burn if the fuel load is critically high.

The area of forest in the south-east of Australia has been expanding even as public support for the necessary hazard reduction burning has been waning. Support has been waning because hazard reduction burning involves active management, which is generally considered incompatible with wilderness values. The concept of wilderness derives from the idea of

---

1   https://www.science.org.au/news-and-events/news-and-media-releases/statement-regarding-australian-bushfires.

a balance of nature, which does not actually have a basis in the ecological sciences. It is a myth: a widely held but possibly false belief or idea.

Excluding appropriate fire regimes from open eucalyptus forests will not keep them in their current state, rather it can have the effect of simply making them more prone to severe fires that will eventually destroy them. Further, if these forests do not receive moderately intense fires within their life span of up to 350 years – for seed germination and seedling establishment – they will eventually be replaced by temperate rainforest species with closed canopies. This is ecological succession.

A rainforest supports a quite different fauna and flora. Koalas do not live in rainforests. Koalas live in eucalyptus forests. To be clear, eucalyptus forests can evolve into rainforests under particular conditions.

For traditional Aboriginal Australians, the concept of wilderness was not a cause for fond nostalgia but was perceived as a land without custodians. The first Australians deliberately burned open eucalyptus forests at times and in ways that kept fuel loads low. This is because they preferred this type of more open forest. Documentary evidence for this conclusion is provided by historian Bill Gammage in his book *The Biggest Estate on Earth: How Aborigines Made Australia* (2011), where he shows that before the arrival of the First Fleet in 1788 – which founded the first European settlement and penal colony – the landmass of Australia was actively managed by Aboriginal people in a sophisticated and systematic manner.

Major fires

Since that first European settlement, Australia has had a long history of major fires that incinerate whole forests, destroy property and kill people.

About 5 million hectares of the state of Victoria reportedly burned at the time of the Black Thursday bushfires in February 1851. Five million hectares is a vast area, and nearly one quarter of the landmass of Victoria. In January 1939, almost 2 million hectares burned, which is a lot less than 5 million, but a lot more than the 1.4 million hectares, which is the extent of the unplanned severe fires across Victoria in the summer of 2019–2020.

More extensive areas burned in southern New South Wales, with the government reporting that it was as much as 5.4 million hectares. It has been claimed that this is unprecedented in extent, except that somewhere in the vicinity of 7.2 million hectares burned during the summer of 1951–1952. A CSIRO report titled the 'Bushfire History of the South Coast Study Area' details fires at Bega, Nowra, Mt Dromelly and other parts of southern New South Wales in 1951 (Duggin 1976). Fires consumed that same coastal strip again in 1994 (Cheney 1995).

In the summer of 2019–2020, 20 million hectares of the land mass of Australia may have burned. This is an extraordinarily vast area considering much of it was in the south east. A similarly vast area of 21 million hectares was lost to unplanned fires as recently as 2012–2013, mostly in Queensland and the Northern Territory ('State of the Forests Report 2018'). However, this is not the largest area burned by uncontrolled fires. In 1974–1975, 117 million hectares burned (Cheney 1995).

It could be argued that the fires in 1974–1975 were predominately burning grasses and forbs (herbaceous flowering plants) and are not comparable to the fires of the summer of 2019–2020. About 25% of the landmass of Australia is tropical savannah, as shown in Figure 15.1. Some would argue much of this area should be burned with low-intensity, early season fires every two years, or at least every five years.

There is no consistent Australia-wide methodology for determining the severity of fires, or even the amount of forest burned by either wildfire or through prescribed burning each summer (Department of Agriculture, 2020). The five-yearly 'State of the Forests Report' (SOFR) series only began reporting on the area burnt Australia-wide in 2001 and has data only until 2016 (Australian Government, 2008; 2013).

Some claim that while the fires this last summer may not have been unprecedented in terms of area burned, they were more ferocious – able to turn day into night.

Through live streaming on social media, we witnessed the seaside town of Mallacoota become as dark as night as a firestorm engulfed the small town in the East Gippsland region of Victoria on 31 December 2019. Few Australians have any direct experience of such phenomena. Yet reports describing past bushfires portray similar scenes. For example, on prominent display at the State Library of Victoria is a painting of the devastating 1851 bushfires by William Strutt showing just this scenario.

Most pertinently the 'Report of the Royal Commission to inquire into the causes of and measures taken to prevent the bush fires of January 1939

Figure 15.1 Map of Australia

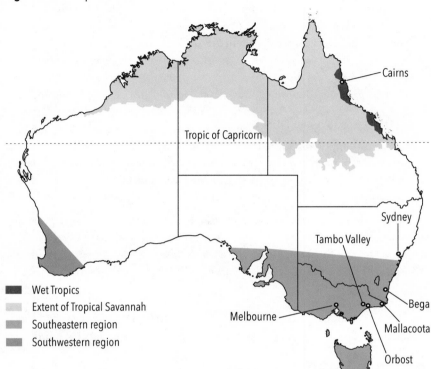

The area of tropical rainforest has historically always been much smaller than the area of tropical savannah across northern Australia. These areas are broadly defined by regional ecosystem mapping. The Southeastern and Southwestern regions are categories defined for calculating regional weather and climate statistics by the Australian Bureau of Meteorology.

Source: Area of savannah as defined by Fox et al. (2001); regions of southwestern and southeastern Australia as defined by the Australian Bureau of Meteorology.

and to protect life and property and the measures to be taken to prevent bush fires in Victoria and to protect life and property in the event of future bush fires' (Victoria 1939) by Judge Leonard Stretton begins:

> In the State of Victoria, the month of January of the year 1939 came towards the end of a long drought which had been aggravated by a severe hot, dry summer season. For more than twenty years the State of Victoria had not seen its countryside and forests in such travail. Creeks and springs ceased to

run. Water storages were depleted. ... The rich plains, denied their benef-
icent rains, lay bare and baking; and the forests, from the foothills to the
alpine heights, were tinder. The soft carpet of the forest floor was gone;
the bone-dry litter crackled underfoot; dry heat and hot dry winds worked
upon a land already dry, to suck from it the last, least drop of moisture.
Men who had lived their lives in the bush went their ways in the shadow of
dread expectancy. But though they felt the imminence of danger they could
not tell that it was to be far greater than they could imagine. They had not
lived long enough. The experience of the past could not guide them to an
understanding of what might, and did, happen. And so it was that, when
millions of acres of the forest were invaded by bushfires which were almost
State-wide, there happened, because of great loss of life and property, the
most disastrous forest calamity the State of Victoria has known.

... it appeared that the whole State was alight. At midday, in many places,
it was dark as night. Men carrying hurricane lamps, worked to make safe
their families and belongings. Travellers on the highways were trapped by
fires or blazing fallen trees and perished. Throughout the land there was
daytime darkness.

It is the case that in 1851, 1939, and also in 2019 at Mallacoota, there
were apocalyptic scenes as day turned to night because of the wildfires.

East Gippsland – a perspective

East Gippsland in Victoria is an area that extends from the seaside town
of Mallacoota, up into the mountainous Kosciusko National Park,
with its snow ski resorts. This is rugged terrain with deep gorges, where
communities were historically dependent on grazing or timber milling.

Many of the fires in East Gippsland during the summer of 2019–2020
were started by lightning strikes, specifically in November 2019, and in the
vicinity of the Tambo Valley. These wildfires joined up, eventually burning
more than 320,000 hectares over a period of 91 days. The media blamed
the fires on 'one of the driest landscapes on record', while the president of
the East Gippsland Wildfire Taskforce Inc., John Mulligan, blamed the
fires on government's failure to achieve sufficient fuel reduction. Mulligan
has claimed this has put at risk not only people and property, but also
the very survival of threatened species and ecological communities within
these forests (Milovanovic 2008; Mulligan 2012; 2018).

Figure 15.2 Forest fire East Gippsland, December 2019

Source: Reprinted with permission from State of Victoria (Department of Environment, Land, Water and Planning).

Mulligan was born in Orbost in 1931; his grandparents came to East Gippsland in the 1880s and ran a trading store and wattle bark business. He claims that until the 1940s this vast area of diverse forest types was resistant to severe bushfires because the first settlers continued the Aboriginal practice of patchwork burning (Mulligan 2012).

The forests of East Gippsland were described by the first European settlers as 'park-like' and 'clean-bottomed' (Gammage 2011). This was attributed to burning by Aborigines. It was also recognised that the exact nature of their fire regime could be difficult to replicate.

According to Mulligan, while much of Victoria burned in 1939, the forests of East Gippsland escaped the inferno because they were more open and had much lower fuel loads relative to West Gippsland. Mulligan recalls that Black Friday (13 January 1939) was a 'terribly hot and windy day'. He remembers a fire that was put out to save the Cabbage Tree Hotel on the road from the Cann River to Orbost (Mulligan 2018).

Mulligan refers to the forests east of Orbost as being under a different management regime until the 1940s – a regime that made them more resistant to uncontrolled wildfires.

A map of the fires in 1939, published by the Forest Commission of Victoria supports this contention to some extent, see Figure 15.3. There were fires that did burn considerable areas of East Gippsland at that time, but they did not join up.

It was not until the 1950s that the Forestry Commission of Victoria became interested in East Gippsland (Mulligan 2018). With new tracks, 4WD vehicles, aircraft and more, the government-employed foresters put a stop to the local burning practices that had been continued by the first European settlers of East Gippsland based on traditional Aboriginal practices. According to Mulligan this resulted in the fuel loads building to dangerous levels, giving rise to the 1983 fires – and many other severe fires – that destroyed tall eucalyptus trees and the unique flora and fauna they supported.

Since 1983, it is generally considered that the area of fuel reduction burning carried out in East Gippsland has been completely inadequate. Six months before the recent summer of bushfires, retired Monash University researcher David Packham told SBS News[2] fuel loads were the heaviest they had been since human occupation of the continent and Aboriginal methods needed to be adopted. Responding to claims before this last summer that climate change posed a risk to the forests, Packham is on the public record stating:

> If there is any global warming, the global warming is so slow and so small that the bushfire event is totally overrun by the fuel state.

> In the Australian context you need fire to keep the bush healthy and safe.

> If we got to 10 per cent (fuel reduction burning) then our area burnt would drop by 90 per cent and our intensity would drop by at least that and undoubtedly more ...

---

2   https://www.sbs.com.au/news/fires-not-due-to-climate-change-expert.

Figure 15.3 Area burned by forest fires in Victoria in 1939

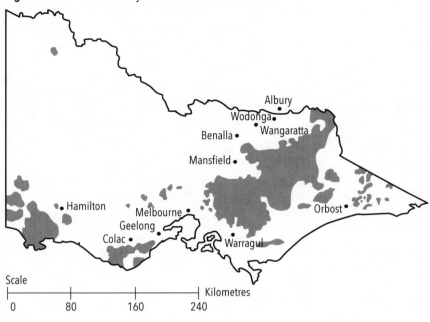

Source: Forest Commission of Victoria.

In the years prior to 2009, only 1.7% of forested land had fuel reduction burning. The Victorian Royal Commission into the 2009 bushfires recommended a target of 5–8% of the public land be burned annually. While some argue an even higher target of 15–20% of the forest burned each year is more appropriate. The point is moot as the actual rates in practice fall far short of that recommended by the Royal Commission. In 2019, for instance, the area burned was just 34% of the 5–8% target recommended by the Royal Commission.[3]

South-western Australia – traditional burning

Before Europeans arrived, an Aboriginal tribe known as the Noongar people managed the dry forests and woodlands in the far south-west of Western Australia. As David Ward, whose doctoral thesis (Ward 2010) was on this topic, has remarked:

---

3 https://www.theaustralian.com.au/nation/politics/bushfires-victoria-wide-of-mark-on-target-burnoffs-in-2019/news-story/67c81e97457bb8de877bd5c2d9a4bacb.

The Noongar lived in these forests without fire trucks, water bombers, helicopters, television journalists, concerned politicians, the Conservation Council, hundreds of firefighters, or the Salvation Army to give them breakfast. They burned frequently, in most places as often as it would carry a mild, creeping fire.

Even where there were no Noongars, Dr Ward explains that much of this landscape would have burned following lightning strikes, and these may have continued to burn for months until put out by the autumn rains. This was stopped with European settlement, leading to a decline in fire frequency as shown in Figure 15.4.

The first ordinance to 'diminish the dangers from bush fires' was established in Western Australia in 1847. Soon after this, Bushfire Acts were also legislated in South Australia and Tasmania. A Conservator of Forests was appointed in Victoria in 1888. The focus was on protecting the states' forests from over-cutting, over-grazing, and fire. Legislation was further strengthened with the lighting of fires prohibited at

Figure 15.4 Reconstructed eucalyptus forest fire history

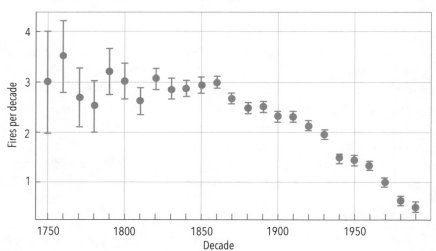

The data matrix used to compile this figure consisted of 1548 records, each record including fields for the number of fires in that decade and vegetation type. There are large standard errors in the earliest decades because of the smaller sample size. The drop-off in fire frequency between 1860 and 1870 coincides with a measles epidemic that decimated the Noongar population.

Source: David Ward 2010.

particular times of the year, while forestry departments and volunteer bushfire brigades were better resourced for fire suppression.

There are many historical references to frequent, widespread burning by the Noongars in Western Australia. In 1975, Mr Frank Thompson was interviewed about his memories of fire near the south coast, before World War I. He said:

> You see, the Natives … they used to burn the country every three or four years … when it was burnt the grass grew and it was nice and fresh and the possums had something to live on and the kangaroos had something to live on and the wallabies and the tamars and boodie rat … It didn't burn very fast because it was only grass and a few leaves here and there and it would burn ahead.

Grasstrees (*Xanthorrhoea spp.*) can live for more than 400 years (Ward & Van Didden 2003). They grow tops with a thatch that is flammable. When they burn in summer, a surrounding ground fire usually results from the thatch dropping to the ground igniting other grasstrees, or at least the short ones. Fire marks on these trees have been used to reconstruct the history of bushfires in the eucalyptus forests of south-western Western Australia as shown in Figure 15.5.

According to Ward, fuel loads, which are a form of leaf litter in these forests, will not rot down to enrich the soil. While there may be some decay in winter, in summer the quantity of dead vegetation (leaves, bark, and capsules) is overwhelming. After twenty years or so, there may be a mulching effect, and build-up ceases. However, according to Ward, by then most wildflowers are smothered, and most of the nutrient is locked up in dead matter.

Northern Australia, woody weeds replace savannah

About 25% of Australia is tropical savannah, which is broadly defined as grassland, or woodland with a grassy understorey. The area of rainforest, even before European settlement, was by comparison small in northern Australia, as shown in Figure 15.1. This is not generally understood; many associate north Queensland with rainforest, when, in fact, it was mostly open grassland or open woodland before European settlement.

Until recently, fire was considered a major factor controlling the rainforest to tropical savanna boundary. Research over the last 30 years, however, has shown that even when eucalyptus woodland is protected

Figure 15.5 After a mild patch burn, south-western Australia

Mild patch burn at the Scott River, south-western Australia, on 30 January 2007 burning the canopy of only the short grass trees.

Source: David Ward.

from fire, only low numbers of rainforest seedlings will establish (Bowman 1988; Fensham & Bowman 1992; Bowman & Panton 1995). Tropical rainforests require higher soil fertility, higher soil moisture and mycorrhizal fungi (Bowman & Panton 1993) as well as protection from fire.

While there is a perception that the area of rainforest was once much greater, in reality it is the area of grassland that has significantly contracted since at least the 1950s. It has not been replaced by rainforest, but rather with what are often referred to as woody weeds – an invasive scrub of mostly *Acacia spp.*, both native and introduced (Burrows 2016). This change to woody, less open vegetation is generally thought to be a consequence of the wider availability and use of 4WD vehicles, graders, water tankers and portable water pumps for fire suppression (Burrows 2016).

Significantly, across northern Australia there does appear to be a better appreciation of the relevance of traditional knowledge in relation to landscape and fire (Cooke 1998). For example, patchwork burning is incorporated into Rural Fire Service Guidelines in Queensland, which emphasises burning not only for conservation, but also for hazard reduction and primary production (grazing) outcomes. There are restrictions, nevertheless, on the use of fire for suppression of woody weeds on many farms. This is in areas classified as remnant vegetation under the *Vegetation Management Act 1999*. An issue for many landholders is that this scrub needs to have only existed for fifteen years for it to be classified as remnant (Neldner et al. 2019). As a consequence, some landholders and some ecologists believe there will be more severe wildfires in Queensland with the encroachment of woody weeds into country that was once more open (Burrows 2016).

Concepts of climate – understanding the statistics

In February 2020, the leader of the Paris Climate Agreement talks, Christiana Figueres, described the bushfires across south-east Australia as the worst disaster to have ever hit the planet.[4] She also claimed the Australian prime minister was denying climate change. In fact, at that time Prime Minister Scott Morrison repeatedly said that summers are becoming hotter and drier.[5] These are the same reasons that some Rural Fire Service chiefs have given for inadequate hazard reduction burning over recent decades; specifically, that weather conditions have prevented it.

Climate is the long-term average of weather, typically averaged over a period of 30 years. The claim that summers are getting hotter and drier should be easy to evaluate in terms of the key variables of temperature and rainfall, respectively. Furthermore, there are statistics for both temperature and rainfall for south-eastern Australia, the region worst affected by the recent fires.

Considering rainfall: since 1900 the wettest summer was in the last decade, in 2010–2011, as shown in Figure 15.6. A mean average record rainfall of 301 millimetres was recorded across south-east Australia during that December, January and February period. This very wet summer was unprecedented for this region since records began in 1900.

---

4   https://www.abc.net.au/triplej/programs/hack/world-top-climate-negotiator-condemns-australian-response-bushfi/11989658.

5   https://www.smh.com.au/politics/federal/pm-s-bushfire-response-must-include-climate-change-experts-20200129-p53vtl.html.

There have been fewer exceptionally dry years when one considers the trend since about 1985. If anything, these official statistics suggest it is getting wetter, as shown in Figure 15.6. Rainfall statistics for the entire Australian continent, available for download from the Australian Bureau of Meteorology, also indicate that more recent years have been wetter, especially the last 50 years (Marohasy 2020).

The temperature data, which is also available for download from the Bureau, is more difficult to interpret because of changes in how temperatures have been recorded over recent years (Marohasy 2018). In order to understand the extent of the problem, let's begin with a single temperature record from the south east of Australia.

The Rutherglen agricultural research station has one of the longest, continuous, temperature records for anywhere in rural Victoria

Figure 15.6 Summer rainfall south-east Australia

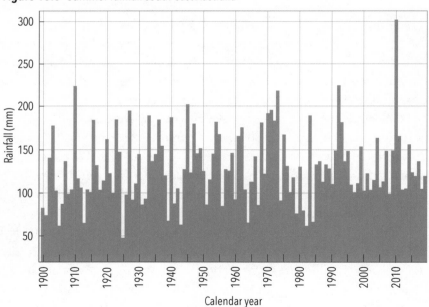

Summer is defined as the months of December, January and February. The area defined as 'Southeastern' by the Australian Bureau of Meteorology is shown in Figure 15.1.

Source: http://www.bom.gov.au/climate/change/#tabs=Tracker&tracker=timeseries&tQ=graph%3Drain%26area%3Dseaus%26season%3D1202%26ave_yr%3D0.

(Marohasy 2016; 2017). Minimum and maximum temperatures were first recorded at the research station using standard and calibrated equipment back in November 1912. Considering the first 85 years of summer maximum temperatures (1910 to 1997), as recorded with a mercury thermometer in a Stevenson screen, they tend to cycle between about 32 ° and 28 °C, with the exception of the summers in 1938–1939 and 1980–1981, which were somewhat hotter, see Figure 15.7.

At Rutherglen, the first equipment change happened on 29 January 1998 (Bureau of Meteorology 2019). That is when the mercury and alcohol thermometers were replaced with an electronic probe – custom built to the

Figure 15.7 Mean maximum summer temperatures at Rutherglen

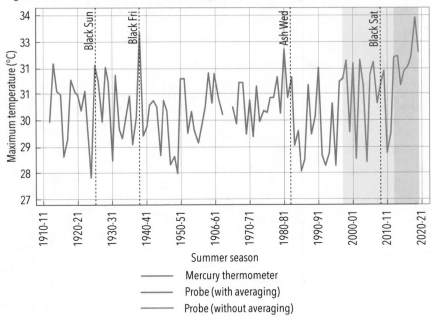

Temperatures recorded with a mercury thermometer are shown by the green line. The change to an electronic probe is shown by the blue line. Then, in 2012, the Bureau stopped numerical averaging, so the hottest temperature recorded each day is now a one-second spot reading, as shown by the red line.

The time of the four deadliest fires (Black Sunday, Black Friday, Ash Wednesday and Black Saturday) are shown by way of a dashed line.

Summer is defined by the months December, January and February.

Source: Monthly values downloaded from http://www.bom.gov.au/climate/data/.

Australian Bureau of Meteorology's own standard, with the specifications still yet to be made public (Marohasy 2018). According to Bureau policy (Trewin 2012, p. 26) and also World Meteorological guidelines, when such a major equipment change occurs there should be at least two years (preferably five) of overlapping (i.e. parallel) temperature recordings. Except the mercury and alcohol thermometers (used to measure maximum and minimum temperatures, respectively) were removed on exactly the same day that the custom-built probe was placed into the Stevenson screen at Rutherglen, in direct contravention of this policy.

In 2011, the Bureau made further changes to how it measures temperatures in that it stopped averaging one-second readings from the probe at Rutherglen over one minute (Marohasy 2018). The maximum temperature as recorded each day at Rutherglen is now the highest one-second spot reading from the custom-built electronic probe.

Figure 15.8 Mean maximum summer temperature for locations with longest continuous records

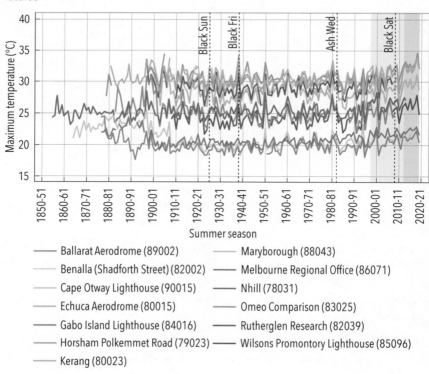

An electronic probe is likely to be much more sensitive to fluctuations in air temperature – as in able to detect transient localised pockets of hot air because of its shorter response time relative to a mercury thermometer – so numerical averaging of one-second readings over at least one minute is standard to ensure equivalence with historical data (Lin & Hubbard 2008). Given the Bureau is no longer averaging, it is unclear whether the uptick in maximum summer temperatures at Rutherglen, as shown in Figure 15.7, is real, or an artefact of the changed measurement method.

Considering other long temperatures series from Victoria, summer temperatures have generally cycled within a narrow band, until about 2011 when there appears to be a step-up in temperatures, as shown in Figure 15.8. Only the Black Friday inferno occurs at the peak in one of these cycles.

Previous page: Overall summer temperatures generally cycled within a band that does not show an increasing trend, until numerical averaging stopped in 2012. The area with the darker orange tint corresponds with measurements from one-second spot readings. Values are as recorded and have *not* been corrected for the urban heat-island effect.

The time of the four deadliest fires - Black Sunday, Black Friday, Ash Wednesday and Black Saturday - are shown by a dashed line.

Summer is defined as the months of December, January and February.

The thirteen stations are

89002. 1908–2018 (111 years: N = 110, M = 1) Ballarat Aerodrome
82002. 1903–2005 (103 years: N = 102, M = 1) Benalla (Shadforth Street)
90015. 1864–2018 (155 years: N = 154, M = 1) Cape Otway Lighthouse
80015. 1881–2018 (138 years: N = 137, M = 1) Echuca Aerodrome
84016. 1877–2018 (142 years: N = 138, M = 4) Gabo Island Lighthouse
79023. 1898–2011 (114 years: N = 101, M = 13) Horsham Polkemmet Rd
80023. 1903–2018 (116 years: N = 108, M = 8) Kerang
88043. 1899–2018 (120 years: N = 111, M = 9) Maryborough
86071. 1855–2013 (159 years: N = 159, M = 0) Melbourne Regional Office
78031. 1897–2007 (111 years: N = 107, M = 4) Nhill
83025. 1879–2008 (130 years: N = 127, M = 3) Omeo Comparison
82039. 1912–2018 (107 years: N = 105, M = 2) Rutherglen Research
85096. 1877–2018 (142 years: N = 141, M = 1) Wilsons Promontory Lighthouse

N is the number of data points/number of summers; M is missing data points/missing summers.

Source: Monthly values downloaded from http://www.bom.gov.au/climate/data/.

Conclusion

The wildfires that burned so much of Australia during the summer of 2019–2020 have been repeatedly described as unprecedented and blamed on climate change. But without supporting statistics.

In reality, areas as extensive and fires as ferocious have burned Australia since at least 1851.

Rainfall data indicates it is getting wetter (not drier) in south-eastern Australia, the region worst hit by the recent fires. While the official statistics show unambiguously that summer rainfall has been increasing (not decreasing as claimed in the popular press), it is difficult to know if air temperatures are increasing, or not, because of all the changes to how the Bureau measures temperature.

The change from mercury thermometers to electronic probes in 1996, and the more recent change from numerical averaging to just taking one-second spot readings since 2012, makes it impossible to construct meaningful continuous historical temperature series. In other parts of the world where electronic probes are used to measure air temperature, the instantaneous one-second readings are averaged over at least one minute, and in the USA over five minutes (Lin & Hubbard 2008).

Vast areas of open eucalyptus forests did burn in south-east Australia this last summer. These wildfires incinerated millions of hectares of important habitat for rare and endangered animal species, and koalas in East Gippsland. The impact on local human communities has also been devastating, with so many lives lost and so much property destroyed.

Across Australia, a better understanding of traditional Aboriginal burning methods, and the types of vegetation that existed before European settlement, could result in more support for the active management of wilderness areas, including for more hazard reduction burning. A focus on hazard reduction burning to keep landscapes generally more open and thus safer for people and wildlife, would be more useful than blaming climate change – at least until there is better quality assurance of actual temperature measurements.

16 Rewriting Australia's Temperature History

Dr Jennifer Marohasy

The Bureau of Meteorology is one of Australia's most important national public institutions; so much depends on our knowledge of past climate and our ability to predict future trends and events. The Bureau's recordings of maximum and minimum temperatures across Australia provide information that is important to our understanding of such trends and patterns in a changing climate. The Bureau does not, however, simply report this information. Rather, complex remodelling is undertaken in the development and compilation of the official climate change statistics. In this chapter, I consider the results of the Bureau's 'homogenisations' of historic temperature measurements through two case studies: Darwin and Rutherglen.

The Bureau undertook some remodelling in 2018 that increased the overall rate of warming by 23% between Versions 1 and 2 of the Australian Climate Observations Reference Network – Surface Air Temperature (ACORN-SAT) database (Trewin 2018). The Bureau claims that this remodelling of temperatures is justified for two reasons: because of changes to the equipment used to record temperatures; and because of the relocation of the weather stations. However, there have been no changes to equipment and no relocations since the release of ACORN-SAT Version 1 for either Darwin or Rutherglen.

Case study: Rutherglen

At an agricultural research station near Rutherglen in south-eastern Australia, maximum and minimum temperatures have been recorded since November 1912 in a Stevenson screen in a paddock, as shown in Figure 16.1. This is an official Bureau of Meteorology weather station with values used to calculate official statistics. The trend in the raw minimum temperatures is for a slight cooling, which is a consequence of land-use change: specifically, the staged introduction of irrigation into the region for cropping, vineyards and orchards (Marohasy 2016).

In 2014, Graham Lloyd, Environmental Reporter at *The Australian*, quoting me, explained how the cooling trend in the minimum temperature record at Rutherglen had been changed into a warming trend by the Bureau by progressively reducing temperatures from 1973 back to 1912, as shown in Figure 16.2. For the year 1913, there was a large difference of 1.8 °C between the mean annual minimum temperature, as measured

Figure 16.1 Location of the temperature recording equipment at Rutherglen

The white circle marks the location of the Rutherglen weather station, with the associated image showing a standard Stevenson screen.

Source: Map data from Google, DigitalGlobe; image of Stevenson screen from Bureau website (http://www.bom.gov.au/climate/cdo/about/airtemp-measure.shtml, viewed 14 May 2017) and reprinted under Creative Commons (CC) Attribution 3.0 Licence.

at Rutherglen using standard equipment at this official weather station, and the remodelled ACORN-SAT Version 1 temperature. Remodelling the data to cooling the past relative to the present in this way has the general effect of making the present appear hotter.

The Bureau responded to Lloyd, claiming the changes were necessary because the weather recording equipment had been moved between paddocks. This is not a logical explanation where the local terrain is flat and, furthermore, the official ACORN-SAT catalogue clearly states that there has never been a site move (Bureau of Meteorology 2012).

Nevertheless, readers might want to give the Bureau the benefit of the doubt and let them make a single set of changes. But just six years

Figure 16.2 Homogenisation of Rutherglen's minimum temperatures

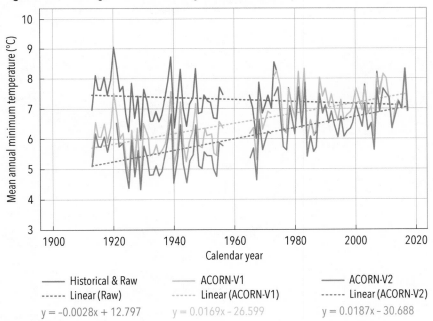

The historical observations (green) show a mild cooling trend of minus −0.28 °C per 100 years. This cooling trend has been changed to warming of 1.7 °C per 100 years in ACORN Version 1 (orange). These temperatures have been further homogenised/remodelled in ACORN Version 2 (red) to give a slightly more dramatic warming, which is now 1.9 °C per century.

Source: Raw data was downloaded as daily minimum from http://bom.gov.au/climate/data/ and the ACORN-SAT data was downloaded from ftp://ftp.bom.gov.au/anon/home/ncc/www/change/ACORN_SAT_daily/.

later, the Bureau again changed the temperature record for Rutherglen. In Version 2 of ACORN-SAT, the minimum temperatures as recorded before 1970 at the Rutherglen Research Station have been further reduced, making the present appear even warmer relative to the past. The warming trend is now 1.9 °C per century.

The Bureau has variously claimed that they need to cool the past at Rutherglen to make the temperature trend more consistent with trends at neighbouring locations. But this claim is not supported by the evidence. For example, the historical recordings at the nearby towns of Deniliquin, Echuca and Benalla also show cooling, as shown in Figure 16.3. The consistent cooling in the minimum temperatures at all these locations

Figure 16.3 Mean annual measured raw minimum temperatures at Ruthergen and nearby locations

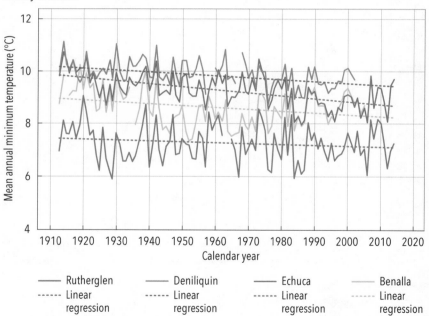

The chart shows the night-time temperatures as recorded at Benalla (1913-2005), Echuca (1913-2014), Deniliquin (1913-2002), and Rutherglen (1913-2014). The shorter records for Benalla and Deniliquin represent the longest complete records for these locations as measured at the same site.

Source: Data sourced from Climate Data Online, Australian Bureau of Meteorology, December 2016 http://www.bom.gov.au/climate/data.

has been caused by land-use change in this region – specifically, the development of water infrastructure and irrigation from the 1930s through to the 1970s (Marohasy 2016).

The changes to the raw data begin with changes to the daily temperatures. For example, on the first day of temperature recordings at Rutherglen, on 8 November 1912, the measured minimum temperature is 10.6 °C, as shown in Table 16.1. This measurement is changed to 7.6 °C in ACORN-SAT Version 1. In Version 2, the already remodelled value is changed again, to 7.4 °C – applying a further cooling of 0.2 °C.

If we consider historically significant events – for example temperatures at Rutherglen during the January 1939 bushfires that devastated large areas of Victoria – the changes made to the historical record are even more significant. The minimum temperature on the hottest day was measured as 28.3 °C at the Rutherglen Research Station. This value was changed to 27.8 °C in ACORN Version 1, a reduction of 0.5 °C. In Version 2, the temperature is reduced by a further 2.6 °C, producing a temperature of 25.7 °C.

This type of remodelling of temperature series will potentially have implications for understanding the relationship between past temperatures and bushfire behaviour. Of course, changing the data in this way

Table 16.1 Daily minimum temperatures Rutherglen, measured and homogenised

Station No.	Date	Raw & Historical	ACORN-V1	ACORN-V2
Beginning of the record				
82039	8/11/1912	10.6	7.6	7.4
82039	9/11/1912	5.6	4.0	3.4
82039	10/11/1912	16.1	14.5	12.7
82039	11/11/1912	5.6	4.0	3.4
During the 1939 bushfire				
82039	12/01/1939	20.9	19.9	17.4
82039	13/01/1939	20.9	19.9	17.4
82039	14/01/1939	28.3	27.8	25.7
82039	15/01/1939	18.9	16.9	15.9

Source: Data sourced from Australian Bureau of Meteorology.

will also affect analysis of climate variability and change into the future. By reducing past temperatures, there is potential for creating new 'record' hottest days for the same weather.

Case study: Darwin

Darwin is a city in northern Australia. Maximum and minimum temperatures were first recorded at Darwin from a thermometer in a modified Greenwich stand in the yard of the Darwin post office in 1882. In 1894, a Stevenson screen in the same yard replaced this modified Greenwich stand as the official standard equipment for the housing of the recording thermometer, as shown in Figure 16.4. In February 1941, a second weather station was established at the Darwin airport. A year later, in February 1942, the post office and its weather station were bombed during Japanese air raids. The airport weather station survived the bombing.

The historical maximum temperature record for Darwin shows a cooling of nearly 2 °C from 1895 to 1941; this is for the period when temperatures were recorded in a Stevenson screen at the post office. This cooling is changed to warming in the creation of the first version of ACORN-SAT. In the creation of ACORN-SAT Version 2 this warming trend of 1.3 °C per 100 years in ACORN-SAT Version 1 is increased to 1.8 °C per 100 years, as shown in Figure 16.5.

As an example of the type of changes made, the daily values at the beginning of the official record are reduced by 1.6 °C. For example, the maximum temperature measured at the Darwin Post Office on 6 January 1910 was 34.6 °C. This value was changed to 34.2 °C in ACORN-SAT Version 1. It is now 33.0 °C in Version 2, as shown in Table 16.2.

In a newspaper article in *The Weekend Australian*, the Bureau claims the changes to the Darwin record in Version 2, of ACORN-SAT are necessary for the following reasons:

> For the case of Darwin, a downward adjustment to older records is applied to account for differences between the older sites and the current site, and differences between older thermometers and the current automated sensor. In other words, the adjustments estimate what historical temperatures would look like if they were recorded with today's equipment at the current site. (Lloyd 2019)

Figure 16.4 Historical photographs showing the Stevenson screen at the Darwin post office

Top photograph taken February 26, 1890 outside the telegraph inspector's residence showing the Stevenson screen in foreground at left (white louvred box on a stand). Second (middle) photograph taken outside the post office circa 1930 does not show shading by trees. Third (bottom) photograph taken outside the post office January 1, 1940.

Source: Northern Territory Library.

Figure 16.5 ACORN-SAT Version 1 and Version 2 temperature trends for Darwin

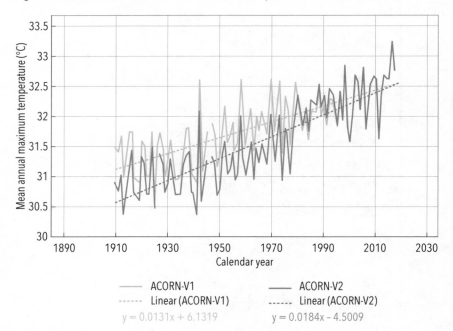

The extent of warming increases from 1.3 °C per 100 years to 1.8 °C in the latest revisions by the Bureau to Australia's temperature history.

Source: Data as daily maximum temperatures downloaded from http://bom.gov.au.

Table 16.2 Daily temperatures measured and homogenised for Darwin, January 1910

Darwin Daily Maximum Temperatures – Depending on Dataset

Date	Raw	ACORN-V1	ACORN-V2	Diff Raw-V2	Diff Raw-V1
1 Jan 10	34.2	33.8	32.8	1.4	0.4
2 Jan 10	32.7	32.3	31.5	1.2	0.4
3 Jan 10	32.7	32.3	31.5	1.2	0.4
4 Jan 10	33.6	33.2	32.4	1.2	0.4
5 Jan 10	34.6	34.2	33.0	1.6	0.4
6 Jan 10	34.6	34.2	33.0	1.6	0.4

The daily maximum temperatures for early 1910 as shown in the three different datasets for Darwin

Source: Raw data was downloaded from http://bom.gov.au/climate/data/and ACORN-SAT was downloaded from ftp://ftp.bom.gov.au/anon/home/ncc/www/change/ACORN_SAT_daily/.

Figure 16.6 Measured maximum temperatures and ACORN-SAT Version 2 temperatures for Darwin

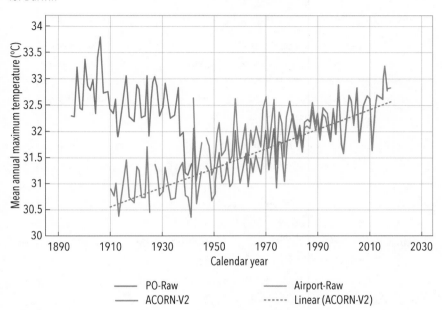

There is a large difference between the ACORN-SAT temperature series (red line) and the historical measurements (green line) from 1937 back to 1910. The Bureau remodels the data on the basis there was shading by trees before 1937. It is more likely there was a cyclone in March 1937 clearing away vegetation that had previously screened the post office from the sea breeze.

Source: Raw data downloaded from http://bom.gov.au/climate/data/and ACORN-SAT data downloaded from ftp://ftp.bom.gov.au/anon/home/ncc/www/change/ACORN_SAT_daily/.

In March 2012, when Version 1 of ACORN-SAT was published, the Bureau was claiming in its catalogue (Bureau of Meteorology 2012) that there was abnormal cooling of the Darwin temperature record before 1 January 1937 because of shading from trees. Somewhat peculiarly, this was then used as a justification for adjusting *all* the temperatures down before this date back to 1910, in effect further dramatically cooling the early record as shown in Figure 16.6. Might it have been more logical to warm that period of the record to correct the 'artificial cooling' caused by the trees?

Regarding actual causes, it could be the case that a cyclone caused the reduction in temperatures in early 1937. According to the *Northern Standard* newspaper reporting on 12 March 1937, there was a cyclone that:

> ... raged and tore to such vicious purpose that hardly a home or business in Darwin did not suffer some damage ... Telephone wires and electric mains were torn down by falling trees and flying sheets of iron, windmills were turned inside out, garden plants and trees were ruined, roads and tracks were obstructed by huge trees ...

This would suggest that rather than shading by trees, there were no trees after the cyclone, possibly allowing cooling from the sea breeze. In a study of modifications to orchard climates in New Zealand, McAneney et al. (1990) showed that screening could increase the maximum temperature by 1 °C for a 10 metre high shelter. Whichever, neither the equipment nor the site changed between ACORN-SAT Version 1 (2012) and Version 2 (2018). To be clear, the weather station has been at the airport since February 1941 and an automated sensor was installed on 1 October 1990. A Stevenson screen was first installed at the post office site in 1894, while a Stevenson screen has always been used at the airport site.

I generated a minimally homogenised temperature record for Darwin, taking into consideration only the move from the post office to the airport based on the available single year of overlapping data. This temperature series shows no overall warming trend. Rather it is consistent with other locations in northern Australia with long high-quality records, for example Richmond in north-western Queensland, that show cooling and then warming, as shown in Figure 16.7.

The calculations of temperature change at Darwin, including for the minimally homogenised series, were submitted for publication in *The International Journal of Climatology* in early 2015. On 25 June 2015, I was informed by the editor that the paper had been assessed by two referees who were leading international experts in this field. Major revisions to the manuscript were requested by the second referee and subsequently made. In an email received from the editor on 7 December 2015, it was stated that both referees were satisfied with the revisions that

Figure 16.7 Annual mean maximum temperatures as measured at Richmond and Darwin

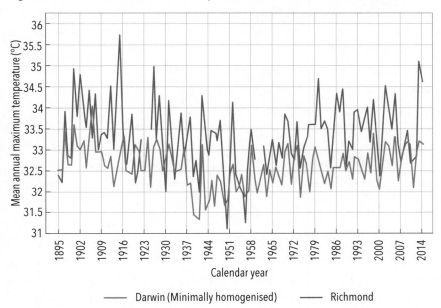

Darwin (Minimally homogenised) ——— Richmond

Richmond, in inland north Queensland, has a continuous temperature time series from 1895 to 2014. Over this period maximum temperatures were recorded in a Stevenson screen using a mercury thermometer at the same site. The annual mean maximum temperatures shown for Richmond (purple line) have not been adjusted in any way, they represent the original historical measurements.

Source: Unpublished manuscript by Jennifer Marohasy and John Abbot, JOC-15-0396.R1.

had been made. The second referee said that the minimally homogenised series for Darwin probably represented the best estimate of what had actually happened to the temperatures at Darwin, since at least 1941. I waited for galley proofs until 9 February 2016, when I received an email from the editor informing me that the manuscript was now rejected and would not be published by the journal; no logically consistent reason was provided. Certainly the manuscript was critical of the Bureau's homogenisation of the historical Darwin temperature record.

Conclusion

Science is a method of study that purports to explain natural phenomena based on direct observation and measurement. Official historical temperature series are assumed to be scientific, and represent actual observations. Yet clearly temperature series are significantly remodelled by the Bureau to the extent that they differ in magnitude and direction from the original. In short, the official series are constructs that accord with popular climate change theory that claims a continual increase in temperatures through the twentieth century, but this has not been the reality in all places. The potential to report this in the peer-reviewed literature is limited, with detailed criticisms denied.

ACORN-SAT Version 2 is the official temperature record for Australia. It was completed just in time for the new remodelled values to be included in the next report of the United Nations Intergovernmental Panel on Climate Change (IPCC). This new version of our temperature history will also underpin annual state-of-the-climate reports that are always widely reported in the mainstream media as the historical temperature record for Australia. The remodelled temperature series are also passed on to universities, CSIRO, and other climate scientists, who base much of their climate research on these 'second-hand' statistics that do not accord with the actual historical measurements.

17 Perspectives on Sea Levels

Dr Arthur Day and Dr Jennifer Marohasy

What is meant by sea level, and how is it measured? A long time ago, an influential Prussian scientist proposed that mean sea level be struck on newly discovered coasts and islands. The idea was taken up by an Antarctic explorer who organised for a mark to be cut deeply into a cliff face just off the coast of Tasmania in 1841 as a permanent reference for the mean sea level of the Southern Ocean. We can now calculate from this mark that sea level has risen 13 cm ±3 cm relative to the land at an average rate of 0.8 ±0.2 mm/year – or 1.0 ±0.3 mm/year if local upward movement of the land surface is taken into account – since 1841. This increase is about half the global rate being promoted by the Intergovernmental Panel on Climate Change (IPCC) for the sea-level rise over the last century.[1]

Climate models – the results of which are compiled and assessed by the IPCC – forecast that one of the consequences of global warming from greenhouse gases will be rising sea levels due to the thermal expansion of the ocean water mass, plus the contribution of water from melting of ice sheets and glaciers residing on land. In 2013, the IPCC concluded (IPCC 2013) that the oceans had already risen 19 cm (17 to 21 cm) between 1901 and 2010, which is an annual rate of 1.7 mm/year

---

1   This chapter was inspired by the work of the late John L. Daly (1943–2004), particularly his work on sea levels. This book chapter is dedicated to John's memory and life's work. Tide gauge plots have been adapted on the basis of the latest data, *without modification*, from The Permanent Service for Mean Sea Level (PSMSL) (https://www.psmsl.org/data/), retrieved 1 June 2020.

(1.5 to 1.9 mm/year). They further predicted the oceans will rise approximately an additional 51 to 98 cm with a substantially accelerated rate of increase during 2081–2100 of 8 to 16 mm/year. There are a few things, however, that are almost never pointed out in discussions about climate change and sea-level rise. For example, the estimates of past and current global rates of sea-level rise, and the future projections, are calculated constructs that are largely the product of extremely complex computer models. We are being asked to simply trust them. However, the success of this modelling is dependent on chains of assumptions. If one assumption turns out to be incorrect then the results produced by the models could be wrong. This applies equally to estimates of past and present rates of global sea-level rise, as well as to future projections. They are hypothetical. Yet these calculated values are broadcast widely with such a sense of confidence that a false impression is created.

None of the global estimates derived from models correspond to directly observed and measurable sea-level change at any point on the open sea, or along any coastline.

Large-scale regional land movements are occurring naturally and causing changes in the sea level, both up and down, depending on the region. Then there are simultaneous pauses, declines, or accelerations in sea level recorded in tide gauges thousands of kilometres apart that can persist from just a few years up to several decades, corresponding with regional changes in atmospheric pressures, winds, or sea-surface temperatures, over the world's ocean basins. Where these persist long enough they may be reflecting natural climatic swings. Their impact on the sea-level trends recorded in tide gauges might hinder efforts to model global sea-level rise.

Accordingly, if we truly seek to understand climate change and hope to one day be able to predict it, then it is perhaps more important that we understand this spatial and temporal variability than to attempt to calculate an elusive global average sea level.

Time and tide, local to regional

We are familiar with 'tides', the twice daily rise and fall of the level of the sea as the fluid mass of the oceans dynamically adjusts to the changing gravity interactions between Earth, Moon and Sun. There are the spring tides and neap tides, the tendency for the tides to be greater during a full moon and a new moon, than they are during a half moon. We are also likely to be aware that the height of tides differs markedly in different places.

The Moon is the primary influence on our tides. As the Earth rotates, the Moon's gravity 'pulls' at the ocean, causing it to rise in a wide wave riding in the wake of its relative passage across the sky. The Sun also raises a similar, though smaller tide, being 46% of the Moon's tide. At any one time, some variables act to reinforce the tide, while others act to dampen it. It takes 18.6 years (Lisitzin 1974) for many permutations and combinations of the astronomical variables to play themselves out at any one location. Only then could a researcher calculate a long-term mean sea level (MSL) with some degree of confidence and exactness.

A global 'zero point of the sea' is an elusive concept. MSL is defined as the 'mean height of the surface of the sea for all stages of the tide cycle over a 19-year period'. Where sea level is measured at only one location, it is sometimes referred to as 'relative sea level' or RSL, since the measurement only specifies the height of the sea in relation to local landmarks. If the land itself is rising or sinking, this would manifest itself as apparent sea-level falls or rises, even if the actual level of the ocean had not changed. Until the advent of satellite-based sea level measurements, which only commenced in 1993, it was not possible to establish a truly global MSL. 'Mean Sea Level' as a descriptive term is often used to mean either the MSL at one location (i.e. its RSL), or over an entire region.

The standard instrument to determine MSL is a 'tide gauge', which measures the height of the sea at regular intervals to record the passage of high and low tides and the harmonics unique to the location where the tide gauge is situated. Most such gauges are installed in the populated areas of the Northern Hemisphere, particularly in Europe and North America. There are relatively few such gauges with long records in the Southern Hemisphere.

Tide gauges, like surface temperatures measured using thermome-ters (Marohasy & Abbot 2015; Marohasy et al. 2014), are subject to several local errors that can distort the data. Just as temperature data are affected by urban heat islands (Watts 2017; Quirk 2017; Vlok 2019), tide gauges located at major cities or ports are subject to urbanisation as well, mainly the tendency of large cities to subside, or sink, due to the weight of the structures and the changes in the underground water table. The larger the city, the greater is the tendency toward subsidence. It is a creeping effect over time, which will manifest itself at a tide gauge as a creeping rise in RSL. Cities located on alluvial low-lying coasts are the most affected. For example, Adelaide in South Australia is showing a strong sea-level 'rise'. The Adelaide anomaly has since been found to be caused by long-term withdrawal of groundwater since European settlement, giving rise to local urban subsidence (Belperio 1993). Another case of subsidence exists in Bangkok, where the sea has 'risen' by 1 m in the last 30 years. In extreme cases, such as Bangkok, the land is sinking far more rapidly than the sea is rising (Schneider 1998). This also applies to Manila, almost a metre in 60 years, and Manhattan (New York) where the sea level has gone up by 50 cm in 160 years. In these cases the tide gauges are located on unstable sinking foundations.

A second subsidence error arises from the fact that most tide gauges are mounted on man-made structures such as piers and docks. Over many decades, these undergo some subsidence, unless they are built on bedrock. Further problems arise when older gauges are replaced or relocated without overlapping data. If there is an unexpected vertical displacement between them a subjectively applied 'correction' is needed to 'synchronise' the data. This is a less than ideal situation.

There are also regional effects that could influence measuring stations thousands of kilometres apart. The tide gauge records plotted in the following figures illustrate some of the challenges involved in trying to estimate long-term changes in 'global' sea levels using the data from tide gauges. Each chart is different but what they show when considered together is that the rate and direction of local sea-level change not only differs from one region to another, but also from one decade to the next, even in the same place.

An example of long-term natural regional influences on historic sea-level records is shown in Figure 17.1. This figure shows the records of two of the oldest tide gauge stations in the world. The oldest is at

Figure 17.1 Annual mean sea level at Brest (France) and Aberdeen (Scotland), 1807 to 1918

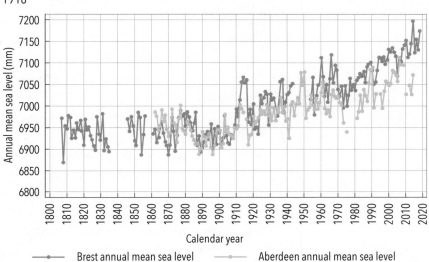

Tide gauge data from the Permanent Service for Mean Sea Level (PSMSL 2020). [2]

This book chapter reproduces charts using data retrieved from www.psmsl.org/data on 1 June 2020 *without modification*.

Sources: Permanent Service for Mean Sea Level: https://www.psmsl.org/data/obtaining/stations/1.php, https://www.psmsl.org/data/obtaining/stations/21.php, https://www.psmsl.org/data/obtaining/stations/361.php.

Brest, in France, facing the Atlantic Ocean at the mouth of the English Channel. Aberdeen is on the east coast of Scotland, facing the North Sea. Brest and Aberdeen are highly significant because their records are almost continuous since 1807 and 1862 respectively. Even though these two stations are more than 1000 km apart, they both record a long period of essentially no sea-level rise up to around 1880, then a small fall in sea level until 1890. Both then commenced an extended period of continuous relative sea-level rise, up to the most recently available

---

2 **Why are sea levels plotted in thousands of millimetres?** To construct comparable local time series of sea-level measurements, the tidal data have to be reduced to a common datum. This reduction is performed by the PSMSL using tide gauge history. The common datum has been arbitrarily set to ~7000 mm below mean sea level to avoid negative numbers in the mean sea-level plots. The adjusted data form the 'Revised Local Reference' (RLR) dataset. For a full definition see http://www.psmsl.org/data/obtaining/rlr.php.

records. Both also record a short dip in sea levels in the early 1970s. The overall relative sea-level rise since around 1890, when it commenced at both stations, is about 20 cm over 130 years in the case of Brest, and about 14 cm for Aberdeen, based on a simple linear trend line. That is an average annual rate of +1.5 mm/year for Brest and +1.1 mm/year in the case of Aberdeen since 1890 – both slower than the calculated global rate reported by the IPCC, although Brest is close.

Elsewhere in the Northern Hemisphere large-scale regional uplift is leading to net falls in sea level over vast areas. For example in Sweden we have a very different record to those of Brest and Aberdeen, as shown for Stockholm in Figure 17.2. The RSL is clearly falling, by about 50 cm over a period of 129 years. That is an average annual rate of decrease of –3.9 mm/year. But Figure 17.3 shows how the IPCC portrays Stockholm in (IPCC 2001). Figure 17.3 is a 'detrended' record to correct for a phenomenon that puts all European and North American tide gauge data in doubt. It is the 'post-glacial rebound' (PGR) effect.

Figure 17.2 Annual mean sea level at Stockholm (Sweden) and Juneau (Alaska), 1889 to 2018

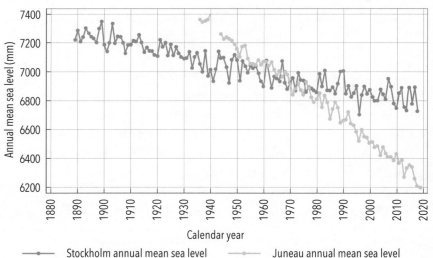

Tide gauge data from the Permanent Service for Mean Sea Level (PSMSL 2020).

Sources: Permanent Service for Mean Sea Level: https://www.psmsl.org/data/obtaining/stations/78.php, https://www.psmsl.org/data/obtaining/stations/405.php.

Figure 17.3 The IPCC-corrected sea-level record at Stockholm (scale in mm)

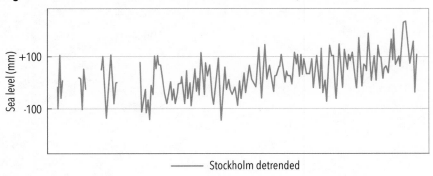

——— Stockholm detrended

This chart was included in the IPCC AR3, IPCC (2001) report in Figure 11.7, on page 662.

Source: IPCC (2001).

The uplift at Juneau, in Alaska, also shown in Figure 17.2 is extreme. In just 80 years the RSL has fallen by 120 cm at a steady rate of decrease of –15 mm/year. Juneau may be undergoing additional tectonic uplift (mountain-building) in addition to PGR.

During the most recent glaciation, Stockholm was buried under several kilometres of ice. The ice retreated from Sweden around 9900 to 10,300 years ago with a rapid melting of ice sheets across Europe and North America, resulting in a rapid rise in sea level. With the ice gone, the plasticity of the mantle below the solid crust of the Earth is allowing rebound of the overlying crust now that the dead weight of the ice is no longer there. (The analogy is the way a floating cork being pushed down into water rebounds when it is released.) This process has been ongoing since the last glacial maximum and will continue well into the future.

PGR is underway all over Europe, North America, and east Asia (Peltier & Tushingham 1991), as shown in Figure 17.4. These are the continents that were most affected by the enormous ice sheets. Those regions enclosed in the solid red line were weighed down the most by ice. They are now rebounding, like Sweden. Peripheral regions around the margins of the former ice masses (bounded by a dashed red line) are subsiding as the continental crust adjusts and rebalances to the new weight redistribution. Tidal stations with more than 75 years of records that show continuously falling relative sea levels due to PGR include,

Figure 17.4 Areas most affected by post-glacial rebound (PGR)

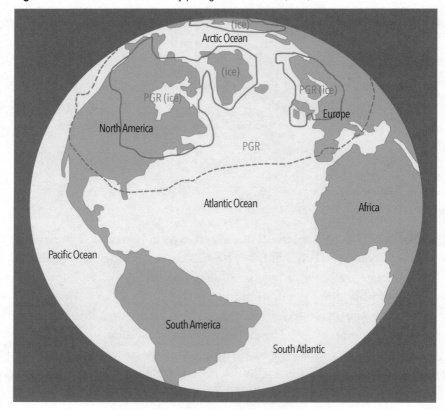

Source: Modified from a map in Microsoft Encarta96 World Atlas.

from Scandinavia: Stockholm, Smögen, Ratan, Furuogrund, Oslo, Barentsberg, Helsinki, Hanko, Mantyluoto, Vaasa, Jakobstad, and Oulu. And from North America: Churchill, Adak Sweeper Cove, Unalaska, Yakutat, Sitka, Juneau, and Skagway.

It is not just the removal of the weight of the ice that can cause uplift or subsidence in the regions directly affected, but the melting of the ice itself has raised global sea levels by more than 120 m. This changes geological stresses on regions far away from the ice due to the added weight of water in ocean basins and over continental shelves, causing a slow readjustment and rebalancing of continental crusts everywhere to compensate for the added weight of water. Unfortunately, many of the tide stations with records that go back into the 19th century originate in places affected by PGR, showing sea-level falls in places like Sweden and equally large rises (3.5 mm/year) in areas like Chesapeake Bay on the US Atlantic coast (Denys & Hannah 1998). In these cases, it is the vertical movement of the land itself that is causing the apparent, or 'relative', sea level to change.

Many other parts of the world, such as in New Zealand, Japan, the Mediterranean, the entire west coast of North and South America, a lot of south-east Asia plus some of the Pacific islands, are tectonically active, leading to vertical land movements, so we cannot simply attribute apparent sea-level changes solely to the sea itself. Indeed, we have to regard tide gauge data from anywhere in the geologically active parts of the world to be potentially compromised – and this is a high proportion of them. We could, of course, disregard all tide gauge data from geologically unstable locations, and places where urban development has corrupted the data, but if we do that, we end up with hardly any data at all (Schneider 1998). So it should at least be recognised that much of the data being analysed is of poor quality and poorly distributed geographically. Consider, for example, there is only one single tide gauge with complete records (that only go back to 1958) for the entire coastline of Antarctica. This is too short a record to be of any real use and yet there is abundant geological evidence that the ice sheet there has undergone vast changes since the most recent deglaciation. PGR is recognised in Antarctica, too, but the lack of tide gauge data can only make its long-term impact on global sea-level rise far more difficult to constrain because of the lack of direct measurements of the Antarctic sea-level response.

To make some sense out of the mass of contradictory sea-level data, a series of 'de-glaciation' models has been developed. These use calculations of the viscosity of the Earth's mantle to determine how the mantle and overlying continental and seabed crusts have reacted to the melting of the great ice sheets, the consequent weight redistribution in the ocean basins, and how these ongoing adjustments will contribute to sea-level rise for many centuries into the future. The trouble is we just don't have the data to properly constrain these models.

A brief world tour of twentieth-century sea levels

The IPCC concluded the oceans have risen around 19 cm at an annual rate of 1.7 mm/year between 1901 and 2010 (IPCC 2013). More recently, the CSIRO (2017) position remains similar – they claim, from 1880 to 2014, globally-averaged sea level rose about 23 cm, with an average rate of rise of about 1.6 mm/year. The wide variability between tide gauges presents a considerable challenge when trying to calibrate the models these calculations depend on. The extent of that challenge is rarely discussed openly, but it is real.

Figure 17.5 shows three stations located along a 200 km stretch of coastline to the north and south of Sydney, Australia. The tide gauge at Sydney's Fort Denison, commencing in January 1886, is the oldest operating tide gauge in the Southern Hemisphere. The Newcastle record is also one of the oldest. The official record at Port Kembla commenced only recently. Fort Denison and Newcastle both show a remarkable similarity in their long-term records. Prior to the commencement of the Newcastle records, Fort Denison showed a slight increase in sea level from the

Figure 17.5 Annual mean sea level at Sydney Fort Denison, Newcastle, and Port Kembla (Australia), 1886 to 2019

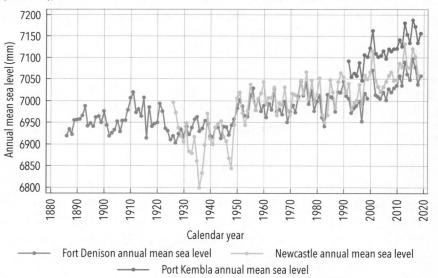

Tide gauge data from the Permanent Service for Mean Sea Level (PSMSL 2020).

Sources: Permanent Service for Mean Sea Level: https://www.psmsl.org/data/obtaining/stations/65.php, https://www.psmsl.org/data/obtaining/stations/196.php, https://www.psmsl.org/data/obtaining/stations/267.php, https://www.psmsl.org/data/obtaining/stations/831.php, https://www.psmsl.org/data/obtaining/stations/837.php.

1880s up to around 1910. Sea level then fell again into the late 1920s when Newcastle also joined the record. Both stations then recorded an ongoing dip in sea levels from the 1920s up to just after 1950 when they both abruptly jumped back up to the previous 1910 level. From that level there was almost no sea-level rise for nearly 50 years, up to just before 2000. From then a trend of increasing sea level commenced to the present, as is also confirmed by the Port Kembla station.

Because it has the oldest record, there has been a lot of debate about the rate of sea-level increase recorded at Fort Denison. But the 'rate' depends on the time frame. It depends on the 'commencement' date and this is arbitrarily imposed by the record itself. So, in the case of Fort Denison, if the beginning of the record is selected there has been an increase of 10.5 cm in the 134 years from 1886 to the present. This is an overall rate of increase of 0.8 mm/year, based on the use of a simple linear trend line. If 1930 is chosen instead, there has been an increase of 11 cm in 90 years at a rate of 1.2 mm/year – a 50% increase by simply selecting a different time frame. The great majority of tide stations have a much shorter record than Fort Denison. Interestingly, Fremantle, about 3500 km to the west, also records the same 50-year hiatus in sea-level rise seen at both Fort Denison and Newcastle from 1950 to almost 2000.

Marseille and Trieste (Figure 17.6) are two of the oldest tide gauges in the Mediterranean. They lie 730 km apart on opposite sides of the Italian Peninsula. Despite being separated by a significant land barrier, they show similar long-term trend behaviour. For example they both show a fairly steady increase in sea level from 1890 up to about 1960, then they both undergo a twenty-year decline in sea level until about 1980, and then both resume a steady increase again up to the present. Since 1890, Marseille has undergone 17 cm of sea-level rise over 130 years at a rate of 1.3 mm/year, and the increase is similar for Trieste.

The Argentine Islands and Mossel Bay stations are more than 6000 km apart, separated by open water on opposite sides of the Southern Ocean. They both have the longest and most complete tidal records in their respective regions, going back to 1958. The Argentine Islands tide gauge is at the Ukrainian Vernadsky Antarctic Base on an island along the western side of the Antarctic Peninsula. It has the only useful tidal

Figure 17.6 Annual mean sea level at Marseille (France) and Trieste (Italy), 1880 to 2016

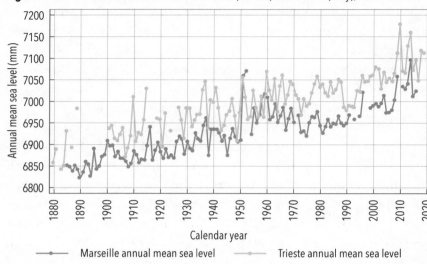

Tide gauge data from the Permanent Service for Mean Sea Level (PSMSL 2020).

Sources: Permanent Service for Mean Sea Level: https://www.psmsl.org/data/obtaining/stations/61.php, https://www.psmsl.org/data/obtaining/stations/154.php.

records for the entire Antarctic continent. The Mossel Bay station is near the southern tip of South Africa and faces into the Southern Ocean.

Despite the incredible distance between them, the two stations show remarkably similar and quite unique synchronous wave-like patterns in their tidal records, as shown in Figure 17.7. The wave-like plots for both stations overlap in frequency and amplitude with a 'wavelength' of about 25 years. The Mossel Bay station drifts upwards slightly after about 2005 but, overall, there is only a very modest rate of long-term sea-level rise recorded at the two stations.

What causes this behaviour? Any possible cause has to be regionally extensive, periodic, and affect the entire Southern Ocean. By far the most likely phenomenon that fits these criteria is a periodic change in atmospheric circulation known as the Southern Annular Mode (SAM), also called the Antarctic Oscillation (AO). This periodic weather phenomenon is characterised by the changing position of the dominant westerly wind belt that circulates around Antarctica over the Southern

Figure 17.7 Monthly mean sea level at Argentine Islands (Antarctica), and Mossel Bay (South Africa), 1958 to 2019

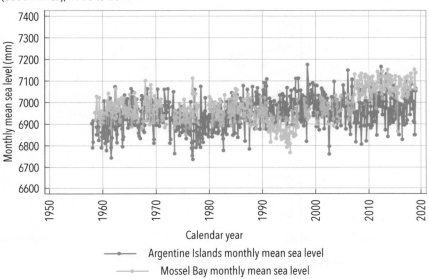

Tide gauge data from the Permanent Service for Mean Sea Level (PSMSL 2020).

Sources: Permanent Service for Mean Sea Level: https://www.psmsl.org/data/obtaining/stations/910.php, https://www.psmsl.org/data/obtaining/stations/913.php.

Ocean. As these winds periodically move towards and away from Antarctica they influence the position and strength of the embedded high- and low-pressure systems. More frequent high-pressure systems over the Southern Ocean will result in a fall in sea levels while, in the opposite phase of the oscillation, more frequent low-pressure systems will cause an increase in sea levels. Hence the synchronous long-term wave-like patterns seen in tide gauges situated half a world apart on opposite sides of an ocean.

The Nauru sea-level record, see Figure 17.8, comprises three successive tide gauges installed since early 1974. There is almost no overlap between the gauges and an almost 200 mm difference in mean sea level between the first, and the second and third gauges. There has been no attempt to resolve this in the data supplied to the PSMSL. Here, the data are presented in their unadulterated form so that readers can decide for themselves how to slide the pieces of the puzzle together and resolve the 'discontinuity'.

Figure 17.8 Monthly mean sea level at Nauru, western Pacific Ocean, 1974 to 2020

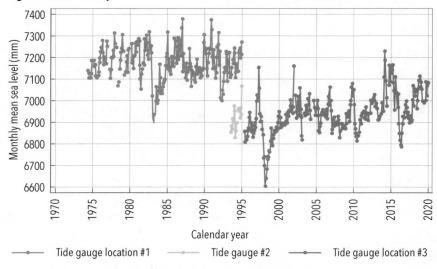

Calendar year

Tide gauge location #1 ——•—— Tide gauge #2 ——•—— Tide gauge location #3

Tide gauge data from the Permanent Service for Mean Sea Level (PSMSL 2020).

Sources: Permanent Service for Mean Sea Level: https://www.psmsl.org/data/obtaining/stations/1374.php, https://www.psmsl.org/data/obtaining/stations/1844.php.

Aside from the discontinuities between the tide gauges, the most striking feature of the Nauru chart is that it clearly demonstrates the impact of the El Niño Southern Oscillation on sea levels in the western Pacific. During three of the big El Niño events in 1982–1983, 1997–1998, and 2015–2016, there was a sharp, but temporary, fall in sea level by up to 30 cm. Despite the large break points between the three gauges, it is clear that sea levels either side of the El Niño events are largely unchanged, although there are large swings.

Betio, in the Republic of Kiribati, lies about 750 km east of Nauru. About 1300 km south-east of Betio lies Funafuti in the island nation of Tuvalu. Both are small low-lying coral atolls in the western Pacific Ocean. They lie only a few metres above sea level. Betio has a growing population at a density of around 15,000 people per km^2. They are heavily dependent on groundwater and, at 2.8 m above sea level, Betio could be vulnerable if excessive groundwater extraction causes subsidence. These two atolls have been grouped together because their tide gauge records, plotted in Figure 17.9, suggest there has been no discernible long-term sea-level rise over

Figure 17.9 Monthly mean sea level at Betio (Kiribati) and Funafuti (Tuvalu), western Pacific Ocean, 1974 to 2020

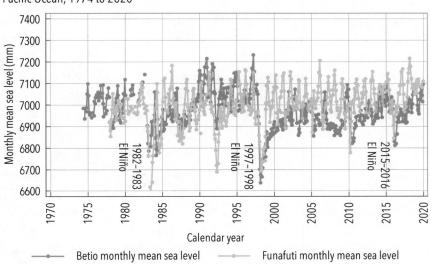

Tide gauge data from the Permanent Service for Mean Sea Level (PSMSL 2020).

Sources: Permanent Service for Mean Sea Level: https://www.psmsl.org/data/obtaining/stations/1381.php, https://www.psmsl.org/data/obtaining/stations/1452.php, https://www.psmsl.org/data/obtaining/stations/1579.php, https://www.psmsl.org/data/obtaining/stations/1739.php, https://www.psmsl.org/data/obtaining/stations/1804.php, https://www.psmsl.org/data/obtaining/stations/1839.php.

the 45 years of records. This is not consistent with claims that such atolls are under immediate threat, at least from this cause. Perhaps there should be more immediate concern about the amount of groundwater extraction rather than climate change. Again, both plots show there has been a sharp temporary fall in sea levels at both atolls due to the three big El Niño events in 1982–1983, 1997–1998, and 2015–2016.

Kwajalein, in the Marshall Islands, lies about 1200 km to the north-west of Betio. About 3000 km south-east of Kwajalein lies Pago Pago in American Samoa. Geologically, these are quite different islands. They also have divergent long-term sea-level outcomes. Kwajalein is one of the world's largest coral atolls, surrounding one of the largest lagoons in the world. Despite its enormous size the average height above sea level for all the small islands ringing the atoll is only about 1.8 metres. In contrast, Pago Pago is a high mountainous island.

Figure 17.10 shows a record of sea level for Kwajalein and Pago Pago going back to the 1940s. Once again, the three big El Niño events from 1982–1983, 1997–1998, and 2015–2016 are visible in both records even though the two islands are more than 3000 km apart. Perhaps for this reason, there is a discernible offset between the El Niño responses at the two localities with a significant delay in the arrival of the sea-level signal at Pago Pago compared to Kwajalein. Earlier events common to both islands can also be seen in the data going right back into the 1940s. The sea-level trends are identical from the late 1940s up to just before 2010 with only a very slight acceleration building over this time. From 2010, however, the rate for Pago Pago began to accelerate. Eventually it decoupled from the long smooth trend still unfolding at Kwajalein.

The acceleration at Pago Pago is not being caused by sea-level rise. On 29 September 2009, a magnitude 8.1 earthquake took place, known as the 'Samoa–Tonga earthquake'. Han et al. (2019) report that before the earthquake occurred, the rate of relative sea-level rise was 2–3 mm/year

Figure 17.10 Monthly mean sea level at Kwajalein (Marshall Islands) and Pago Pago (American Samoa), western Pacific Ocean, 1946 to 2020

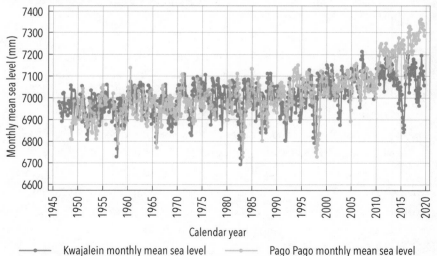

Tide gauge data from the Permanent Service for Mean Sea Level (PSMSL 2020).

Sources: Permanent Service for Mean Sea Level: https://www.psmsl.org/data/obtaining/stations/513.php, https://www.psmsl.org/data/obtaining/stations/539.php.

but afterwards it rapidly accelerated to 8–16 mm/year, as shown in Figure 17.10. The phenomenon is known as 'post-seismic subsidence'. It occurs as the solid Earth responds to the stress change caused by an earthquake. A combination of GPS displacement and satellite gravity measurements have determined that the local rate of relative sea-level rise is approximately five to nine times faster than the IPCC's global average. Han et al. (2019) believe this is likely to continue for decades and result in total relative sea-level rise of 30–40 cm. This is approaching the IPCC's projection for global sea levels over the rest of this century, except that it is not being caused by 'climate change'. Pago Pago is a mountainous island and the worst to be expected is that there will be some flooding. However, if such an event were to occur again at a low coral atoll like Kwajalein the result would be catastrophic, potentially rendering such an atoll uninhabitable. So, perhaps we should worry more about earthquakes than climate change.

The parallel behaviour in the long-term trends in tide gauge records of stations thousands of kilometres apart, as shown in Figures 17.1, 17.5, 17.6, and 17.7, cannot be a function of localised site-specific factors. They also can't be artefacts connected with the tide gauges themselves. In the case of the Pacific Ocean tide gauges, there is no question the synchronous sharp falls in sea levels shown in Figures 17.8, 17.9 and 17.10 – some more than 3000 km apart – are linked to the onset of the El Niño Southern Oscillation, a climate index that can be used for long-range weather forecasting (i.e. Abbot & Marohasy 2012). The onset of an El Niño is accompanied by high air pressure over the western Pacific. During an El Niño, an Australia-sized mass of heavy high-pressure air descends over the surface of the western Pacific Ocean. The weight of this air depresses the sea surface by as much as 30 cm over millions of km^2 of open ocean.

An historic tidemark on the 'Isle of the Dead'

The 'Isle of the Dead' MSL benchmark is not mentioned by the IPCC in any of its reports, despite it being one of the oldest of its kind in the world. The site chosen was a small island adjacent to Port Arthur, 50 km south-east of Hobart (see Figure 17.11). The benchmark was located on the island because it was deemed more secure than the mainland. The 1841 benchmark lies approximately at today's tidal high water. It is the first scientific attempt to record a fixed reference point relative to the 'mean sea level', or the 'zero point of the sea', in the Southern Ocean. Antarctic explorer Captain Sir James Ross explained why at the time:

> My principal object in visiting Port Arthur was to afford a comparison of our standard barometer with that which had been employed for several years by Mr. Lempriere, the Deputy Assistant Commissary General, in accordance with my instructions, and also to establish a permanent mark at the zero point, or general mean level of the sea as determined by the tidal observations which Mr. Lempriere had conducted with perseverance and exactness for some time: by which means any secular variation in the relative level of the land and sea, which is known to occur on some coasts, might at any future period be detected, and its amount determined. The point chosen for this purpose was the perpendicular cliff of the small islet off Point Puer, which, being near to the tide register, rendered the operation more simple and exact; the Governor, whom I had accompanied on an official visit to the settlement, gave directions to afford Mr. Lempriere every assistance of labourers he required, to have the mark cut deeply in the rock in the exact spot which his tidal observations indicated as the mean level of the ocean. (Ross 1847)

A little further on though, Ross also explained quite clearly it wasn't essential the mark be made exactly at the mean level of the ocean. In fact he felt it was actually more desirable to place it out of the water with the exact distance of the mark above MSL to be recorded nearby on a plate of copper. Ross continues:

> I may here observe, that it is not essential that the mark be made exactly at the mean level of the ocean, indeed it is more desirable that it should be rather above the reach of the highest tide: it is, however, important that it be made on some part of a solid cliff, not liable to rapid disintegration, and the exact distance above the mean level (which may also be marked more slightly) recorded on a plate of copper, well protected from the weather, by placing a flat stone with cement between, upon the plane surface or platform which should constitute the mark from which the level of mean tide should be measured.

Tasmania has a long history of earthquake activity. Between 1883 and 1892 south-eastern Tasmania was rocked by a swarm of more than 2500 earthquakes (McCue 2015). In the midst of this earthquake swarm, in 1888, Captain Shortt paid the first scientifically reported visit to the benchmark in 47 years (Shortt 1889) to ascertain if there had been any associated land movement:

> During the years 1883, 1884, 1885, and 1886, or immediately prior to the eruption at Tarawera, this island, and the South-Eastern portion of mainland Australia, were frequently shaken by Earth tremors; and as such disturbances are often known to be associated with local changes to sea and land, it appeared to me to be of great importance to ascertain whether any recent change could be traced along the coast-line of this island.

Shortt found Lempriere's benchmark and the accompanying copper plate:

> Mr. Lempriere, it is evident, carefully carried out these directions, for on a tablet still existing a little above the tide mark in question is the following record. 'On the rock fronting this stone a line denoting the height of the tide now struck on the 1st July, 1841, mean time 4h. 44m p.m.; Moon's age 12 days; height of water in tide gauge 6 ft. 1 in.'

But after some tidal calculations, he concluded:

> At this low water level the mark was found to be 2½ft. above. This very closely corresponds with the normal difference between these levels of low and high water, and would therefore indicate that there has been practically no alteration of the relative levels of sea and land during the last 47 years.

The 'permanent mark' that Lempriere installed at Ross's instigation is still there today, and clearly visible, although it is believed the accompanying copper plate disappeared sometime around 1913 (Lord 1995). The line and arrow mark that can be seen is a standard British Ordnance Survey benchmark. It is about 40 cm across at its base (Figure 17.11).

The conundrum of this historic benchmark and the possibility of establishing the Southern Hemisphere's longest record of sea level has attracted scientific interest including a team of scientists from the CSIRO, the Antarctic CRC at the University of Tasmania, and the Southampton Oceanography Centre in the UK. They installed a new state-of-the-art

Figure 17.11 Isle of the Dead and sea-level benchmark

View of the Isle of the Dead from the north-east, and a map of Port Arthur showing its location (top). Photographs of the Ross–Lempriere sea-level benchmark, located on the northern shoreline of the Isle of the Dead, taken at mean tide on August 29, 1999 (bottom). The mark is a standard British Ordnance Survey benchmark, carved on 1 July, 1841. It consists of a vertical arrow pointing downward to a horizontal line. The line is 40 cm long.

Source: Photograph of the Isle from Anagoria, Created: 15 February 2018, https://upload.wikimedia. org/wikipedia/commons/c/cf/2018-02-15_111355_Port_Arthur_Isle_of_the_Dead_anagoria.JPG and photographs of the benchmark (bottom) taken by John L. Daly.

acoustic tide gauge at the Port Arthur jetty in 1999 on the opposite shore about 1.2 km away from the benchmark, and within 300 m of the probable location of Lempriere's original gauge. The new gauge recorded a 'parallel' tidal history for comparison with the original 1841–1842 observations recently recovered from the archives of the Royal Society in London, and the official Australian archives in Hobart. They also set up a GPS network around the harbour in an attempt to construct a time-series of vertical land movements.

South-eastern Tasmania is in a state of slow uplift as part of a long-term trend widespread along the eastern margin of Australia. As a consequence, vertical movements in the land surface need to be taken into account.

Calculation of the record of sea-level change at Port Arthur is subject to three critical requirements. The first is the existence of the carved bench-mark, the second is the availability of the original 1837–1842 tide gauge records, and the third is the record of the water level in the 1841 tide gauge at the time the benchmark was struck near to high water. Both the time and the water level had to be correctly inscribed on the copper plate mounted alongside the benchmark. If either the historic tidal records or the inscription on the copper plate had been lost, the task of reconstruct-ing the sea-level history would have been impossible. An accurate result also depends on two tasks having been undertaken with considerable care back in 1841. The first is that the benchmark on the island was properly positioned and cut at the water line across the rock face at the same time as the water level was recorded in the tide gauge on the mainland over a kilometre away – these two critical pieces of data having been recorded on the copper plate formerly positioned near the benchmark.[3] The second is that the water between the location of the tide gauge and the benchmark was properly performing its expected function as a 'spirit level'. This requires the assumption that the sea surface was perfectly 'level' at the time.[4] The procedure employed in using these data alongside

---

3   So far, nobody has been able to work out exactly how Lempriere was able to do this, but one suggestion is that he and an associate signalled each other – perhaps by using a musket shot.

4   However any current, or wind passing over the water surface at the time, could have led to a 'slope' on the sea surface – causing a levelling error.

the modern tide gauge data is complex. It also depends on successfully 'levelling' the new tide gauge to the same benchmark so that its readings conform to the same 'datum', or reference height, as the first gauge 160 years previously. This time the 'levelling' was done using modern techniques borrowed from surveying, which has greater accuracy than using the sea surface as a spirit level. In detail, the procedure is complex, particularly the estimations of errors and uncertainties.

The method and results were published in Pugh et al. (2002) and Hunter et al. (2003). Mean tidal level was found to lie 44.5 cm below the benchmark in 1841–1842 and 31.5 cm below it in 1999–2002. From these findings, the estimated sea-level rise, relative to the land, was 13 ±3 cm,[5] rising at a rate of 0.8 ± 0.2 mm/year over the 160-year period from 1841–1842 to 1999–2002. Correction for vertical land movement was not possible using GPS because of insufficient observation time. Instead, an indirect estimate was made based on a 125,000 year-old shell bed 24 m above sea level, about 42 km away. This yielded an estimated uplift rate of 0.19 mm/year. This very long-term average may not apply from 1841 but, if it does, the inferred uplift rate since then is in the range from 0 to 0.2 ± 0.2 mm/yr. This adds to the relative rate to get a true rate of sea-level rise of up to 1.0 ± 0.3 mm/year, or about 16 ±3 cm over 160 years from 1841–1842 to 1999–2002.

The two closest currently active tide gauges at Spring Bay and Hobart (Figure 17.12) confirm sea levels are rising, although their records only go back about 30 years.

The Ross–Lempriere sea-level benchmark has allowed scientists to observe the longest time span of any sea-level observations in the Southern Hemisphere, and one of the longest in the world. It shows that long-term sea-level rise in Tasmania is only just more than half the global rate reported by the IPCC over the last century. Confidence is

---

5   Note the authors are careful to state the wide uncertainties involved. One important source of uncertainty is that the local barometric pressure was unknown on the afternoon in 1841 when the tide was recorded in Lempriere's gauge. No attempt was made to correct for this because of uncertainties in the observations at the time. Another source of uncertainty is that, because the tidal records only spanned two or three years, the measurements will not have accommodated the full 18.6 year tidal cycle that is normally considered to be a minimum requirement for the calculation of MSL.

Figure 17.12 Monthly mean sea level at Spring Bay and Hobart, Tasmania (Australia), 1987 to 2020

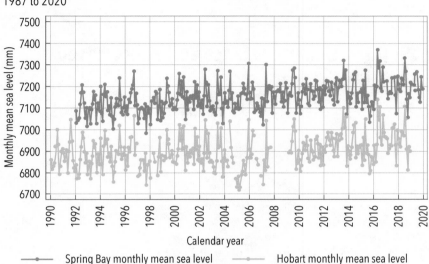

Tide gauge data from the Permanent Service for Mean Sea Level (PSMSL 2020).

Sources: Permanent Service for Mean Sea Level: https://www.psmsl.org/data/obtaining/stations/838.php, https://www.psmsl.org/data/obtaining/stations/1216.php.

enhanced because the sea-level determinations over time are referenced to a single physical benchmark, which still exists. New scientific techniques have enhanced the value of the benchmark, allowing its story to be told. It is a particularly important contribution to the study of past sea-level rise because there is a lack of such long-term data for the Southern Hemisphere.

The mystery of the historic 1841 benchmark on the Isle of the Dead has been solved! An old benchmark also exists at Fort Denison in Sydney Harbour. Are there any other long forgotten benchmarks elsewhere on more remote coastlines? In the words of Captain Shortt:

> It would be desirable also in the interests of Navigation to have such marks carefully made on various parts of our coast line. (Shortt 1889)

If others exist, where are they, and what could we learn from them about ever-changing sea levels?

18 Climate Science and Policy-based Evidence

Professor Aynsley Kellow

Science is the belief in the ignorance of experts.
Richard Feynman

In 1974, two bombs exploded in the city of Birmingham killing 21 people. Shortly afterwards, five men departed the city, travelling by train from Birmingham to attend the funeral of an Irish Republican Army (IRA) member in Belfast. When they arrived in Heysham, from where they were to catch the ferry to Northern Ireland, they were detained by members of the Police Special Branch. While they were being searched, the police officers learned of the bombings back in Birmingham and detained the men who agreed to be taken for forensic testing. A positive Griess test for the presence of nitrite ions on the hands of two of the men convinced the police that they had handled explosives. The men and an associate were subjected to 'vigorous' inter-rogation. Four confessed, and all were convicted at their trial in 1975. Eventually, an appeal in 1991 freed them and they were paid substantial compensation.

At the heart of this case was the forensic evidence that 'proved' the two men had handled explosives. But the Griess test could also detect nitrites from other sources, such as the coating on playing cards – and they had been playing cards on the train. This was a case of bringing evidence to the theory – the theory police formed that the men were responsible for the bombing. They failed to consider alternative explanations for the positive test.

Police in the 'Birmingham Six' case were guilty of 'noble cause corruption' and, thinking they knew the men were guilty, sought evidence that would secure a conviction.

There are other examples from forensic science and these show that there does not need to be conscious bias. In a famous Australian case, a forensic scientist found 'foetal blood' in the car of Lindy Chamberlain, who was convicted of killing her baby, Azaria. Police were convinced of her guilt and were uninclined to accept the (correct) alternative explanation that a dingo had killed Azaria. The stain that was thought to have been blood was later determined to most likely be a sound-deadening compound from a manufacturing overspray.[1]

Judging good science

Bringing evidence to the theory is essentially an example of a confirmation bias; the advance of science relies upon a number of methodological features that guard against flaws such as these leading to wrong conclusions. Legal processes also have features that try to minimise the chances of wrongful convictions – not always immediately successfully, as these examples show. A key feature of the judicial process is that it is adversarial, so evidence presented by either side can be subjected to critical scrutiny by the other. While errors are still possible, they are often corrected by subsequent appeals.

Climate science is replete with examples of research that conclude that the data are 'consistent with model results', where model results themselves are often treated as data; however, all climate models failed to predict a flattening of the temperature record during the past twenty years. This flattening between at least 2002 and 2014 as shown in Figure 18.1 is now known as the 'pause' or 'hiatus'.

Science is inherently an error-ridden undertaking, which advances through the detection and correction of error. Unfortunately, such learning in climate science is inhibited by politics, including by

---

1   Reference Under S.433A of the Criminal Code by the Attorney-General for the Northern Territory of Australia of Convictions of Alice Lynne Chamberlain and Michael Leigh Chamberlain, Supreme Court of the Northern Territory of Australia, No. CA2, 1988. (Acquittal decision.)

Figure 18.1 Global temperature as measured by satellites

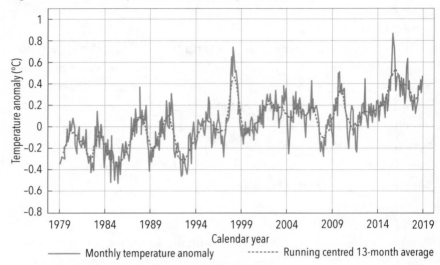

Source: https://www.nsstc.uah.edu.

smearing sceptical voices, calling these people 'deniers', and likening them to Holocaust deniers. Science is more a journey than a destination, and a journey that has many wrong turns and dead ends. John Ioannidis (2005) estimates that more than half of the published research findings in medicine turn out to be false. This is not just because the monetary stakes are high, but because the emotional stakes are also high and confirmation bias can distort the interpretations. As Carl Sagan once remarked, 'Where we have strong emotions, we're liable to fool ourselves'. Sagan also pointed to the antidote to such problems, in another widely-quoted statement: 'Sceptical scrutiny is the means, in both science and religion, by which deep insights can be winnowed from deep nonsense.'

For this reason, philosopher of science Karl Popper (1963; 1968) advocated liberalism in both politics and the scientific method. Thomas Kuhn (1962) saw science involving groups of scientists defending paradigms until they no longer could do so. Groups, of course, are prone to groupthink. As Irving Janis put it:

The more amiability and *esprit de corps* there is among the members of a policy-making ingroup, the greater the danger that independent critical thinking will be replaced by groupthink, which is likely to result in irrational and dehumanizing actions directed against outgroups. (Janis 1971)

Entrenched views are often difficult to dislodge because of the status of their proponents and the widespread acceptance of the concepts, leading the German physicist Max Planck to remark that science advances one funeral at a time. But neither the reputation of scientists, nor the number of adherents a theory has are reliable tests of truth. Popper emphasised that we should accept theories only tentatively and have greater faith in those that have withstood numerous attempts at falsification. Science involves 'Conjectures and Refutations', to quote the title of one of his books. (The frequency of the appearance of the word 'could' in climate science abstracts suggests it emphasises conjectures rather than refutations.) Falsifiability is, therefore, a hallmark of science; propositions that are not falsifiable – at least in theory – are not scientific propositions. The United States (US) Supreme Court has adopted essentially Popperian criteria for admissible scientific evidence (Daubert 1993).

Numerous attempts at falsification are thus preferred to numerous papers *confirming* results, but they are only possible when the data and methods used are made transparent. Unfortunately, climate science is replete with examples of noble cause corruption – or what Sonja Boehmer-Christiansen (1994) called 'policy-based evidence' – and falls well short of the standards we expect for scientific excellence. Moreover, with climate science, the corrective mechanisms of the scientific method have been weakened because of the strong emotions Sagan warned against, and the empowerment of the climate science enterprise – a harnessing of science to state power in a kind of 'official science' that philosopher Paul Feyerabend (1975) warned against.

The strong moral sense that accompanies climate science, seeking to 'save the world' from catastrophic anthropogenic climate change, is part of the problem, because 'high moral purposes' can lead to questionable interpretive practices (QIPs) as moral purpose affects the psychology of the scientists. Established QIPs include: *blind spots* (overlooking data inconsistent with one's moral agenda); *selective preference* (accepting research supporting one's agenda, but subjecting opposing research of comparable or greater quality to criticism); and *phantom facts* (drawing implications without evidence) (Jussim et al. 2016).

Some examples of the noble cause corruption of climate science

In this chapter, I will show how climate science falls short of the criteria for good science, and far too frequently has demonstrated noble cause corruption in producing evidence intended to have a political effect. This has been the case since at least 1988, when James Hansen gave a bravura performance before a committee of the US Congress – carefully stage-managed during an El Niño to give the appearance of uncomfortable heat.[2] And which was then encouraged by a commitment made by the United Nations Framework Convention on Climate Change (UNFCCC) to tighten commitments in response to evidence. It continued as the Kyoto Protocol was being developed, with the US State Department through Timothy Wirth and Rafe Pomerance (both involved in stage-managing Hansen's testimony in 1988) engineering a conclusion in the Second Assessment Report of the Intergovernmental Panel on Climate Change in 1996 that climate change could be attributed to human agency (Santer et al. 1996).

The problem was, the lead author of the paper on which this conclusion was based (Benjamin Santer) was also the co-author of another paper that concluded it was premature to draw such a conclusion (Lewin 2017). Critics were also quick to point out that the Santer et al. paper was based on data collected between 1963 and 1987, whereas data were actually available from 1958 to 1992. When this last dataset was used the effect that had been claimed disappeared. Santer et al. claimed they had 'model-based results' that were 'qualitatively similar to the observations' shown in a graph presented by Michaels and Knappenberger (1996) showing the result vanished when all the data available were used.

---

2   Wirth later boasted that they chose a date when the Washington temperature would likely be at its steamiest, and then had the windows opened so the air conditioning would be overwhelmed, creating dramatic television.

An important feature of good scientific conduct is transparency of data and of methods. If the data used in research cannot be seen and the methods employed made clear, it is not possible to subject the research to critical scrutiny – to attempt to falsify it or, if failing to do so, to lend support to its findings. A further important feature is publication *after* an *anonymous* peer review. Adherence to both these features has been less than ideal in climate science.

Globalisation has undermined the integrity of all science by changing the nature of the anonymous peer review, which is for it to be double blind, so that the identity of neither the authors nor the reviewers is disclosed to the other.

One manifestation of globalisation, electronic communication, has undermined anonymous peer review, especially because knowledge has become more specialised and the experts in each particular field of knowledge are likely to be known to each other. They will likely be in touch via email, but the wide-bodied jet has also made global travel cheaper. They are more likely to have met at conferences and other gatherings than when air travel was the preserve of the affluent. In climate science, the formation of the Intergovernmental Panel on Climate Change (IPCC) has accentuated this, by ensuring scientists are in touch with each other. An email in 2010, for example, seeking to coordinate a response to a *Newsweek* article critical of climate science showed just how close were the leading *dramatis personae* of climate science.[3] This makes it easier for defenders of paradigms to circle the wagons in the face of an attack – or, to use another metaphor, to swarm to the site of the 'infectious' idea or of research finding. And if a smaller, more specialised scientific environment does not allow identification, Google will soon allow the author(s) to be identified (Kellow 2007). Additionally, many journals now ask authors to nominate suitable reviewers.

The Climategate emails released from the Climate Research Unit (CRU) at the University of East Anglia in 2009 showed multiple breakdowns in the peer-review process, in both scientific publication and in

---

3   See: https://eelegal.org/wp-content/uploads/2015/08/Horner-response-Redacted-2-2010-copy-2.pdf.

the IPCC reports, with correspondents conspiring to exclude research challenging the prevailing warming orthodoxy, seeking to arrange a favourable 'pal' review, and even seeking to undermine unhelpful editors (Montford 2012).

The corrupted state of peer review in climate science was evident in a case involving Garth Paltridge, a distinguished climate scientist who, with co-authors, explored a key assumption in climate models: that modest climate forcing from increased CO_2 (about 1.2 °C for a doubling of CO_2) would be amplified by positive feedback because this modest warming increased water vapour, the dominant forcing agent in the atmosphere.[4] Paltridge et al. analysed the available data on tropospheric humidity for the period 1973–2007, and submitted a paper to the *Journal of Climate*, with a conclusion that held significant implications for the positive feedback assumption. Paltridge summarised the paper as follows:

> … if (repeat if) one could believe the NCEP [National Centers for Environ-mental Prediction] data 'as is', water vapour feedback over the last 35 years has been negative. And if the pattern were to continue into the future, one would expect water vapour feedback in the climate system to halve rather than double the temperature rise due to increasing CO_2. (*Climate Audit* 2009)

The paper was rejected, with one reviewer making the rather remark-able statement that:

> … the only object I can see for this paper is for the authors to get something in the peer-reviewed literature which the ignorant can cite as supporting lower climate sensitivity than the standard IPCC range. (*Climate Audit* 2009)

Paltridge and colleagues questioned why the editor paid any atten-tion to such a review, when he should have sent the paper to another referee. Unable to extract a guarantee that any resubmission would not be assigned to the same reviewer, they submitted the paper to *Theoretical and Applied Climatology*, where it was accepted (Paltridge et al. 2009). The editor of the *Journal of Climate* was Andrew Weaver, the leader of the Green Party in British Columbia – politics and science in harmony.

4   These examples are discussed in greater detail in Kellow (2018).

This conduct was shown by the Climategate emails (Climategate, 2012) to be typical, with the *Journal of Climate* again implicated. Climategate revealed that editors were actively seeking negative reviews. For example, Keith Briffa stated: 'Confidentially I now need a hard and if required extensive case for rejecting – to support Dave Stahle's and really as soon as you can.' There was evidence of ensuring papers were rejected and evidence of regret that some 'slipped through the cracks'.

This from Michael Mann:

> While it was easy to make sure that the worst papers, perhaps including certain ones Tom refers to, didn't see the light of day at J. Climate, it was inevitable that such papers might slip through the cracks at e.g. GRL ...'

Mann also demonstrated a desire to counter research that might not be 'helpful', seeking early access to an advance copy from the leading journal *Nature*:

> ... we think it will be important for us to do something on the Thompson et al paper as soon as it appears, since its [sic] likely that naysayers are going to do their best to put a contrarian slant on this in the blogosphere. Would you mind giving us an advance copy. We promise to fully respect Nature's embargo (i.e., we wouldn't post any article until the paper goes public) and we don't expect to in any way be critical of the paper. We simply want to do our best to help make sure that the right message is emphasized.

Mann had suggestions as to how non-compliant journals should be treated after *Climate Research* published a paper critical of Mann's 'Hockey Stick':

> I think that the community should, as Mike H has previously suggested in this eventuality, terminate its involvement with this journal at all levels – reviewing, editing, and submitting, and leave it to wither way into oblivion and disrepute ...

Tom Wigley went further:

> Mike's idea to get editorial board members to resign will probably not work – must get rid of [Editor-in-Chief] von Storch too, otherwise holes will eventually fill up with people like Legates, Balling, Lindzen, Michaels, Singer, etc. I have heard that the publishers are not happy with von Storch, so the above approach might remove that hurdle too.

This 'Community' also played a gatekeeping role in IPCC reports, where its members occupied key positions as authors and editors. For example, Phil Jones wrote to Michael Mann, promising 'I can't see either of these papers being in the next IPCC report. Kevin and I will keep them out somehow – even if we have to redefine what the peer-review literature is!'

They also helped shape media reporting of climate research by assisting journalists – from the BBC to the *New York Times* – by providing helpful interpretations. The close relationship between environment non-government organisations (NGOs) and climate scientists demonstrated by Donna Laframboise (2011), was also apparent, with Barrie Pittock (a CSIRO scientist in Australia) complaining to Mike Hulme at CRU that material for a WWF leaflet was insufficiently alarmist:

> I would be very concerned if the material comes out under WWF auspices in a way that can be interpreted as saying that 'even a greenie group like WWF' thinks large areas of the world will have negligible climate change.

Climategate revealed that members of the 'community' themselves had reservations about controversial research such as the Hockey Stick, a reinterpretation of the temperature record of the past millennium using tree rings and other proxies used in IPCC Third Assessment Report to support conclusions of alarm. For example, Ed Cook wrote: 'I have my doubts about the MBH camp [Mann, Bradley and Hughes – authors of the Hockey Stick paper] … They tend to work in their own somewhat agenda-filled ways.' He also wrote: 'I do find the dismissal of the Medieval Warm Period as a meaningful global event to be grossly premature and probably wrong.'

He was not alone in these thoughts. Tom Wigley wrote: 'At the very least MBH is a very sloppy piece of work – an opinion I have held for some time.'

Keith Briffa was also sceptical: 'Between you and I, I believe there may be problems with the analysis of the Bristlecone data [a key element of the Hockey Stick paper]. We can talk by phone about this.' These problems had been identified by sceptics Steve McIntyre and Ross McKittrick, but the 'community' did not support them publicly.

An important issue surrounding the Hockey Stick controversy was the refusal of Mann to release data to those who wished to scrutinise his research, a refusal hardly consistent with good scientific practice (Montford 2010). This was not an isolated example, and complaints about the lack of data archiving have led to an improvement in this practice by some journals, including greater expectation of the scrutiny of data by referees.

Colleagues at CRU repeatedly discussed tactics to avoid releasing data to outsiders under freedom of information laws. While they provided data to 'trusted' scientists, they gave numerous excuses for concealing the data on which their findings and temperature records were based. Scientists were even advised to delete data, which when done after a freedom of information (FOI) request was a criminal offence, Jones (BBC 2010) later stating he sent this email 'out of frustration'. It emerged subsequently that much of the original raw data upon which its research and its temperature series were based had been destroyed.

Given the centrality of data transparency to the scientific method, the response by Jones to a request from Australian sceptic Warwick Hughes for access to temperature data was remarkable. After initially suggesting an arrangement with the World Meteorological Organization (WMO) precluded this, Jones wrote to Hughes:

> Even if WMO agrees, I will still not pass on the data. We have 25 or so years invested in the work. Why should I make the data available to you, when your aim is to try and find something wrong with it. (Hughes 2005)

As it happened, another Australian, John McLean, later undertook an audit of CRU's global temperature record (HadCRUT, data driving the climate models) and found numerous embarrassing mistakes, acknowledged by the UK Met Office and IPCC (McLean, 2017). In some cases, ships reported ocean temperatures from locations up to 100 km inland. Reporting of ocean temperatures by ships is important because the surface temperature of the 70% of the Earth's surface that is ocean is taken as a proxy for the temperature of the air above it. (The correspondence of the two has, understandably, been questioned.) Many of these readings were made from water in buckets dropped over the side of ships, and adjusted because of the replacement of canvas buckets by steel buckets (Kent et al. 2007). Latterly, most readings came from the continuous recording of temperature at varying heights at the intakes of ships' cooling systems. Problems with these historical data persist and they are still being adjusted (Chan et al. 2019). (One wonders if policy-makers are aware that the 'settled science' of climate change is based upon models tuned to data adjusted to account for canvas versus steel buckets.)

The CRU notes that the use of marine air temperatures (MAT) would be preferable when combining with land temperatures, 'but they involve more complex problems with homogeneity than sea surface tempera-ture (SST)' (CRU 2016). In using SSTs, they acknowledge that 'we are tacitly assuming that the anomalies of SST are in agreement with those of MAT.' This is the equivalent of looking for one's lost keys under a street lamp because the light is better there. More reliable near-global readings have been available since 2005 from the extensive system of about 4000 Argo floats that have produced more observational oceanic data than were collected in the entire 20th century. These errors suggest major quality assurance issues.

A problem for the negotiation of the Paris Agreement in 2015 was that the IPCC's *Fifth Assessment Report* acknowledged 'a hiatus' or 'pause' in temperature increase, which scientists struggled to explain (Stocker et al. 2013, pp. 61–2). This lasted from 2002 through until 2014, a period of twelve years, as shown in Figure 18.1.

This was not a convenient truth in the lead-up to a major negotiating meeting, and Karl et al. 'conveniently' removed the problem, deciding that 'residual data biases in the modern era could well have muted recent warming, and as stated by IPCC, the trend period itself was short and commenced with a strong El Niño in 1998' (Karl et al. 2015, pp. 1469–70). Citing recent improvements in the observational record and a couple of additional years of global data 'including a record-warm 2014' (Karl et al. 2015, pp. 1470), they re-examined the observational evidence related to the 'hiatus'.

Controversy erupted in February 2017 when John Bates, a former colleague of Karl's at NOAA, detailed how the Karl et al. (2015) paper had been mishandled, accusing Karl of influencing the results and release of the crucial paper, stating that 'in every aspect of the preparation and release of the datasets ... we find Tom Karl's thumb on the scale pushing for, and often insisting on, decisions that maximize warming' (Bates 2017). Bates was dumbstruck that Karl, in charge of NOAA's climate data archive, 'would not follow the policy of his own Agency nor the guidelines in *Science* magazine for dataset archival and documentation.'

Bates noted that the Chairman of the House Science Committee, Lamar Smith, had questioned the timing of the paper; it was published just prior to the Obama Administration's Clean Power Plan submission to the Paris Climate Conference in 2015. But Smith had also sponsored a Bill in 2014 and 2015, proposing a *Secret Science Reform Act* – which Obama vowed to veto. This bill would have prohibited the EPA from creating regulations based on science that was 'not transparent or reproducible'. After Donald Trump's election, the same provisions were proposed in an *Honest and Open New EPA Science Treatment Act*, which did not pass, but Scott Pruitt, as EPA Administrator, effectively implemented the requirements of Smith's proposed legislation by orders within the EPA.

These were very much in line with principles of good scientific practice (and the *Daubert* rule), but scientific organisations and environment groups claimed this would make it more difficult for the EPA to create rules at all, and craft them based on the best available science. Their main objections seemed to be that environmental changes often made replication of exactly the same experiment impossible (a very narrow interpretation of reproducibility, essentially an argument that you can't step into the same river twice) and confidentiality of medical records. Somehow testing drugs for efficacy and risk manages to overcome this difficulty in supporting regulatory decisions, but research supporting environmental regulations would necessarily breach patient confidentiality.

John Bates stated that the issue was about the 'timing of a release of a paper that had not properly disclosed everything it was' (Cornwall & Voosen 2017). In his blog post on *Climate, etc* he stated: 'Karl failed to disclose critical information to NOAA, *Science Magazine*, and Chairman Smith regarding the datasets used in K15.'

Karl himself stated that 'there wasn't a lot of new science in it'. He added that they just assembled the data they had and paired it with a published, non-operational dataset of land surface temperatures. 'We said, let's just put it together, and that's what made it newsworthy and important.' Bates's complaint was about the failure of Karl et al. (2015) to wait until the non-operational dataset had met quality assurance requirements, and it was quite clear that Karl et al. (2015) were in a rush to produce scientific evidence that was not only timely, but 'newsworthy and important' – especially politically important.

The Karl et al. (2015) paper was helpful for Paris, but in 2016 Donald Trump was elected president, threatening to withdraw the US from the Paris Agreement. There was one final, futile, attempt to drive US policy on this issue with a paper by Ben Santer and others rushed into e-print on 24 May 2017 attempting to disprove, not a piece of published science, but – remarkably – a statement by the incoming Environmental Protection Agency (EPA) Administrator Scott Pruitt at his Senate confirmation hearing that satellite data indicated there had been a 'levelling off' of warming. Santer et al. (2017) reasserted the trend of 0.01 °C per annum,[5] and concluded that this was inconsistent with Pruitt's 'levelling off' statement. The paper enjoyed rapid passage though peer review at *Nature Scientific Reports* in only 29 days (received 6 March, accepted 4 April, published May 24), possibly with the aim of influencing the G7 meeting in Sicily commencing on 26 May, but certainly before the looming decision on the Paris Agreement by the President. Of the last, approximately, 500 articles published in *Nature Scientific Reports*, fewer than 1% had been accepted within 30 days from submission and for all such articles, except for Santer et al., supporting materials were available (Goldstein 2017). Sceptical climate scientist Roy Spencer (2017) summed it up thus: 'It's sad to see how far peer-reviewed climate science has fallen.'

Climate science presents too many examples of evidence being conveniently provided with the intention of having a political impact – from Santer et al. intending to affect the development of the Kyoto Protocol in 1996, to Karl et al. (2015) conveniently removing 'the hiatus' that had

---

5   This is equivalent to 1° C per hundred years.

detracted from the scientific narrative behind the Paris negotiations, and Santer (again) and colleagues (remarkably) taking issue with Congressional testimony in a scientific journal in an attempt to influence President Trump's decision on whether to withdraw from the Paris Agreement. The Climategate emails show that this behaviour is endemic among the 'Community'. The nobility of the cause has trumped adherence to the scientific method; the generation of policy-based evidence has limited the ability of policy makers to respond on the basis of reliable scientific evidence.

Conclusion

These few examples of what might be considered noble cause corruption should concern both citizens and governments. Some have characterised climate science as 'post-normal science', which is pursued when facts are uncertain, values are in dispute, stakes are high, and decisions are urgent (Funtowicz & Ravetz 1993). But the inherent uncertainty in climate science should be embraced, because it changes the nature of the debate over policy responses; if we don't know what size the cap should be, other measures should be preferred to 'cap and trade' – a point acknowledged in relation to climate change by the inventors of this policy (Hilsenrath 2009).

Philosophers David Coady and Richard Corry (2013, p. 62) admit that climate models have been 'practically falsified', but rather strangely assert that, rather than this being evidence that 'climate science is bunk', we should accept that the criterion of falsifiability is bunk, because 'Falsifiability is too much to require of a complex scientific theory since no theory that is part of an interconnected set of theories will be falsifiable' (Coady & Corry 2013, p. 67). Garth Paltridge had difficulty getting published data that falsify a core component of the global warming theory propped up by the models, with an editor supporting a reviewer who correctly identified, not a problem with Paltridge's data or logic, but with its implications for the politics and policies supported by climate science.

Aside from leaving us with no guidance as to how we might judge the veracity of climate science, this ignores the point (for example) that there was a very recent pause in warming as measured by global satellite and it was denied.

The US Supreme Court, in the Daubert decision adopted essentially Popperian criteria for assessing scientific information because they must rule on conflicting scientific claims. Policy makers are in the same position, and we have little option but to ask that climate science adheres to Popperian standards: open disclosure of data and methods; genuine anonymous peer review; specification of what might falsify hypotheses and attempts to do so.

Complexity and uncertainty do not excuse climate science from Popperian standards. As was eloquently written in the blog *The Hockey Schtick* (2018) put it, 'If you can't explain the pause, you can't explain the cause'. Nor does it exempt adherence to research ethics, but there are too many breaches in evidence. A recent story in *Nature* questioned whether the net effect of forests (emissions and other factors considered) was positive or negative for global warming. One scientist stated, 'I have heard scientists say that if we found forest loss cooled the planet, we wouldn't publish it' (Popkin 2019, p. 282).

Climate science is in urgent need of repair.

19 Of Blind Beetles and Shapeshifting Birds: Research Grants and Climate Catastrophism

Dr Paul McFadyen, Scott Hargreaves and
Dr Bella d'Abrera

Public policy in government is not made based simply on reason and evidence. In Westminster systems it is made based on the number of supporters the policy has in Cabinet and the authority of those supporters. The rules are distinct from those traditionally applied to science.

In a democracy, *successful* government public policy is what the government or parliament determines it is. *Good* public policy, on the other hand, is a subjective value judgement made by voters at elections. This is self-evident and is a fundamental tension within democracy. While bad science and bad public policy may be overturned in a democracy in the long run, it can inflict a great deal of harm in the short term.

With climate change, the key determinants in the development of national and international climate change public policy are:

- the ideologies arising out of the cultural *zeitgeist* of the time, which place the anthropogenic global warming (AGW) hypothesis as somehow above the realm of contestable science and policy
- the reliance on expert knowledge of scientists in government scientific organisations and universities, even when the issue is one of policy rather than science
- the methodology for the funding of science, which creates perverse incentives.

We shall look at each in turn, and then outline the results of an audit of government-funded climate-related research in Australian universities.

Universities have attained a special status on policy making both as centres of expertise and as vehicles to pursue public-benefit research; therefore, close examination is required of the institutional environment in which they operate, and uses to which they put research funds.

How public policy is made

The belief that carbon dioxide (CO_2) emissions by humanity will cause catastrophic global warming is the dominant paradigm worldwide. The escalation of the AGW hypothesis from national governments to the United Nations (UN) increases its authoritative impact on government public policy. Most countries have signed the UN's 2015 Paris Agreement to limit CO_2 emissions. The only significant exception is the United States of America (US), which indicated on 1 June 2017 that it would withdraw from the Paris Agreement.

Most countries in the UN are in the developing world and the AGW hypothesis offers the prospect of a massive wealth transfer of hundreds of billions of dollars from rich countries to poor countries both in the Green Climate Fund and in international carbon credits.

National governments are advised directly by government scientific institutions and by the UN through the Intergovernmental Panel on Climate Change (IPCC), which itself draws on those institutions for its consolidated analysis. However, scientists employed in government institutions and universities are partly grant funded. Given that scare campaigns promoted by environmental activists and amplified in the media work to unlock research funds from government, furthering the man-made global warming hypothesis aligns with the long-term economic interest of researchers.

These dynamics check the influence of the many individual scientists who strongly disagree with the AGW hypothesis, particularly in its catastrophic forms. They publish articles in scientific journals and newspapers, host climate blogs and talk and write to politicians stating that the evidence to support such a hypothesis is not there. However, independent scientists – no matter the weight of their findings in purely scientific terms – *are not* the formal scientific advisers on climate-change policy to government.

That so many individual scientists have doubts but no scientific organisation or association has publicly stated them appears to be an ethical failure of the collective scientific body politic.

There is no formal scientific academy, organisation or association in Australia that has publicly opposed the catastrophic man-made global warming hypothesis. This appears to be the case in most countries. Montford (2012) noted the capitulation of the Royal Society (the very first such body) some years ago. While consensus is not a determinant in science, it is in the making of government policy. This means that statements made by presidents of the Royal Society, or equivalents in other countries and on behalf of different specialisations, carry great weight. Individual politicians may agree with such independent scientists, yet overwhelmingly government climate policy remains based on the catastrophic AGW hypothesis.

Expert scientific knowledge and economics

In climate science, the range of issues are so complex and cover so many scientific disciplines that it is extremely difficult not only for much of the public to understand, but also for many scientists. Politicians and policy bureaucrats are dependent on the advice from government scientific organisations.

The assumption is that the scientific advice from government scientific organisations is disinterested and objective, yet every economist knows that funding arrangements can change behaviour. They are a matter of economics, not science, but they do affect science – and in economics, like science, theories change.

In the post-war period, economic arguments held greater sway, and two separate but related ideas took hold, with deleterious consequences. The first was that the power of the state could be harnessed to boost science, and that basic research leads to technology, which in turn leads to economic growth. Davidson (2005) points out this 'linear model' can be traced to Vannevar Bush in his 1945 report 'Science, the endless frontier'. However, Davidson considered:

> The benefits of publicly financed R&D are oversold … Indeed many studies show basic research often follows technology, and not vice versa. This model

is so discredited an otherwise self-serving report into publicly funded research in the UK described it as being 'dead'.

Even more damning is an OECD report into the sources of economic growth. This 2003 report found a positive relationship between R&D and economic growth. Yet when it disaggregated R&D into public and private R&D the differences were very stark. Publicly funded R&D had a large and very significant negative relationship to growth. This is a particular challenge for Australia that has a relatively large public component in R&D expenditure. (Davidson 2005)

The second idea was that the competition inherent in market economics was more efficient, and therefore its application to the world of research would necessarily improve productivity and lead to even more of the benefits promised by Bush. If removing socialist practices and implementing competitive market economics improved productivity in other areas of society, then it seemed logical that it should be applied to improve productivity in science.

Increasing competition in science could be achieved by replacing ongoing base-funded scientific research with a competitive arrangement of government 'buying' research outputs from scientists with time limited grants. This is the well-known purchaser–provider arrangement.

Government would be the purchaser and scientists would be the providers. The purchaser–provider model was embraced by government for appearing to improve both economic efficiency and administrative flexibility compared to the old fixed-base funding system.

But there were a few problems the policy bureaucrats had overlooked. The purchaser–provider economic model requires an *informed* purchaser for the model to work effectively. In this transaction, however, the most informed person is the scientist and researcher – the provider – rather than the purchaser, or government. This gives the provider the capacity to manipulate the purchaser and can lead to adverse outcomes that may not improve economic efficiency and effectiveness.

The second problem is that the purchaser in democratic countries is an elected government. Scientists know that the way to get government to fund their research is by harnessing scare campaigns using the green pressure groups and the media.

Figure 19.1 The Iron Triangle

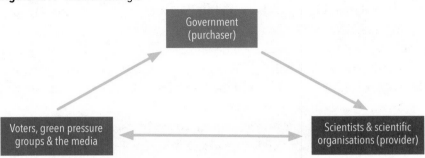

Thus was established a self-reinforcing 'Iron Triangle' (see Figure 19.1, above) between government (as the purchaser), scientists (as the provider), and the electorate, media and green pressure groups.

So instead of government having control over the direction of scientific research, the application of competitive market economics by grant funding has influenced the establishment of a science funding model of interacting and reinforcing self-interest groups. This has reduced government flexibility and made government policy hostage to the scientific organisations and pressure groups. This model is now largely the way science, including climate science, is funded in government scientific institutions and universities. This can have adverse outcomes for the economy.

The third problem is the inefficiency of grant funding. For a typical three-year grant, scientists spend two years doing research and much of the third year lobbying and writing proposals for the next three-year grant. This leads to a significant decline in productivity compared to the equivalent base-funded tenured scientist.

The Australian Research Council

Yet bad as the Iron Triangle is, the arrangements for competitive grant funding through the Australian Government's Australian Research Council (ARC) are even worse. At least under the traditional 'Iron Triangle' – with government as the purchaser – there is full democratic accountability.

The ARC was established in 2001 as an independent body under the *Australian Research Council Act* 2001, reporting to the Australian

Government via the responsible minister. It claims that 'the outcomes of ARC-funded research delivers cultural, economic, social and environmental benefits to all Australians'.

Under the Act the minister's flexibility is restricted. The legislation specifically states, 'The Minister must not direct the CEO to recommend that a particular proposal should, or should not, be approved as deserving financial assistance'.

The ARC legislation restricts the freedom of their minister regarding grant proposals and reduces the government's flexibility to direct grants to areas of greatest benefit to society. He does, for instance, sign off on high-level 'Science and Research Priorities' (such as energy, food, or environmental change), but implementation is the preserve of the ARC.

It is difficult to say whether recent changes, which assert some notional authority to the views of the minister (Tehan, 2018), increase or reduce confidence in the system. The minister now requires that each applicant spell out how the research will 'advance the national interest', and reserves the right to veto particular grants (as a previous minister had done, on occasion). However, the minister did not, in 2018, veto any grants at all (Nott, 2018).

ARC funding is substantial, with $3 billion in grants for individual university research projects budgeted over four years (Tehan, 2018). Ideally, taxpayer-funded grants would reflect the direction of the democratically elected government rather than the ARC bureaucracy, and legislative limitations on the minister's prerogative should be removed. This would, however, not necessarily fix the fundamental issue inherent in the purchaser–provider model, which can only be overcome by a fundamental change to the way base funding is provided.

Before we reveal the results of our audit of ARC grants, it is necessary to look more deeply at the universities that are the major recipients of the grants, as they have their own governance issues.

Universities

The combination of the Iron Triangle, the ARC and universities as they are currently governed in Australia is diabolical.

In Australia virtually all universities (other than the Australian National University) are established by state government legislation. While state government universities are state government entities, they are mainly funded by the Australian Commonwealth Government, not by state governments.

The root cause is the flawed financial arrangements that underpin the Australian Federation. The Commonwealth Government raises most of the tax revenue while the states have most of the expenditure responsibilities. This 'vertical fiscal imbalance' increases central control by the Commonwealth Government, but reduces the good governance principle of subsidiarity – which requires that the level of government raising the revenue to be the same as the one that spends that revenue, providing a direct feedback loop for voters between taxes raised and taxes spent. This is a major fiscal and political dysfunction of the Australian federation. As a result, neither the states nor the Commonwealth are properly accountable for Australia's universities.

The states can take the view that as the universities are federally funded, the management of the university is nothing to do with them, while the Commonwealth Government can take the view that as they are state institutions, they are not a Commonwealth responsibility. Consequently, billions of taxpayer dollars are allocated to universities with little accountability to government.

Table 19.1 shows the dominance of Federal funding mechanisms for our two oldest universities, both of which predate the formation of the Commonwealth of Australia by some decades. Federal funds dwarf by an order of magnitude funds from state governments, and revenue from international studies is reliant on Federal measures, such as special visa arrangements.

This separation of formal accountability and funding conflicts with the purposes of a university. Most of them were established and funded primarily to serve society by training people in professions necessary for society. This is no different from schools, just at a higher level. Research activity – allowing teachers to pursue their academic interests and to ensure their teaching is up to date – was seen as a part of academic life. In the Australian system, dedicated research was mostly the preserve of

Table 19.1 Revenue from continuing operations (2018–19)

Revenue ($ million)	University of Melbourne	University of Sydney
Australian Government grants	772.4	683.4
HELP - Australian Government payments	288.2	257.4
HECS-HELP - student payments	31.1	33.1
Fees and charges – international students	904.7	1,061.9
Fees and charges – other	202.7	162.2
State and Local Government financial assistance	39.5	33.0
Investment revenue	226.8	119.4
Consultancy and contracts	123.6	111.3
Other revenue	218.1	184.3
TOTAL	2,807.2	2,646.1

Source: Annual Reports.

government research organisations such as the Commonwealth Scientific and Industrial Research Organisation (CSIRO).

The more recent emphasis on research in universities creates a tension with the historic mission, because the fundamental function of training in public universities is important for the economy. Australia has a skilled migration program partly because the educational institutions are not producing the requisite number of skilled workers.

The Senate of each state public university should have a senior official from the state treasury as an *ex officio* member. State treasuries are the appropriate agency because they operate at the highest level of government, are economically and employment focused, and have intimate ongoing connections to all state government line agencies. This would provide a high-level conduit between the state government and the university senate for information transfer on skill needs and governance.

Some may have a philosophical view that public universities should be independent of government, yet, while public universities are taxpayer funded and established by statute, government has not only a right, but a duty to oversee the governance of such institutions. Peter Ridd's case at James Cook University (JCU) illustrated the failure of the hands-off view. In the absence of government intervention, JCU is understood to have spent in the order of $1 million of taxpayer funds to fight a test case in which academic freedom is under threat.

Audit of ARC grants for climate change

The ARC administers the National Competitive Grants Program (NCP), which comprises two programs, Discovery and Linkage. The Discovery program schemes include Australian Laureate Fellowships; Discovery Early Career Researcher Award; Discovery Indigenous; Discovery Projects, and Future Fellowships.

The Linkage Schemes are ARC Centres of Excellence; Industrial Transformation Research Program; Learned Academies Special Projects; Linkage Infrastructure, Equipment and Facilities; Linkage Projects; and Special Research Initiatives. The ARC has developed a grants search data portal that contains all humanities-funded projects from 2002–2019. The data used in this chapter is drawn from the grants search data portal.

To determine the number of projects that refer, or have referred, to climate change, a number of key words were identified, and the occurrence of those keywords was identified in the descriptions of the projects' content. The key words employed were 'climate change', 'global warming', and 'greenhouse gases'.

While sympathising with them for needing to do so, we excluded from analysis those research grants – and there were many – where the researcher had described a coherent rationale and methodology, and then more or less said 'and this is somehow relevant to global warming' in the hope of increasing their chance of success.

Between 2002 and 2019, 85 ARC-funded humanities research projects were identified as addressing the theme of climate change. The total amount of funding was $32.7 million, distributed across a wide range of fields, including historical studies, philosophy and religious studies, anthropology, human geography, social work and ethics. The outcomes are shown in Table 19.2, below.

Table 19.2 ARC grants for climate-change research in the humanities, 2002–2019 ($'000)

	Number of Projects	Amount ($'000)
Historical studies	11	$2,997.4
Philosophy and religious studies	6	$2,148.4
Anthropology, criminology, demography, human geography	28	$12,681.4
Political science	11	$3,995.2
Social work	5	$1,708.2
Creative arts	2	$250.4
Cultural Studies, language studies, literary studies, linguistics	7	$2,221.8
Archaeology	13	$5,931.5
Ethics	2	$805.8
TOTAL SPEND	85	$32,740.0

Source: ARC Data Portal, Authors.

There are eleven historical studies research projects that also discuss climate change. Unsurprisingly, a great many of these seek to employ historical examples in order to make the case that by looking at the past, society will be more prepared for the effects of climate change. Other fields have also received generous support for projects which concern climate change, such as philosophy and religion. For example, in 2018, the University of New South Wales was given $367,770 for a project entitled 'Ethics, responsibility and the carbon budget' which states that 'in order to avoid climate change the world must drastically limit its emissions of greenhouse gases'. Meanwhile:

- Researchers at the University of Sydney were granted $892,179 for 'Climate change the history of environmental determinism'.
- At the University of Melbourne a project examining 'Climate justice' was awarded $646,823, which would 'offer significant insights into the effects of climate change and adaption policy on the key area of rural well-being and energy use'.
- The University of Adelaide was given $772,454 in the field of demography to study 'Climate change and migration in China: theoretical,

empirical and policy dimensions', stating that researchers would 'analyse the complex relationship between climate change and migration by focussing in depth on two areas in China anticipated being major hotspots of Climate change impact'.

• At the University of Melbourne, human geographers were granted $751,985 for 'Climate Change and Security in the South Pacific', based on the assumption that 'Climate change is dangerous to many Pacific societies and countries, which will in turn generate problems for Australian aid, diplomatic, immigration and security policy'.

In the sciences we limited ourselves to six key areas where climate change was most likely to be relevant. The results are presented in Table 19.3, below.

It is important to note that while the outcomes of research projects are often reported to confirm the factual basis of climate-change models and predictions, anthropogenic climate change is overwhelmingly the assumption on which the program was framed. Take the University of Technology Sydney's 'The Coal Rush and Beyond: Climate Change, Coal Reliance and Contested Futures', which received $540,000 in 2014. According to the proposal, 'Globally, coal extraction and burning is booming' but 'the burning of coal has released unprecedented quantities of CO_2 into the atmosphere and exacerbated anthropogenic climate change'. Another example is a project from the University of Queensland (UQ) in 2007, which yielded $707,000, and left behind no cliché

Table 19.3 ARC grants for climate-change research in the sciences, 2002–2019 ($'000)

	Number of grants	Amount ($'000)
Physical sciences	8	$2,423.9
Chemical sciences	25	$11,214.8
Atmospheric sciences	240	$141,547.5
Environmental sciences	111	$50,278.3
Biological sciences	190	$47,257.6
Engineering	72	$33,847.2
TOTAL	646	$286,569.3

of catastrophism: 'The recent anthropogenic global warming is causing polar icecap melting, sea level rise, reef coral bleaching and degradation, and increased frequency and intensity of severe droughts, floods, tropical cyclones/hurricanes/typhoons.'

It is often said by activists that to combat climate change we need to consider bold new technologies. It appears the ARC endorses this approach by backing proposals that test the boundaries of science, engineering and – one might say – common sense. The engineers at UQ received $713,000 to develop a:

> Floating Forest: a breakwater for protecting the Australian coastline. This project aims to develop structural, materials and foundation solutions for a large floating forest that will act as a mega breakwater and windbreaker to protect the Australian coastline from strong waves and winds caused by climate change.

This preliminary design for such a structure has already been published by UQ (2019), and we'll let readers form their own judgement (see Figure 19.2, below), while noting that the coastline of mainland Australia is more than 35,000 km long.

While some proposals are specially related to the AGW hypothesis, in a significant number of the studies it appears the researchers have had to adapt a proposal that may have had some intrinsic scientific merit, in order to justify it on the grounds of climate change. A notable sub-genre in this respect is projects where researchers wish to study particular flora or fauna, but have to frame their studies in terms of current or potential 'impacts' of AGW. Examples include:

- The southern hairy-nosed wombat ('Predators and climate change threaten Australia's arid-zone wildlife.')
- Blind beetles ('Climate change is having drastic effects on animal biology, threatening many species. Recent data suggest that changes in body shape (the size of appendages) is one such effect.')
- Frogs ('Are frogs in fragmented lowland rainforest especially susceptible to both disease and climate change?')
- Koalas ('This project ... will also predict the effect of future climate change on western koala populations.')

Figure 19.2 Floating forest design aiming to reduce wind and wave damage

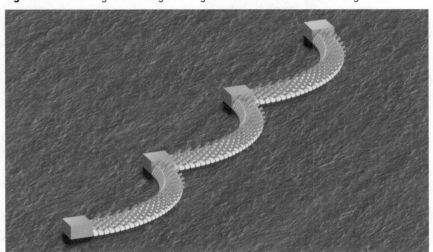

The University of Queensland. The researchers say theirs is the first design that places a windbreak on top of a floating breakwater structure. The structure consists of a concrete deck tilted upstream to allow a wave run-up, which will dissipate the wave's energy, similar to the way a truck safety ramp can slow a speeding truck. Several arrays of hollow column tubes are placed on top of the breakwater deck to form the 'trees' of the floating forest, which will ultimately reduce wind speed. [1]

Source: https://www.uq.edu.au/news/article/2019/05/floating-forest'-could-protect-vulnerable-shorelines.

- Emus ('This project investigates how decline of a key seed disperser, the emu, due to global environmental ... affects the persistence and migration potential of endemic SW Australian forest plant species.')
- Fish ('Why are fish shrinking as the climate warms?')
- Fiddler crabs ('Climate change is already affecting many Australian animals, including fiddler crabs ...')
- Cuckoos ('This project will assess the impact of climate change on interactions between parasitic cuckoos, hosts and prey and formulate predictions.')
- Adélie penguins ('How will animals respond to climate change? A genomic approach ... The research will help us understand the biology of climate adaptation.')

Alternatives

The ARC funding mechanism may be beyond reform, as it embodies the worst aspects of the Iron Triangle while undermining the accountability of the minister–purchaser and the university–provider. It may do less harm to society if the $800 million in grants were simply divided up between the universities as untied research grants rather than the current *faux* competitive grants system, which promotes support for trendy but dysfunctional ideologies.

It would ensure that accountability for disbursement of funds within the university is clearly identified, and that it sits where it belongs with the vice chancellor and ultimately the governing senate of the university. Moreover, it would save up to $40 million in administration. Even if the research was on something totally useless it would be better than on something that harms Australia.

This allocation to universities could be done on a formulaic basis, for example dollars per student per university (per capita) with the amount for each university further divided on a faculty basis. This could mirror the existing faculty split for the current ARC grants (the latter would be the easiest way, or alternatively it could be adjusted if desired if some universities specialised in particular areas). Or it could be left entirely to the discretion of the vice chancellor.

Universities would be required to report annually (to the Treasury) outlining how these funds were spent on various grant projects, and as an assurance that the funds weren't being used to provide increased salaries, for example – which is the same reason capital funds to government agencies are allocated separately to ongoing funds. This would be no different to the current acquittals process for grants so it would not require extra work.

There are precedents for arrangements in which a budget allocation is distributed according to some formula. For example, the Commonwealth Grants Commission grants to states are made on a formulaic basis – essentially on a per capita basis, per state, weighted for a range of disability factors, for example, dispersion of the population, demographics, and so on. The formula is updated every five years. What is more, federal funding allocations for health to the states are also allocated on a formulaic per capita basis weighted for various factors. Subsequently, state governments also allocate these health funds on a formulaic basis to their public hospitals either on a casework (i.e. workload) or another basis.

Conclusion – climate science, a trust misplaced

Leaders in government rely on the authority of the scientific institutions to guide and justify their decisions. True science does not rely on the authority of the speaker, but on the evidence provided. However, this is not the way that public policy in government is made. Ministers and policy bureaucrats do not have the capacity or the time to assess the scientific evidence, so they rely on the authority of scientific institutions. It is a matter of trust and if this trust is based on a self-interested funding methodology, it is a trust misplaced. We must seek a better way.

20 A Descent into Sceptics' Hell
Scott Hargreaves

> To enter the lost city, go through me.
> Through me you go to meet a suffering
> Unceasing and eternal. You will be
> With people who, through me, lost everything.

No one who wants a quiet or sociable life would be a climate change sceptic. I like dinner parties as much as the next man, but expressing doubts as to the coming climate Armageddon can really ruin what might otherwise be a fine evening.

One recommendation (Marohasy 2017a) is to arm oneself with some arguments, and a genteel manner, and seek some kind of dialogue. Sadly, this rarely works, as too often the interlocutors are zealots who will brook no discussion, or aren't really that interested in climate change, but, rather, are reciting standard opinions in order to make conversation.

My last experience of dinner party popularity was when I worked for an energy company that dominated the Australian market in 'green energy' (renewables only), and where a major part of my job was securing permits for its wind farms. When networking in my professional life, and also at dinner parties, all of this was greeted with admiration; it was clear to everyone that I was living a virtuous life and was most likely a similarly virtuous person. When I told them I was a manager of 'Sustainability', the approbation was a wonder to behold.

Sadly, for my professional life and my reputation in polite society, I was progressing from holding a genuine interest in global warming and the need for carbon-reduction strategies, to daily becoming more uncomfortable with the outlandish claims and the failed predictions of the global-warming movement.

My first tilt towards scepticism involved renewable energy. As many of my colleagues at the energy company understood, wind and solar are terrible vehicles for accomplishing carbon reduction. They combined a belief in anthropogenic global warming (AGW) with a commitment to rigorous public policy analysis, and an expectation that when a price is put on carbon the subsidy schemes for renewables will just wither away, being no longer necessary. I, on the other hand, was much closer to the renewable energy lobby and its political milieu, and so was much less surprised when a carbon price was introduced in Australia and yet the subsidy schemes remained. The rent-seekers clothe their arguments in the language of carbon reduction, but their actual positions when lobbying are based on naked self-interest.

Over time, and once I examined the critiques of the institutions and research that supports the AGW industry, I could not help but descend further into sceptics' hell. This naturally accelerated when I joined the staff of the Institute of Public Affairs.

I will describe my descent into scepticism by taking inspiration from the epic poem, the foundational work of Italian literature, the *Divina Commedia (Divine Comedy)* of Dante Alighieri (1265–1321). In the first of three books, *Inferno,* Dante describes his descent through the nine circles of hell. On his journey, Dante was protected by the ghost of the Roman poet Virgil and could indulge old grudges by including his enemies in various places and have them suffering ever-more-inventive torments. For me, I have a succession of guides rather than a lone poet, and being increasingly cut off from mainstream opinion feels like it is me not my enemies who are in hell. As in Dante's journey, the further the descent, the further from the stars.

In *Climate Change The Facts 2017* (Marohasy 2017a) the great Clive James (1939–2019) skewered the catastrophist tone in the media that cherry picks the worst-case scenarios from official bodies, such as the Intergovernmental Panel on Climate Change (IPCC), and which through a combination of mendacity, stupidity and concern for repeat

Figure 20.1 Dante's circles of hell

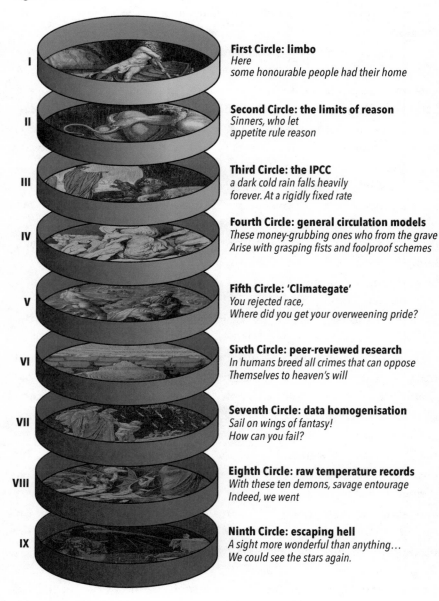

I **First Circle: limbo**
Here
some honourable people had their home

II **Second Circle: the limits of reason**
Sinners, who let
appetite rule reason

III **Third Circle: the IPCC**
a dark cold rain falls heavily
forever. At a rigidly fixed rate

IV **Fourth Circle: general circulation models**
These money-grubbing ones who from the grave
Arise with grasping fists and foolproof schemes

V **Fifth Circle: 'Climategate'**
You rejected race,
Where did you get your overweening pride?

VI **Sixth Circle: peer-reviewed research**
In humans breed all crimes that can oppose
Themselves to heaven's will

VII **Seventh Circle: data homogenisation**
Sail on wings of fantasy!
How can you fail?

VIII **Eighth Circle: raw temperature records**
With these ten demons, savage entourage
Indeed, we went

IX **Ninth Circle: escaping hell**
A sight more wonderful than anything…
We could see the stars again.

Source: Image adapted from Dante's Inferno, Cantos VI thru XI by Manny, Ashes from Burnt Roses blog: http://ashesfromburntroses.blogspot.com/2018/02/dantes-inferno-cantos-vi-thru-xi.html.

sales aims to keep readers and viewers in a permanent state of fear. It is only fitting, then, that in adapting Dante's structure of hell I can draw upon the brilliant translation by James (Alighieri 2015), in which the great man renders an 800-year-old Italian poem into English that is both accessible and beautiful, and which has echoes of the original's rhymes, metaphors and puns. To the extent that he started people on a journey of inquiry into the whole AGW project, James should have included in his climate-change writing the same warning that appears above the gates of hell, and which appears at the start of this chapter. (His wit and erudition are greatly missed.)

First Circle: limbo

> Ahead of us, beginning as a bar
> Of light, it swelled to form a hemisphere
> Of fire that through the shadows burned a dome –
> And from a good way off I guessed that here
> Some honourable people had their home.

For Dante this is the home for eternity of the great poets and thinkers of the past who had the misfortune to be born before the Christian saviour arrived on the scene. Thus Homer, Plato, Aristotle, Cicero and other giants of Graeco-Roman civilisation are here condemned to eternity, not suffering, but yet without the gift of grace.

This is the realm reserved for anyone who is not necessarily a sceptic, and who has a life. They may not actively deny AGW, but they are more interested in other social and economic problems or – to their shame – they are more concerned with what Aristotle called human flourishing (eudemonia), which involves learning, virtue, work, society and family. To Extinction Rebellion[1] and others protesting 'climate breakdown', someone who does not share their apocalyptic sense of urgency may not (immediately) deserve punishment, but they are certainly cast out of decision making.

---

1   A socio-political movement with the stated aim of using civil disobedience and nonviolent resistance to protest against climate breakdown, biodiversity loss, and the risk of social and ecological collapse. Wikipedia.

My guide through limbo was former Australian prime minister John Howard, a champion of the centre-right who was goaded into proposing a domestic carbon cap-and-trade scheme in 2007. His technocratic solution was cloaked in a neutral language of risk mitigation and economic considerations, and was blocked in the Senate by his Australian Labor Party and Australian Greens political opponents. They much preferred the formulation that climate change was 'the great moral challenge of our time' (Rudd 2007) and held out for something much more dramatic. Howard and his legislation were in limbo, the first circle of sceptics' hell.

Second Circle: the limits of reason

> I understood this was the punishment
> For carnal sinners, who let appetite
> Rule reason, and who, once drawn, are now sent –
> Like winter starlings by their wings in flight –
> Across the bleak sky in a broad, thick flock.

In 1988 it would have been hard to argue (successfully) against the establishment of the IPCC (although the Institute of Public Affairs was doubtful from the start). Its model was that commonplace in public policy – an expert panel called to prepare a report for policy makers. This was reason in action, science at its best, replacing the ignorance, self-interest and prejudice of the political class. It was more Woodrow Wilson and the technocrats than the Jacobin approach.

My guide through the second circle of hell was Bjorn Lomborg. His radical approach is to take as gospel all the conclusions of the IPCC, and to build on the promise of reason made at its foundation. Importantly, the IPCC's climate predictions are typically less catastrophist than portrayed by activists and used in the media (where journalists tend to cherry pick the outliers from the model runs). Swedish doctor and humanitarian Hans Rosling was a man committed to evidence-based policy making and so – despite being concerned about AGW – he was appalled by a conversation with Al Gore in which the latter said 'We need to create fear!' To his credit, Rosling replied, 'No!' (Rosling et al. 2018).

In Dante's second circle, those who let passion overwhelm reason (like the story of the tragic lovers Paolo and Fransesca related by Dante) suffer by being blown about by endless winds ('Here, there, now up, now down, the winds dictate their track').

Lomborg also uses the output of the IPCC climate models as inputs to his cost–benefit analysis (Lomborg 2017). This determines that investments in low-emissions technology should be supported (to produce technology better than what we have now), but economy destroying measures should not:

> … remember that any realistic policy that we are going to embark on in the next 50 years will have a trivial impact on climate change and hence also a trivial impact on the risks of these tail events. So, in reality, a lot of people seem to be saying, 'I really, really, really worry about this far out thing that could happen, like extinction of some sort. And therefore, I'm going to pursue very costly but incredibly ineffective policies.' (Roberts 2019).

My colleague Daniel Wild (2018) adapted Lomborg's approach to economic impacts of the Paris Accord on Australia, and calculated that the 'economic cost of Australia meeting its emissions reduction requirement under the agreement (was) estimated to be $52 billion in net present value terms, over the period 2018–2030'.

Third Circle: the IPCC

And here there was no fire, and nothing burned.
Instead, a dark cold rain falls heavily
Forever. At a rigidly fixed rate …

The descent into the third circle comes with the onset of doubts about the veracity of the IPCC. This might arise by following the systematic critiques of the IPCC's work, and there is no shortage of those. Aynsley Kellow (2018) has, for instance, pointed out that the IPCC's 'Summary for Policy Makers' – which is edited by committee rather than by the technical specialists who assemble the research – is the result of a deeply politicised process.

Evidence abounds of IPCC perfidy. Simon Breheny (2017) pointed out that between its first (1991) and third (2001) reports (ten years apart) the IPCC literally 'ironed out' the natural ups and downs in temperature, instead preferring neat straight lines and thus eliminating both the Mediaeval Warm Period beginning around AD 985 when the Vikings settled southern Greenland, and the Little Ice Age beginning around AD 1200 when Greenland became too cold for human habitation.

Beyond such analysis, it is good to doubt an organisation like the IPCC from first principles. 'Mainline economics' (Mitchell & Boettke 2016) tells us that principals in institutions have incentives just like actors in a marketplace, and we can make predictions based on those incentives. Call me cynical (and people at dinner parties do, in this context), but the IPCC is an example of the Shirky Principle, which is that 'Institutions will try to preserve the problem to which they are the solution'.

Also, even with a less cynical reading, there is flaw built into the very concept that the IPCC is responsible for 'government sanctioned assessments', as Judith Curry (2015) pointed out. These assessments do not account for, or admit:

> ... the very substantial disagreement about climate change that arises from:
> - Insufficient observational evidence
> - Disagreement about the value of different classes of evidence (e.g. simulation model output versus observations from satellites)
> - Disagreement about the appropriate logical framework for linking and assessing the evidence
> - Assessments of areas of ambiguity and ignorance
> - Belief polarization as a result of politicization of the science.

Out of such systemic exclusion of evidence is built the apparent consensus.

So far reason is our guide, but Dante's third circle was reserved for the gluttons, and I think his insight into the nature of sin is applicable. It is more than *ad hominem* to observe that the inaugural head of the IPCC, Rajendra Pachauri (1940–2020), was forced to resign under a cloud of allegations around governance and sexual harassment. He denied them all (Worrall 2016) but, like Caesar's wife, the head of such an important body should be above suspicion. Yet he served for thirteen years, basking in the afterglow of the Nobel Peace Prize he shared with Al Gore.

Fourth Circle: general circulation models

These money-grubbing ones who from the grave
Arise with grasping fists and foolproof schemes,
And these with hair close-cropped to show they gave
As those took – a two-sided robbery …

I barely require a guide through the fourth circle, which is scepticism about modelling. The credulity with which the latest outcomes of model runs is greeted never ceases to astound. I wouldn't know one end of a *climate* model from the other, but my own training in economics and finance has predisposed me to doubt.

In my Master of Business Administration (MBA) degree, undertaken in the late 1990s, we learnt the relatively new and fashionable theories that had revolutionised corporate finance. The Black-Scholes-Merton models promised a way to price options (call options, put options) using historic data. This would enable a market of perfect information in which risk could be calibrated (as indeed it can, so long as the future looks like the past). The markets were so impressed by the cleverness of the model that in 1994 they backed a hedge fund known as Long Term Capital Management, in which Robert C. Merton and Myron Scholes were directors and principal advisers on the modelling used to guide investments. As it happened, the fund operated for only three years before the Asian Financial Crisis sent it into a tailspin. Models are only as good as the assumptions and the data fed into them, which don't necessarily scale or interact in predictable ways.

So, Judith Curry was really waiting for me at the entrance to the fourth circle when she published *Climate Models for the Layman* (2017).

There is growing evidence that climate models are running too hot and that climate sensitivity to carbon dioxide [CO_2] is at the lower end of the range provided by the IPCC. Nevertheless, these lower values of climate sensitivity are not accounted for in IPCC climate model projections of temperature at the end of the 21st century or in estimates of the impact on temperatures of reducing carbon dioxide emissions.

How can the decision makers of the world mortgage our futures off the back of a mode of analysis with inherent limitations, undertaken by people with an institutional bias toward demonstrating a warming trend, whose models can't predict the past let alone the future?

The modellers also have an interest in proclaiming that big data can provide the answers, when they do not have the expertise in physics and actual climate science to generate or test falsifiable hypotheses. As the great Richard Lindzen (Linzen 2019, pers. comm. 21 July) has observed:

> ... almost all reports of model results are from people who played no role in developing models, and have no idea of their properties and shortcomings. They are pretty much used as Ouija boards selected for their ability to produce ominous results.

Fifth Circle: 'Climategate'

You outcasts from the sky,' the angel cried
On the ghastly threshold. 'You rejected race,
Where did you get your overweening pride?
Why fight against the will none can outface
And which so often has increased your pain?
By flouting the decrees of providence
Some call the fates, what can you hope to gain?

Any sceptic who has descended below Lomborg's second circle knows the experience of being branded a conspiracist as well as a 'science denier' and/or a 'shill for oil companies' (in fact the company I worked for made money from renewables, and even more money supplying the gas to fast-response power plants necessary to compensate for their intermittency; I chose my current job to reflect my beliefs, not the other way around).

The charge of conspiracist is often first launched when I've deployed an institutional argument from the third circle, and the charge is made by those with a touching belief in national and international climate bodies that they have only the 'public interest' at heart. The heirs of Plato (380 BC), they believe in the wisdom of the guardians.

But how was it possible to avoid the charge of being a conspiracist when the 'Climategate' emails revealed that there really was a world-wide conspiracy to (a) bring to the fore only worst-case predictions,

(b) suppress research which did not fit the narrative, and (c) seek to destroy the careers of those who wished to research and publish genuine but non-conforming research?

It's hard to imagine a reader of this book unfamiliar with the Climategate emails, a leaked trove of emails, but if you are such I commend to you the compilation published by the Australian-based Lavoisier Group, co-founded by the indefatigable Ray Evans (Costella 2010). Here is a sample email from Keith Briffa, as he looks over some data provided in good faith by an American:

> The data is of course interesting but I would have to see it and the board would want the larger implications of the statistics clearly phrased in general and widely understandable (by the ignorant masses) terms before they would consider it not too specialised.

'Larger implications' being code for 'we're all gonna fry!' – an approach deemed necessary for the 'ignorant masses'.

Hugh Morgan, a *bona fide* businessman and sceptic, puts it well in his introduction:

> … a very small cabal of climate scientists, based at the University of East Anglia and at Penn State University, were able to control the temperature record fed into the IPCC reports and which comprised the foundation on which the whole global warming structure was based. The only data base which they could not influence was the satellite measured temperature data which John Christy and Roy Spencer, from the University of Alabama, had established from 1979 on.

> That this was a real conspiracy is beyond argument. The word 'conspiracy' is used by the players themselves. In any conspiracy there is a tight inner core and then successive rings of collaborators, who accept the leadership of the central core.

And so I found myself descended to the fifth circle, rejected and labelled a conspiracy theorist for pointing to an actual, admitted, conspiracy.

Sixth Circle: peer-reviewed research

> Don't you recall how Aristotle shows,
> In the *Ethics*, that three different kinds of state
> In humans breed all crimes that can oppose
> Themselves to heaven's will? Incontinence,
> Malice and brutishness?

At the hypothetical dinner party, it is somewhere between the mains and desserts that you'll be told you are a 'science denier' because 'the science is settled'. There are of course strong methodological arguments against the notion that science is, or can be, settled. But it is usually hopeless trying to point out that the 'consensus' of '97% of scientists' is nothing more than a meme, artfully established and repeated *ad nauseam*.

The claim was originated by Cook:

> Among abstracts (in peer-reviewed literature) expressing a position on AGW, 97.1% endorsed the consensus position that humans are causing global warming. (Cook et al 2013)

My objection is always that even if true, this does not establish the key question we might ask those scientists, which is, 'what do you mean by "causing"?'. As Andrew Montford found:

> ... the consensus referred to is trivial:
> - that carbon dioxide (CO_2) is a greenhouse gas;
> - that human activities have warmed the planet to some unspecified extent.
> (Montford 2014)

In any event, no sane person puts complete faith in peer-reviewed research. As Dr Peter Ridd – who was sacked by James Cook University for questioning some of the more alarmist claims of his colleagues – said on my podcast:

> Scientists will go on about peer review as though it's a dozen scientists (who) pore over the work for a month on end and they repeat the experiences ...

> [In fact] you read the work maybe for a couple of hours. You might spend a whole day on it, [but] usually not ... You can't do the experiments. You're not being paid for it.

And that's peer review ...

[T]his is not proper quality assurance, and they're finding that when you actually try to replicate peer-reviewed literature, it tends to be wrong around about half the time. (Berg & Hargreaves 2019)

John Ioannidis (2005) started something big with his paper on research in medicine, 'Why most published research findings are false', and now there is a full-blown debate (although perhaps not in Australia, or in climate science) about the replication crisis in science, described as a 'methodological crisis in which it has been found that many scientific studies are difficult or impossible to replicate or reproduce'.

With my own openness to institutional explanations, and the application of the public choice theory embedded in mainline economics, I was always going to be concerned with the potential of big government to skew big science.

Seventh Circle: data homogenisation

Your star, if always by your star you steer,
You can't fail to make glorious harbour. Had
I sensed this rightly in your life so fair,
And not too soon died, seeing Heaven glad
To help you would have made me take more care
To aid your work.

Dante referred to the last three circles of hell as the *City of Dis* – a dark world of imaginative torments ruled by Satan, a city surrounded by steep walls.

These walls deter many, even among the group of sceptics who might have reached the sixth circle. My colleague Jennifer Marohasy has lamented that most sceptics doing their own critiques of 'consensus' climate models will use the official temperature records while reluctantly acknowledging they may be an invention of the very small cabal of climate scientists, based at the University of East Anglia, and their colleagues.

It has been a small but dedicated group of scientists and journalists – such as Jennifer Marohasy, Tony Heller, Jaco Vlok, Jo Nova, Anthony Watts and Tom Quirk – who have delved deep into the mysteries of how raw temperature data is 'homogenised' into the official temperature records, and the warming bias that results (Vlok 2019; Nova 2017; Quirk 2017; Marohasy & Abbot 2016; Marohasy & Abbot 2015; Marohasy 2017b).

The necessity for some form of 'homogenisation' is not contested; no one could argue that, say, moving a weather station is not a break in the series, or that some account shouldn't be taken of the urban heat island (UHI) effect (Quirk 2017). But they remodel in the wrong direction, as Anthony Watts explains:

> All the key institutions apply adjustments to the actual measurements, which has the effect of exaggerating, rather than correcting for, the urban heat island (UHI) effect. (Watts 2017)

Further, these same institutions keep cooling the past even when there have been no further changes to the equipment or location of the weather station. This is the situation at Rutherglen and Darwin in Australia, for example, where a cooling trend in the record before 1950 was changed to a warming trend in the first official homogenisation, and then this warming trend further increased with each new iteration of the Australian Climate Observations Reference Network for Surface Air Temperatures (ACORN-SAT), as Marohasy explains in Chapter 16 of this book.

I know firsthand how difficult it is to engage with the critique. For my own part, I first read Marohasy's material in 2016 and found it difficult to follow. At this stage she had been working on it for five years and was still having trouble convincing even members of the global sceptics' conspiracy to acknowledge the problem, lest they would have to begin with the raw temperature data themselves and engage in much more tedious analysis.

The walls of Dis are steep, too steep even for many sceptics who would rather walk with Lomborg in the second circle, or perhaps with Peter Ridd and John Ioannidis in the sixth circle, but it rewards the effort. Once inside, you'll never again trust the official climate record, which is homogenised to accord with the outputs of the climate models. I commend to you the work of Marohasy and her collaborators as detailed in Chapters 5 to 10 of the previous book in this *Climate Change The Facts* series (Marohasy 2017a).

Eighth Circle: raw temperature records

With these ten demons, savage entourage
Indeed, we went: but you know how things are –
Pray with the saints, drink with the sots.

There I was, comfortably in the seventh circle, thinking there could be nowhere further to slide. We couldn't trust the homogenised data, but at least the raw temperature data was clean, wasn't it?

In September 2017, the Australian Bureau of Meteorology published a 71-page internal report entitled *Review of the Bureau of Meteorology's Automatic Weather Stations* (Bureau of Meteorology 2017). The report begins by explaining that the Bureau has 695 automatic weather stations spread across Australia, and that the data from this network underpins all of the services the Bureau delivers. The report then goes on to admit that two of these weather stations are 'not fit for purpose' – Goulburn Airport and Thredbo. The Bureau subsequently accepted that many weather stations across Tasmania, Victoria and the ACT were also not relaying temperatures below –10 °C (Lloyd 2017a).

The review (Bureau of Meteorology 2017) was forced by Marohasy who had been getting up in the early hours of the morning to compare the temperatures measured at the coldest part of the day versus the values archived by the Bureau in the Australian Data Archive for Meteorology (Lloyd 2017b). In the process she had evidence to prove that a strict limit of –10 °C had been set (Lloyd 2017b). There would be no more 'record' cold days, because there were now processes to correct *even* the raw temperature data (Marohasy 2017c). That was until this limit was lifted, and, as a consequence, the temperatures plunged at Thredbo (Lloyd 2017c).

Since 1996, the Bureau has been transitioning from manual recordings of daily temperatures to the automated system using electronic probes, with the probes more responsive to fluctuations in temperatures and, therefore, likely to record both hotter and colder for the same weather.

You might think it prudent to have maintained the two methods of measurement alongside each other for some period of time, to determine whether any biases were being introduced that might, say, exaggerate a warming trend. This was done at 22 of the 695 locations with new Automatic Weather Stations (Bureau of Meteorology 2012) generating parallel measurements. That is, an alcohol in glass thermometer used for manual measurement of minimum temperatures and a mercury in glass thermometer used for manual measurement of maximum temperatures were recording temperatures in the same Stevenson screen as the new electronic temperature probes. But the equivalence, or otherwise, of temperatures from these different measuring systems appears to have never been established by way of any report or technical paper published by the Australian Bureau of Meteorology. Furthermore, the reams of manually recorded data have never been digitised or analysed.

On 25 August 2015, Marohasy first asked for access to the parallel measurements for Wilsons Promontory lighthouse. On 16 October 2017, John Abbot made a formal Freedom of Information (FOI) request for the parallel data for not only Wilsons Promontory, but also the desert location of Giles for where there should be parallel data for the last 25 years, and also Mildura (an agricultural region in western Victoria) for which 34 years of parallel daily data exists.

It took the intervention of the then Federal Minister for the Environment, Josh Frydenberg, in order for the Mildura parallel data to be released, and only then because the latter was under pressure from the highly influential (and informed) broadcaster, Alan Jones. It is telling that the information released was in the form of 10,000 separate scanned records, which at no time have been tabulated by the Bureau. That task fell to Marohasy, who has established that the current electronic probe at Mildura often records 0.4 °C hotter for the same weather and that first electronic probe recorded cooler, by a statistically significant 0.3 °C (Marohasy 2018). The parallel data for Wilsons Promontory, Giles and the other eighteen locations for which it must exist according to one of the Bureau's own technical reports (Bureau of Meteorology 2012) is still being refused, most recently in correspondence to John Abbot dated 14 April 2020.

While I am not qualified in any of the relevant technical disciplines, I *am* a former Manager of Sustainability acquainted with the concept of data assurance. For the sustainability reports of the type I assisted in producing, the gold standard was external assurance by a reputable authority. In this case, we observe that the Bureau has chosen to use its own standard rather than the global one, and there is no independent verification that it actually applies that standard. This is acknowledged on page 11 of the AWS review report:

> To address measurement quality, the WMO recommends that national meteorological agencies have ISO 17025 accreditation of key measurement processes. Accreditation covers the technical procedures and processes that ensure the traceability and integrity of measurements, as well as the technical competence of the staff making the measurements. While the Bureau does not currently hold ISO 17025 accreditation, it has internal processes, technical procedures, and measurement traceability and integrity that are largely in accordance with ISO 17025 requirements. (Bureau of Meteorology 2017)

In the absence of such quality assurance and real research into the different results arising from switching from analogue to digital temperature measurement, no reliability can be placed on any claim to 'record hot days', or on an accelerated warming trend or anything similar.

Ninth Circle: escaping hell

Ah, Genoese, you that know all the ropes
Of deep corruption yet know not the first
Thing of good custom, how are you not flung
Out of this world?

What you've read so far is pretty much my descent into climate scepticism. It is true that I have made new friends along the way, but there is no doubt it has contributed to a form of social and professional hell. Nothing like what the sacked Professor Peter Ridd, formerly of James Cook University, may have suffered, to be sure, but not fun either. Last year, it was with some regret that I had no choice but to walk out of a dinner with a friend of 30 years' standing who not only thought I was wrong (his prerogative) but that I was exercising bad faith and was, essentially, bad.

But the new friends can also be problematic. Inside the City of Dis, in the ninth circle, I find those determined to prove to me that CO_2 cannot possibly be a greenhouse gas, and that any theory that it does so violates the second law of thermodynamics, or something. Some are labelled, or label themselves, 'Dragon Slayers', for a reason I learnt once but can't seem to remember.

My physics and chemistry don't extend beyond first-year university, so I have difficulty challenging them, but my lack of qualifications also means I can't agree with them, either. They see me as a 'luke warmist' – someone who acknowledges that the Earth may be warming (probably is, very slightly, at the moment) and that CO_2 *could* have something to do with that (most likely, but a small contribution), and so I am. And I further believe catastrophist arguments are built on shoddy data and even shoddier modelling, and that nearly every idea that has been proposed for radical action is very bad indeed.

So, just as Dante moved past the stationary figure of Satan, moored at the bottom of hell, I will do the same and reach towards mere *Purgatorio*, the middle place that provides the opportunity for redemption found in the next book of Dante's *Divine Comedy*. There, real data informing real science and real policy analysis will lead to much better decisions about the path forwards. I hope for what Dante experienced:

And up we went, her first, I second, to
The point where I could see an opening.
And it was there I saw, when I looked through,
A sight more wonderful than anything –
Some of the loveliness revealed to men
By Heaven. We could see the stars again.

Appendices

Appendix 12.1: Calculating the tropical mid-atmospheric radiation flux E_R and the relationship between deep convection altitude and the emissions level.

From Riehl and Simpson (1979) Figures 5 and 12, the energy balance for the sub-500 hPa level is

$$S_b + Q_{SE} + E_i = E_R + E_C$$

where S_b = 1.8 PW is the solar radiation absorbed in the atmosphere below 500 hPa, Q_{se} = 6.4 PW is the latent and sensible heat input, E_i = 5.1 PW is the import of (mostly latent) heat into the trough zone, E_c = 8.7 PW from which E_R = 4.6 PW.[1]

A similar balance for the 100 hPa (15 km for the tropics) level yields a value for E_s = 7.9 PW, which is equivalent to 182 W/m². This implies that the non-spectral window component of emissions crossing the 100 hPa level has a blackbody temperature of 238 K which places the emissions level at around the 300 hPa (10 km) altitude in the tropics. Deep convection must reach at least this level to satisfy the requirement that for the heat engine to operate that heat is rejected at the cold part

---

1   Riehl and Simpson state that 73 W/m² of short-wave radiation is absorbed below 100 hPa. Partitioning the absorption according to the masses of air in the 1000–500 hPa (41 W/m²) level and the 500 hPa to 100 hPa (32 W/m²) level, then for the half-equatorial trough S_b = 1.8 PW, and S_T = 1.4 PW (insolation absorbed in the 100–500 hPa layer).

of the cycle. Deep convection regularly approaches 15 km. In reality, the emissions level covers a broad range of altitudes, but it is still true that the engine will run more efficiently if deep convection reaches a higher altitude due to more effective heat rejection by radiation.

Appendix 12.2: Variation of E_c and E_R with humidity

It is assumed that for modest perturbations from present conditions, the work output, W, from the tropical atmospheric heat engine is proportional to the energy flow rate through the engine in the deep convection, E_c, $W = \eta E_C$ where η is the efficiency of the engine.[2] The work done by the engine produces the wind circulation including much of the trade winds and is ultimately dissipated as friction at a rate proportional to the cube of the characteristic velocity of the circulation V (Peixoto & Oort 1992), i.e.

$$W = \beta V^3 \tag{1}$$

where β is a constant. Thus

$$E_c = \frac{\beta V^3}{\eta} \tag{2}$$

The energy added to the heat engine E_c at the hot part of the cycle is the latent heat released in the updraft. We may thus assume that

$$E_c = \beta_2 \dot{M} q^* \tag{3}$$

where $\dot{M}$ is the mass flow rate through the deep convection and q^* is the saturation water vapour mixing ratio of the air entering the cell. β_2 is a constant. Here it is assumed that the rainfall efficiency and the relative humidity will not change with temperature. By mass conservation, we also assume that the net mass flow rate of the updraft in the deep convection is proportional to V

$$\dot{M} = \beta_3 V \tag{4}$$

where β_3 is a constant.

---

2   Estimated to be about 1%.

Combining equations 2 to 4 results in

$$E_c = Kq^{*\ 3/2} \tag{5}$$

where K is a constant. Because q^* varies by around 7% per °C, from equation 5, the value of E_c varies by around 10% per °C.

In this work, we have assumed that the Rosseland approximation describes the vertical transport of radiation to the emissions level E_R, i.e.[3]

$$E_R = \frac{16}{3} \sigma \frac{T^3}{a} \frac{\Delta T}{\Delta z} \tag{6}$$

where a is the absorption coefficient of the air due to greenhouse gases and ΔT is the difference in temperature over a vertical distance Δz. $\sigma = 5.67 \times 10^{-8}$ W·m⁻²·K⁻⁴. Assuming that a will increase linearly with greenhouse gas concentration that is dominated by water, and again assuming constant relative humidity, from equation 6, E_R is inversely proportional to q^*. It is notable that the convection pathways E_c increases with the $q^{*3/2}$ which converts to about 10% per °C, whereas the radiation pathway reduces with $1/q^*$ and thus reduces at 7% per °C. The effect of higher absolute humidity due to higher temperature is to increase the convection pathway at a faster rate than the radiation pathway reduces. This is a very powerful negative feedback mechanism.

---

3   http://jullio.pe.kr/fluent6.1/help/html/ug/node486.htm.

Appendix 13.1

Feedback analysis

By 'feedback' we mean some mechanism whereby a change in temperature causes a further change in some greenhouse substance (water vapour, clouds, etc.) that, in turn, contributes to the climate forcing. This is illustrated in Figure 1 (taken from Lindzen et al. 2001). The top panel shows the situation in the absence of feedbacks, while the bottom panel shows the effect of feedbacks. As we see from this figure, the response, once feedbacks are included is given by the expression:

$$\Delta T 5 \frac{\Delta T_0}{1 - G_0 F}$$

The quantity $G_0 F$ is sometimes referred to as the feedback factor, f it is simply the response of the climate system to the fed-back flux (non-dimensionalised by 1 °C) resulting from $\Delta T = 1C$. If there are several feedbacks, then f is replaced by Σf_i. If f is negative, then we have a negative feedback and ΔT is smaller than ΔT_0. However, if f is positive, ΔT is greater than ΔT_0, and the behaviour is far more dramatic since ΔT blows up for $f=1$. Thus, if the alleged water-vapour feedback contributes 0.5 to f, thus doubling ΔT_0, another 0.5 brings ΔT to infinity. Of course, other processes will prevent such an unphysical response.

Figure 1 Schematic illustrating operation of feedbacks

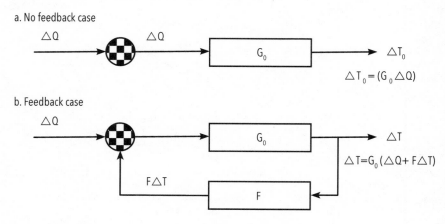

a. No feedback case

$\Delta T_0 = (G_0 \Delta Q)$

b. Feedback case

$\Delta T = G_0 (\Delta Q + F \Delta T)$

Moreover, the greater ΔT is, the longer the time it will take to reach the final value. It should be noted that the above relation is not built into models; rather, it describes how the models behave. A more detailed discussion can be found in both Lindzen et al. (2001), and in Roe and Baker (2007). This feedback analysis is essentially taken from electrical engineering.

References

1: Walruses, Polar Bears and the Fired Professor

9 News 2014, 'Melting sea ice forces 35,000 walruses to swarm onto Alaskan beach', *9 News*, viewed 2 November 2019, https://web.archive.org/save/https://www.9news.com.au/world/thousands-of-walruses-leave-alaska/cc91d021-dfa0-4596-8679-89de45c0732b.

Aars, J 2018, 'Population changes in polar bears: protected, but quickly losing habitat', *Fram Forum Newsletter 2018*, Tromso, January 1, https://framsenteret.no/2018/12/population-changes-in-polar-bears-protected-but-quickly-losing-habitat/.

Amstrup, SC 2019, 'Written Testimony before the Subcommittee on Energy and Mineral Resources of the Committee on Natural Resources United States House of Representatives legislative hearing on "The Need to Protect the Arctic National Wildlife Refuge Coastal Plain"', 26 March 2019, https://naturalresources.house.gov/download/testimony-polar-bears-international-amstrup.

Amstrup, SC, Marcot, BG & Douglas, DC 2007, 'Forecasting the rangewide status of polar bears at selected times in the 21st century', US Geological Survey, Reston, VA, https://www.plexusowls.com/PDFs/forecasting_polar_bears_amstrup_etal_lowres.pdf.

Atwood, TC, Peacock, E, McKinney, MA, Lillie, K, Wilson, R, Douglas, DC, Miller, S & Terletzky, P 2016, 'Rapid Environmental Change Drives Increased Land Use by an Arctic Marine Predator', *PLoS ONE*, vol. 11, iss. 6, https://journals.plos.org/plosone/article?id=10.1371%2Fjournal.pone.0155932.

BBC 2019a, 'Walrus living on the edge' (Episode 2), *Seven Worlds, One Planet: Asia*, 3 November 2019, https://www.bbc.co.uk/programmes/p07sly3r [viewable in the UK only].

BBC 2019b, 'Walrus on the Edge', 3 November 2019, https://www.bbc.co.uk/programmes/articles/4zh2Dd3JC8gprNZcGY6BbHB/walrus-on-the-edge.

REFERENCES

Bernard, JF 1923, 'Local Walrus Protection in Northeast Siberia', *Journal of Mammalogy*, vol. 4, iss. 4, pp. 224–227, https://nofrakkingconsensus.com/wp-content/uploads/2019/12/1923_Joseph_Bernard_Local-walrus-protection-in-Northeast-Siberia-6-pp.pdf.

Bromaghin, JF, McDonald, TL, Stirling, I, Derocher, AE, Richardson, ES, Regehr, EV, Douglas, DC, Attwood, T & Amstrup, SC 2015, 'Polar bear population dynamics in the southern Beaufort Sea during a period of sea ice decline', *Ecological Applications 25*, iss. 3, pp. 634–651, https://esajournals.onlinelibrary.wiley.com/doi/abs/10.1890/14-1129.1.

Castro de la Guardia, L, Myers, PG, Derocher, AE, Lunn, NJ & Terwisscha van Scheltinga, AD 2017, 'Sea ice cycle in western Hudson Bay, Canada, from a polar bear perspective', *Marine Ecology Progress Series*, vol. 564, pp. 225–233, http://www.int-res.com/abstracts/meps/v564/p225-233/.

Cherry, SG, Derocher, AE, Thiemann, GW & Lunn, NJ 2013, 'Migration phenology and seasonal fidelity of an Arctic marine predator in relation to sea ice dynamics', *Journal of Animal Ecology*, vol. 82, pp. 912–921, http://onlinelibrary.wiley.com/doi/10.1111/1365-2656.12050/abstract.

Crockford, SJ 2014, 'On the Beach: Walrus Haul-outs are Nothing New', *Global Warming Policy Foundation Briefing Paper 11*, https://polarbearscience.files.wordpress.com/2015/08/walrus_on-the-beach_crockford-2014.pdf. Video: http://www.thegwpf.org/susan-crockford-on-the-beach-2/.

Crockford, SJ 2015, 'The Arctic Fallacy: sea ice stability and the polar bear', *Global Warming Policy Foundation Briefing Paper 16*, https://www.thegwpf.org/content/uploads/2015/06/Arctic-Fallacy2.pdf.

Crockford, SJ 2017, 'Testing the hypothesis that routine sea ice coverage of 3–5 mkm^2 results in a greater than 30% decline in population size of polar bears (*Ursus maritimus*)', *PeerJ Preprints*, DOI: 10.7287/peerj.preprints.2737v3.

Crockford, SJ 2018a, 'The real story behind the famous starving polar bear video reveals more manipulation', *Financial Post*, 29 August, https://business.financialpost.com/opinion/the-real-story-behind-the-famous-starving-polar-bear-video-reveals-more-manipulation.

Crockford, SJ 2018b, State of the Polar Bear Report 2017. 'Global Warming Policy Foundation Report' #29. London.

Crockford, SJ 2019a, 'Attenborough's tragedy porn of walruses plunging to their deaths because of climate change is contrived nonsense', *PolarBearScience.com*, 7 April, https://polarbearscience.com/2019/04/07/attenboroughs-tragedy-porn-of-walruses-plunging-to-their-deaths-because-of-climate-change-is-contrived-nonsense/.

Crockford, SJ 2019b, 'Netflix is lying about those falling walruses. It's another "tragedy porn" climate hoax', *National Post*, 29 April, https://web.archive.org/

web/20190428091331/https:/business.financialpost.com/opinion/netflix-is-lying-about-those-falling-walruses-its-another-tragedy-porn-climate-hoax.

Crockford, SJ 2019c, *The Polar Bear Catastrophe That Never Happened*, Global Warming Policy Foundation, London.

Crockford, SJ 2019d, 'Fabulous lectures in Holland done: Munich conference forced to relocate', *PolarBearScience.com*, 20 November, https://polarbear science.com/2019/11/20/fabulous-lectures-in-holland-done-munich-conference-forced-to-relocate/.

deMarban, A 2006, 'Fence erected on Alaskan cliff keeps walrus from taking plunge', *Wenatchee World*, 30 June, https://web.archive.org/web/20191205221455/https://www.wenatcheeworld.com/news/local/fence-erected-on-alaskan-cliff-keeps-walrus-from-taking-plunge/article_48d2461c-2770-5387-839e-824b164987d9.html.

Delingpole, J 2019, 'Walrusgate – Attenborough exposed in #fakenews Netflix eco documentary scandal', Breitbart, 17 April, https://www.breitbart.com/europe/2019/04/17/walrusgate-the-netflix-attenborough-scandal-gets-worse/.

Dyck, M, Campbell, M, Lee, D, Boulanger, J & Hedman, D 2017, '2016 Aerial survey of the Western Hudson Bay polar bear subpopulation', Final report, Nunavut Department of Environment, Wildlife Research Section, Igloolik, NU. https://www.gov.nu.ca/sites/default/files/pb_wh_2016_population_assessment_gn_report_27_june_2017.pdf.

EIKE 2019, 'Program – 13th International Conference on Climate and Energy', Munich, 22–23 November, https://www.eike-klima-energie.eu/13thconference/program/.

Fay, FH & Kelly, BP 1980, 'Mass Natural Mortality of Walruses (*Odobenus rosmarus*) at St. Lawrence Island, Bering Sea, Autumn 1978', *Arctic*, vol. 33, no. 2, pp. 226–245, http://pubs.aina.ucalgary.ca/arctic/Arctic33-2-226.pdf.

Fischbach, AS, Kochnev, AA, Garlich-Miller, JL & Jay, CV 2016, 'Pacific walrus coastal haul-out database, 1852–2016 – Background report', *U.S. Geological Survey Open-File Report 2016–1108*. http://pubs.aina.ucalgary.ca/arctic/Arctic65-4-377.pdf.

Gosselin, P 2019, 'Radicals Bully NH Munich Conference Center ... Force Cancellation of 13th Skeptic Climate Conference!' *NoTricksZone.com*, 19 November, https://notrickszone.com/2019/11/19/radicals-bully-nh-munich-conference-center-force-cancellation-of-13th-skeptic-climate-conference/.

Harwood, LA, Smith, TG, Melling, H, Alikamik, J & Kingsley, MCS 2012, 'Ringed seals and sea ice in Canada's western Arctic: harvest-based monitoring 1992-2011', *Arctic*, vol. 65, pp. 377–390, http://pubs.aina.ucalgary.ca/arctic/Arctic65-4-377.pdf.

Joling, D 2009, 'Trampling blamed for Alaska walrus deaths', Associated Press in *Seattle Times*, 1 October, https://www.seattletimes.com/nation-world/trampling-blamed-for-alaska-walrus-deaths/.

Linshi, J 2014, 'Here's Why Thousands of Walruses Are Gathering on Alaska's Shore', *Time*, 1 October, https://web.archive.org/web/20141205121458/https://time.com/3450896/walrus-alaska/.

Laframboise, D 2019a, 'Was this zoologist punished for telling school kids politically incorrect facts about polar bears?' *National Post*, 16 October 2019, https://web.archive.org/web/20191017195343/https://business.financialpost.com/opinion/was-this-zoologist-punished-for-telling-school-kids-politically-incorrect-facts-about-polar-bears

Laframboise, D 2019b, The UVic, Susan Crockford Story Continues, BigPicNews.com, 30 October 2019, https://nofrakkingconsensus.com/2019/10/30/the-uvic-susan-crockford-story-continues/

Lunn, NJ, Servanty, S, Regehr, EV, Converse, SJ, Richardson, E & Stirling, I 2016, 'Demography of an apex predator at the edge of its range – impacts of changing sea ice on polar bears in Hudson Bay', *Ecological Applications*, vol. 26, pp. 1302–1320, https://esajournals.onlinelibrary.wiley.com/doi/abs/10.1890/15-1256.

Mittermeier, C 2018, 'Starving-Polar-Bear Photographer Recalls What Went Wrong', *NationalGeographic.com*, August 2018, https://www.nationalgeographic.com/magazine/2018/08/explore-through-the-lens-starving-polar-bear-photo/.

NSIDC 2019, 'Falling up' Figure 3b. September monthly mean trends for 1979–2019, showing overall trend and trends for the most recent 13 years, and the steepest 13 years in the 41-year record, W. Meier, NSIDC, https://nsidc.org/arcticseaicenews/2019/10/falling-up/.

New York Times 1996. '60 Walruses Plunge to Their Death, and Alaskans Wonder Why', 31 August, https://web.archive.org/web/20190409042832/https:/www.nytimes.com/1996/08/31/us/60-walruses-plunge-to-their-death-and-alaskans-wonder-why.html.

Obbard, ME, Stapleton, S, Szor, G, Middel, KR, Jutras, C & Dyck, M 2018, 'Re-assessing abundance of Southern Hudson Bay polar bears by aerial survey: effects of climate change at the southern edge of the range', *Arctic Science*, vol. 4, iss. 4, pp. 634–655, https://www.nrcresearchpress.com/doi/10.1139/as-2018-0004.

Polar Bears International 2019, 'Press Release: Scientist Ian Stirling Receives Ice Bear Lifetime Achievement Award from Polar Bears International', 12 November. https://www.newswire.ca/news-releases/scientist-ian-stirling-receives-ice-bear-lifetime-achievement-award-from-polar-bears-international-859185523.html.

Regehr, EV, Laidre, KL, Akçakaya, HR, Amstrup, SC, Atwood, TC, Lunn, NJ, Obbard, M, Stern, H, Thiemann, GW & Wiig, Ø 2016, 'Conservation status of polar bears (*Ursus maritimus*) in relation to projected sea-ice declines', *Biology Letters*, vol. 12, iss. 12, https://doi.org/10.1098/rsbl.2016.0556.

Regehr, EV, Hostetter, NJ, Wilson, RR, Rode, KD, St. Martin, M & Converse, SJ 2018, 'Integrated population modeling provides the first empirical estimates of vital rates and abundance for polar bears in the Chukchi Sea', *Scientific Reports*, vol. 8, article 16780, https://www.nature.com/articles/s41598-018-34824-7.

Rode, KD, Wilson, RR, Regehr, EV, St. Martin, M, Douglas, DC & Olson, J 2015, 'Increased land use by Chukchi Sea polar bears in relation to changing sea ice conditions', *PLoS One* 10 e0142213.

Rode, KD, Wilson, RR, Douglas, DC, Muhlenbruch, V, Atwood, TC, Regehr, EV, Richardson, ES, Pilfold, NW, Derocher, AE, Durner, GM, Stirling, I, Amstrup, SC, St Martin, M, Pagano, AM & Simac, K 2018, 'Spring fasting behavior in a marine apex predator provides an index of ecosystem productivity', *Global Change Biology*, vol. 24, iss. 1, pp. 410–423, http://onlinelibrary.wiley.com/doi/10.1111/gcb.13933/full.

Siberian Times 2017, 'Village besieged by polar bears as hundreds of terrorised walruses fall 38 metres to their deaths', 19 October, https://web.archive.org/web/20171024175051/http://siberiantimes.com/ecology/others/news/village-besieged-by-polar-bears-as-hundreds-of-terrorised-walruses-fall-38-metres-to-their-deaths/.

Steele, J 2014, 'Hijacking Successful Walrus Conservation', *LandscapesAndCyles.net*, viewed 20 November 2019, http://landscapesandcycles.net/hijacking-successful-walrus-conservation.html.

Stirling, I 2002, 'Polar bears and seals in the eastern Beaufort Sea and Amundsen Gulf: a synthesis of population trends and ecological relationships over three decades', *Arctic*, vol. 55, supp. 1, pp. 59–76, http://pubs.aina.ucalgary.ca/arctic/Arctic55-S-59.pdf.

Stone, M 2017, 'Polar Bears Drive Hundreds of Walruses Off Cliff in Siberian Bloodbath', *Gizmodo*, 25 October, https://earther.gizmodo.com/a-freak-walrus-slaughter-in-siberia-reminds-us-how-clim-1819802967.

Stroeve, J, Holland, MM, Meier, W, Scambos, T & Serreze, M 2007, 'Arctic sea ice decline: Faster than forecast', *Geophysical Research Letters*, vol. 34, https://agupubs.onlinelibrary.wiley.com/doi/10.1029/2007GL029703.

Taylor, J 2019, 'Climate Scientists Reduced to Hiding from Climate Thuggery in Germany', *Townhall*, 22 November, https://townhall.com/columnists/jamestaylor/2019/11/22/climate-scientists-reduced-to-hiding-from-climate-thuggery-in-germany-n2556941?2311.

USFWS 2005, 'Togiak: Refuge Finds Unusual Walrus Deaths', Region 7 Fieldnotes, 1 December, https://web.archive.org/web/20170716082620/https:/www.fws.gov/FieldNotes/print/print_report.cfm?arskey=17504.

USFWS 2008, 'Polar Bear Listing: Frequently Asked Questions', USFWS Support Document 037257, 11 pages, https://www.fws.gov/home/feature/2008/polarbear012308/pdf/037257PolarBearQAFINAL.pdf.

USFWS 2017, 'Endangered and Threatened Wildlife and Plants; 12-Month Findings on Petitions to List 25 Species as Endangered or Threatened Species', 4 October, https://polarbearscience.files.wordpress.com/2017/10/walrus-decision-usfws-announcement_for-federal-register-publication_4-oct-2017.pdf.

USFWS 2018, 'Regulations of drones, unmanned aircraft systems (UAS)', viewed 1 December 2019, https://www.fws.gov/refuge/buenos_aires/visit/drone_rules.html [See also 50 CFR 27.34].

USFWS 2019, 'Notice to Public: Cinder River Critical Habitat Area', US Fish & Wildlife Service/Alaska Department of Fish & Game Special Notice, 17 July 2019, viewed 1 December 2019, http://www.adfg.alaska.gov/index.cfm?adfg=bristolbay.permits.

WWF 2019, 'Walruses, the new symbol of climate change. They'd be on ice if they could.' *WWF_UK* on Twitter, 12 April, https://twitter.com/wwf_uk/status/1116804504089968640.

Wang, M & Overland, JE 2009, 'A sea ice free summer Arctic within 30 years?', *Geophysical Research Letters*, vol. 36, https://agupubs.onlinelibrary.wiley.com/doi/pdf/10.1029/2009GL037820https://agupubs.onlinelibrary.wiley.com/doi/pdf/10.1029/2009GL037820.

West, A 2019, 'David Attenborough's 'Our Planet' leaves fans traumatised with "heartbreaking" walrus scene', *Yahoo News*, 11 April, https://uk.news.yahoo.com/david-attenboroughs-planet-leaves-fans-traumatised-heartbreaking-walrus-scene-095709472.html.

Whitworth, D 2019, 'David Attenborough's Our Planet: Walruses plunging to deaths become new symbol of climate change', *The Times*, 5 April, https://www.thetimes.co.uk/article/david-attenborough-s-our-planet-walruses-plunging-to-deaths-become-new-symbol-of-climate-change-23sbkwzlt.

Wiig, Ø, Born, EW & Garner, GW (eds.) 1995, 'Polar Bears: Proceedings of the 11th working meeting of the IUCN/SSC Polar Bear Specialists Group', 25–27 January 1993, in Copenhagen, Denmark. Gland, Switzerland and Cambridge UK, IUCN. http://pbsg.npolar.no/en/meetings/.

Wiig, Ø, Amstrup, S, Atwood, T, Laidre, K, Lunn, N, Obbard, M & others 2015, 'Ursus maritimus', The IUCN Red List of Threatened Species 2015: e.T22823A14871490, viewed 28 November 2015, http://www.iucnredlist.org/details/22823/0.

Williams, TD 2019, 'NH Hotel Group Cancels 'Alternative Climate Conference' After Protests', *Breitbart*, 19 November, https://www.breitbart.com/environment/2019/11/19/nh-hotel-group-cancels-alternative-climate-conference-after-protests/.

York, J, Dowsley, M, Cornwell, A, Kuc, M & Taylor, M 2016, 'Demographic and traditional knowledge perspectives on the current status of Canadian polar bear subpopulations', *Ecology and Evolution*, vol. 6, iss. 9, pp. 2897–2924, https://onlinelibrary.wiley.com/doi/full/10.1002/ece3.2030.

2: Climate Change in the Polar Regions: A Perspective

Bar-Sever, Y, Haines, B, Bertiger, W, Desai, S & Wu, S 2012, 'Geodetic reference antenna in space (GRASP) – a mission to enhance space-based geodesy', *Jet Propulsion Laboratory Papers*, Caltech.

Caltech Jet Propulsion Laboratory 2016, 'Greenland ice loss 2002-2016', viewed April 2019, https://gracefo.jpl.nasa.gov/resources/33/greenland-ice-loss-2002-2016/.

Church, JA, Gregory, JM, Huybrechts, P, Kuhn, M, Lambeck, K, Nhuan, MT, Qin, D & Woodworth, PL 2001, *Changes in Sea Level. In: Climate Change 2001: The Scientific Basis. Contribution of Working Group 1 to the Third Assessment Report of the Intergovernmental Panel on Climate Change*, Houghton, JT, Ding, Y, Griggs, DJ, Noguer, M, van der Linden, PJ, Dai, X, Maskell, K & Johnson, CA (eds.), Cambridge University Press, Cambridge, United Kingdom and New York, NY, USA, pp. 639–693.

Chylek, P, Dubey, M & Lesins, G 2006, 'Greenland Warming of 1920–1930 and 1995–2005', *Geophysical Research Letters*, vol. 33, iss. 11.

Chýlek, P, Dubey, M, McCabe, M & Dozier, J 2007, 'Remote sensing of Greenland ice sheet using multispectral near-infrared and visible radiances', *Journal of Geophysical Research*, vol. 112, iss. D24.

Danish Meteorological Institute 2018, 'Arctic Sea Ice Charts 1893–1965', viewed April 2019, https://nsidc.org/data/g02203.

Fasullo, J, Nerem, R, Hamlington, B 2016, 'Is the detection of accelerated sea level rise imminent', *Scientific Reports*, Vol.6, Article number 31245.

Folland, CK, Karl, T & Vinnikov, KYA 1990, *Observed Climate Variations and Change. In: CLIMATE CHANGE The IPCC Scientific Assessment. Report Prepared for IPCC by Working Group 1*, Houghton, JT, Jenkins, GJ & Ephraums, JJ (eds.), Cambridge University Press, Cambridge, United Kingdom and New York, NY, USA, pp. 195–238.

Gwyther, D, O'Kane, T, Galton-Fenzi, B, Monselesan, D & Greenbaum, J 2018, 'Intrinsic Processes drive variability in the basal melting of the Totten Glacier Ice Shelf', *Nature Communications*, vol. 9, no. 3141.

Hollingsworth, B 2013, 'Wrong: Al Gore predicted Arctic Summer Ice Could Disappear In 2013', *CNS News*, 13 September, 2013.

Houston, J & Dean, R 2011, 'Sea-level Acceleration Based on U.S. Tide Gauges and Extensions of Previous Global-Gauge Analyses', *Journal of Coastal Research*, vol. 27, iss. 3, pp. 409–417.

Johnson, J, Bentley, M, Smith, J, Finkel, R, Rood, D, Gohl, K, Balco, G, Larter, R & Schaefer, J 2014, 'Rapid Thinning of Pine Island Glacier in the Early Holocene', *Science*, vol. 343, iss. 6174, pp. 999–1001.

Joughin, I, Smith, B & Medley, B 2014, 'Marine ice sheet collapse potentially underway for the Thwaites Basin, West Antarctica', *Science*, vol. 344, iss. 6185, pp. 735–738.

Nerem, RS, Beckley, BD, Fasullo, JT, Hamlington, B, Masters, D & Mitchum, G 2018, 'Climate-change–driven accelerated sea-level rise detected in the altimeter era'. *Proceedings of the National Academy of Sciences*.

NOAA 2020, 'NOAA Tides & Currents; Sea Level Trends', https://tidesandcurrents.noaa.gov/sltrends/sltrends_us.html

Przybylak, R 2000, 'Temporal and Spatial Variation of Surface Air Temperature over the Period of Instrumental Observations in the Arctic', *International Journal of Climatology*, vol. 20, pp. 587–614.

Rignot, E, Mouginot, J, Morlighem, M, Seroussi, H & Scheuchl, B 2014, 'Widespread, rapid grounding line retreat of Pine Island, Thwaites, Smith and Kohler glaciers, West Antarctica, from 1992 to 2011', *Geophysical Research Letters*, vol. 41, iss. 10, pp. 3502–3509.

Rignot, E, Mouginot, J, Scheuchl, B, van den Broeke, M, van Wessem, M & Morlighem, M 2019, 'Four decades of Antarctic Ice Sheet mass balance from 1979–2017', *Proceedings of the National Academy of Science,* vol. 116, pp. 1095–1103.

Solly, M 2018, 'The Wreck of a WWII Fighter Plane Will Be Unearthed from a Greenland Glacier', Smithsonianmag.com, 28 August 2018.

Thomas, ER, Bracegirdle, TJ, Turner, J & Wolff, EW 2013, 'A 308 year record of climate variability in West Antarctica', *Geophysical Research Letters*, vol. 40, no. 20, pp. 5492–5496.

Trenberth, KE, Jones, PD, Ambenje, P, Bojariu, R, Easterling, D, Klein Tank, A, Parker, D, Rahimzadeh, F, Renwick, JA, Rusticucci, M, Soden, B & Zhai, P 2007, *Observations: Surface and Atmospheric Climate Change. In: Climate Change 2007: The Physical Science Basis. Contribution of Working Group I to the Fourth Assessment Report of the Intergovernmental Panel on Climate Change*, Solomon, S, Qin, D, Manning, M, Chen, Z, Marquis, M, Averyt, KB, Tignor, M & Miller, HL (eds.), Cambridge University Press, Cambridge, United Kingdom and New York, NY, USA, pp. 235–335.

Wang, Y, Thomas, E, Hou, S, Huai, B, Wu, A, Sun, W, Qi, S, Ding, M & Zhang, Y 2017, 'Snow Accumulation Variability Over the West Antarctic Ice Sheet Since 1900', *Geophysical Research Letters,* vol. 44, iss. 22, pp. 11,482–11,490.

Zwally, H, Li, J, Robbins, J & Saba, J 2015, 'Mass gains of the Antarctic ice sheet exceed losses', *Journal of Glaciology,* vol. 61, iss. 230, pp. 1019–1036.

3: Counting Emperor Penguins

Ainley, D, Russell, J, Jenouvrier, S, Woehler, E, Lyver, P, Fraser, E & Kooyman, G 2010, 'Antarctic penguin response to habitat change as Earth's troposphere reaches 2 °C above preindustrial levels', *Ecological Monographs,* vol. 80, pp. 49–66.

Barbraud, C & Weimerskirch, H 2001, 'Emperor penguins and climate change', *Nature,* vol. 411, pp. 183–186.

BirdLife International 2018, 'Emperor Penguin, Aptenodytes forsteri' The IUCN Red List of Threatened Species, viewed 26 April 2019, <http://dx.doi. org/10.2305/IUCN.UK.2018-2.RLTS.T22697752A132600320.en>.

Coria, N & Montalti, D 2000, 'A newly discovered breeding colony of emperor penguins *Aptenodytes forsteri*', *Marine Ornithology,* vol. 28, pp. 119–120.

Dugger, K, Ballard, G, Ainley, D & Barton, K 2006, 'Effects of flipper bands on foraging behavior and survival of Adélie Penguins (*Pygoscelis adeliae*)', *The Auk,* vol. 123, pp. 858–869.

Fraser, A, Massom, R, Michael, K, Galton-Fenzi, B & Lieser, J 2012, 'East Antarctic Landfast Sea Ice Distribution and Variability, 2000-08', *Journal of Climate,* vol. 25, pp. 1137–1156.

Fretwell, PT & Trathan, PN 2019, 'Emperors on thin ice: three years of breeding failure at Halley Bay', *Antarctic Science,* viewed 26 April 2019, <https://www. cambridge.org/core>.

Jenouvrier, S, Caswell, H, Barbraud, C, Holland, M, Stroeve, J & Weimerskirch, H 2009, 'Demographic models and IPCC climate projections predict the decline of an emperor penguin population', *PNAS,* vol. 106, pp. 1844–1847.

Jenouvrier, S, Holland, M, Stroeve, J, Serreze, M, Barbraud, C, Weimerskirch, H & Caswell, H 2014, 'Projected continent-wide declines of the emperor penguin under climate change', *Nature Climate Change,* vol. 4, pp. 715–718.

Kooyman, G, Siniff, D, Stirling, I & Bengtson, J 2004, 'Moult habitat, pre- and post-moult diet and post-moult travel of Ross Sea emperor penguins', *Marine Ecology Progress Series,* vol. 267, pp. 281–290.

Kooyman, GL, Ainley, DG, Ballard, G & Ponganis, P 2007, 'Effects of giant icebergs on two emperor penguin colonies in the Ross Sea, Antarctica', *Antarctic Science,* vol. 19, pp. 31–38.

Kooyman, GL & Ponganis, P 2017, 'Rise and fall of Ross Sea emperor penguin colony populations: 2000 to 2012', *Antarctic Science,* vol. 29, pp. 201–208.

REFERENCES

Landrum, L, Holland, MM, Schneider, DP & Hunke, E 2012, 'Antarctic Sea Ice Climatology, Variability, and Late Twentieth-Century Change in CCSM4', *Journal of Climate*, vol. 25, pp. 4817–4838.

LaRue, MA, Kooyman, G, Lynch, HJ & Fretwell, P 2015, 'Emigration in emperor penguins: implications for interpretation of long-term studies', *Ecography*, vol. 38, pp. 114–120.

Libertelli, MM & Coria, NR 2017, 'Monitoring the northernmost colony of Emperor penguin *Aptenodytes forsteri* at Snow Hill Island, Weddell Sea, Antarctica', in Van de Putte (ed), *Book of abstracts XIIth SCAR Biology Symposium Cambridge*, pp.132, Scientific Committee on Antarctic Research, Polar Research Institute.

Massom, R, Hill, K, Barbraud, C, Adams, N, Ancel, A, Emmerson, L & Pook, MJ 2009, 'Fast ice distribution in Adélie Land, East Antarctica: interannual variability and implications for emperor penguins *Aptenodytes forsteri*', *Marine Ecology Progress Series*, vol. 374, pp. 243–257.

New York Times 2019, 'An Emperor Penguin Colony in Antarctica Vanishes', viewed 26 April 2019, https://www.nytimes.com/2019/04/25/science/emperor-penguins-antarctica.html.

Saraux, C, Le Bohec, C, Durant, JM, Viblanc, VA, Gauthier-Clerc, M, Beaune, D, Park, YH, Yoccoz, NG, Stenseth, NC & Le Maho, Y 2011, 'Reliability of flipper-banded penguins as indicators of climate change', *Nature*, vol. 469, pp. 203–206.

Steele, J 2013, *Landscapes and Cycles; An Environmentalists Journey to Climate Skepticism*, CreateSpace Independent Publishing Platform.

Todd, FS, Adie, S & Splettstoesser, JF 2004, 'First ground visit to the emperor penguin *Aptenodytes forsteri* colony at Snow Hill island, Weddell sea, Antarctica', *Marine Ornithology*, vol. 32, pp. 193–194.

Trathan, PN, Fretwell, PT & Stonehouse, B 2011, 'First Recorded Loss of an Emperor Penguin Colony in the Recent Period of Antarctic Regional Warming: Implications for Other Colonies', *PLOS ONE*, vol. 6.

Turner, J, Lu, H, White, I, King, JC, Phillips, T, Hosking, JS, Bracegirdle, TJ, Marshall, GJ, Mulvaney, R & Pranab, D 2016, 'Absence of 21st century warming on Antarctic Peninsula consistent with natural variability', *Nature*, vol. 535, pp. 411–415.

Wang, G, Hendon, HH, Arblaster, JM, Lim, E-P, Abhik, S & van Rensch, P 2019, 'Compounding tropical and stratospheric forcing of the record low Antarctic sea-ice in 2016', *Nature Communications*, vol. 10.

Wienecke, B 2011, 'Review of historical population information of emperor penguins', *Polar Biology* vol. 34, pp. 153–167.

Younger, JL, Clucas, GV, Kao, D, Rogers, AD, Gharbi, K, Hart, T & Miller, KJ 2017, 'The challenges of detecting subtle population structure and its importance for the conservation of emperor penguins', *Molecular Ecology*, vol. 26, pp. 3883–3897.

4: Fire and Ice, Volcanoes at Antarctica

Birch, WD, Gleadow, AJW, Nottle, BW, Ross, JA & Whately, R 1978, 'Geology and structural development of the Cerberean Cauldron, Central Victoria', *Memoirs of Museum Victoria*, vol. 39, pp. 1–17.

Clemens, JD & Birch, WD 2012, 'Assembly of a zoned volcanic magma chamber from multiple magma batches: The Cerberean Cauldron, Marysville Igneous Complex, Australia', *Lithos*, vol. 155, pp. 272–288.

Corr, HFJ & Vaughan, DG 2008, 'A recent volcanic eruption beneath the West Antarctic ice sheet', *Nature Geoscience*, vol. 1, pp. 122–125.

IPCC 2013, *Climate Change 2013: The Physical Science Basis. Contribution of Working Group I to the Fifth Assessment Report of the Intergovernmental Panel on Climate Change*, Stocker, TF, Qin, D, Plattner, G-K, Tignor, M, Allen, SK, Boschung, J, Nauels, A, Xia, Y, Bex, V & Midgley, PM (eds.), Cambridge University Press, Cambridge, United Kingdom and New York, NY, USA, p. 1535.

Loose, BN, Garabato, AC, Schlosser, P, Jenkins, WJ, Vaughan, D & Heywood, KJ 2018, 'Evidence of an active volcanic heat source beneath the Pine Island Glacier', *Nature Communication*, vol. 9, p. 2431.

Naish, T, Powell, R, Levy, R & Wilson, GS 2009, 'Obliquity-paced Pliocene West Antarctic ice sheet oscillations', *Nature*, vol. 458, pp. 322–328.

Pedersen, GBM & Grosse, P 2014, 'Morphometry of subaerial shield volcanoes and glaciovolcanoes from Reykjanes Peninsula, Iceland: effects of eruption environment', *Journal of Volcanology and Geothermal Research*, vol. 282, pp. 115–133.

Pollard, D & De Conto, RM 2009, 'Modelling West Antarctic ice sheet growth and collapse through the past five million years', *Nature*, vol. 458, pp. 329–333.

Ritz, C, Edwards, TL, Gaël, D, Payne, AJ, Peyaud, V & Hindmarsh, RCA 2015, 'Potential sea-level rise from Antarctic ice-sheet instability constrained by observations', *Nature*, vol. 528, pp. 115–118.

Scherer, RP 1991, 'Quaternary and Tertiary microfossils from beneath Ice Stream B: Evidence for a dynamic West Antarctic Ice Sheet history', *Palaeogeography, Palaeoclimatology, Palaeoecology*, vol. 90, pp. 395–412.

Stone, JO, Balco, GA, Sugden, DE, Caffee, MW, Sass, LC, Cowdery, SG & Siddoway, C 2003, 'Holocene Deglaciation of Marie Byrd Land, West Antarctica', *Science*, vol. 299, pp. 99–102.

Van Wyk de Vries, M, Bingham, RG & Hein, AS 2017, 'A new volcanic province: an inventory of sub-glacial volcanoes in West Antarctica', *Geological Society, London, Special Publications*, vol. 461, pp. 231–248.

Winberry, JP & Anandakrishnan, S 2003, 'Seismicity and neotectonics of West Antarctica', *Geophysical Research Letters*, vol. 30, 1931.

5: Sacred bubbles in ice cores

Alley, RB 2010, 'Reliability of ice-core science: historical insights', *Journal of Glaciology*, vol. 56, no. 200, 2010.

Amos, J 2019, 'Climate change: European team to drill for "oldest ice" in Antarctica', BBC News, https://www.bbc.com/news/science-environment-47848344.

Beerling, DJ & Royer, DL 2011, 'Convergent Cenozoic CO_2 history', *Nature Geoscience*, vol. 4, p. 418.

Berner, W, Oeschger, H & Stauffer, B 1980, 'Information on the CO_2 cycle from ice core studies', *Radiocarbon*, vol. 22, no. 2, pp. 227–235.

Finsinger, W & Wagner-Cremer, F 2009, 'Stomatal-based inference models for reconstruction of atmospheric CO_2 concentration: A method assessment using a calibration and validation approach', *The Holocene*, vol. pp. 757–764.

Gore, A 2006, *An Inconvenient Truth*, Lawrence Bender Productions, online video, viewed 17 January 2020, https://www.youtube.com/watch?v=-JIuKjaY3r4.

Gribbin, J 1989, 'The end of the ice ages?' *New Scientist*, 17 June, https://www.newscientist.com/article/mg12216694-400-the-end-of-the-ice-ages/.

IPCC 1990, *Policymakers Summary. In: CLIMATE CHANGE The IPCC Scientific Assessment. Report Prepared for IPCC by Working Group 1*, Houghton, JT, Jenkins, GJ & Ephraums, JJ (eds.), Cambridge University Press, Cambridge, United Kingdom and New York, NY, USA, pp. xi–xxxiii.

Kouwenberg, LLR, Kürschner, WM, Wagner-Cremer, F & Visscher, H 2005, 'Atmospheric CO_2 fluctuations during the last millennium reconstructed by stomatal frequency analysis of Tsuga heterophylla needles', *Geology*, January.

Mulvaney, R, Wolff, EW & Oates, K 1988, 'Sulphuric acid at grain boundaries in Antarctic ice', *Nature*, vol. 331, pp. 247–249, https://www.nature.com/articles/331247a0.

Nealon, S 2019, 'Two million-year-old ice provides snapshot of Earth's greenhouse gas history', *Phys Org*, 30 October, https://phys.org/news/2019-10-million-year-old-ice-snapshot-earth-greenhouse.html.

Rundgren, M, Björck, S & Hammarlund 2005, 'Last interglacial atmospheric CO_2 changes from stomatal index data and their relation to climate variations', *Global and Planetary Change*, vol. 49, pp. 47–62.

Stauffer, B & Tschumi, J 2000, 'Reconstruction Of Past Atmospheric CO_2 Concentrations By Ice Core Analyses, in Physics Of Ice Core Records', Hokkaido University Press, Sapporo, pp. 217–241 http://eprints.lib.hokudai.ac.jp/dspace/bitstream/2115/32470/1/P217-241.pdf.

Wagner, F, Bohncke, SJP, Dilcher, DL, Kürschner, WM, van Geel, B & Visscher, H 1999, 'Century-Scale Shifts in Early Holocene Atmospheric CO_2 Concentration', *Science*, vol. 284, no. 5422, pp. 1971–1973.

Woodward, FI 1987, 'Stomatal numbers are sensitive to increases in CO_2 from pre-industrial levels', *Nature*, vol. 327, pp. 617–618, https://www.nature.com/articles/327617a0.

6: Monitoring Temperatures and Sea Ice with Satellites

Cabedo-Sanz, P, Belt, S, Jennings, A, Andrews, J & Geirsdóttir, A 2016, 'Variability in drift ice export from the Arctic Ocean to the North Icelandic Shelf over the last 8000 years: A multi-proxy evaluation', *Quaternary Science Reviews*, vol. 146, pp. 99–115.

Cheng, L, Abraham, J, Hausfather, Z & Trenberth, KE 2019, 'How fast are the oceans warming?' *Science*, vol. 363 (6423), pp. 128–129.

Christy, JR, Spencer, RW, Braswell, WD & Junod, RW 2018, 'Examination of space-based bulk atmospheric temperatures used in climate research', *International Journal of Remote Sensing*, vol. 39, pp. 3580–3607.

Earth System Science Center, *Global Temperature Report*, viewed 13 March 2020, https://www.nsstc.uah.edu/climate/.

IPCC 2014, *Climate Change 2014: Synthesis Report. Contribution of Working Groups I, II and III to the Fifth Assessment Report of the Intergovernmental Panel on Climate Change*, Pachauri, RK & Meyer, LA (eds.), IPCC, Geneva, Switzerland, p. 151.

KNMI Climate Explorer a, *European Climate Assessment & Data*, viewed 13 March 2020, https://climexp.knmi.nl/selectfield_cmip5.cgi?id=someone@somewhere.

KNMI Climate Explorer b, *European Climate Assessment & Data*, 'Starting Point', viewed 13 March 2020, https://climexp.knmi.nl/start.cgi.

National Snow and Ice Data Center, Data and Image Archive, viewed 13 March 2020, https://nsidc.org/data/seaice_index/archives.

Remote Sensing Systems, *Upper Air Temperature*, viewed 13 March 2020, http://www.remss.com/measurements/upper-air-temperature.

Spencer, RW & Christy, JR 1990, 'Precise monitoring of global temperature trends from satellites', *Science*, vol. 247, pp. 1558–1562.

7: Winter Temperature Trends in Antarctica

Arrhenius, S 1896, 'On the Influence of Carbonic Acid in the Air upon the Temperature of the Ground', *Philosophical Magazine and Journal of Science*, Series 5, vol. 41.

Hansen, J, Sato, M, Ruedy, R, Nazarenko, L, Lacis, A, Schmidt, GA, Russell, G, Aleinov, I, Bauer, M, Bauer, S, Bell, N, Cairns, B, Canuto, V, Chandler, M, Cheng, Y, Del Genio, A, Faluvegi, G, Fleming, E, Friend, A, Hall, T, Jackman, C, Kelley, M, Kiang, N, Koch, D, Lean, J, Lerner, J, Lo, K,

Menon, S, Miller, R, Minnis, P, Novakov, T, Oinas, V, Perlwitz, Ja, Perlwitz, Ju, Rind, D, Romanou, A, Shindell, D, Stone, P, Sun, S, Tausnev, N, Thresher, D, Wielicki, B, Wong, T, Yao, M & Zhang, S 2005, 'Efficacy of climate forcings', *Journal of Geophysical Research*, vol. 110, iss. D18.

Hellmer, HH, Kauker, F, Timmerman, R, Determann, J & Rae, J 2012, 'Twenty-first-century warming of a large Antarctic ice-shelf cavity by a redirected coastal current', *Nature*, vol. 485, pp. 225–228.

Holland, MM & Bitz, CM 2003, 'Polar amplification of climate change in coupled models', *Climate Dynamics*, vol. 21, pp. 221–232.

IPCC 2013, *Climate Change 2013: The Physical Science Basis. Contribution of Working Group I to the Fifth Assessment Report of the Intergovernmental Panel on Climate Change*, Stocker, TF, Qin, D, Plattner, G-K, Tignor, M, Allen, SK, Boschung, J, Nauels, A, Xia, Y, Bex, V & Midgley, PM (eds.), Cambridge University Press, Cambridge, United Kingdom and New York, NY, USA, p. 1535

Kekesi, A 2005, NASA/Goddard Space Flight Center Scientific Visualization Studio. Data provided by Larry Stock. https://svs.gsfc.nasa.gov/vis/a000000/a003100/a003188/index.html

Manabe, S & Wetherald, R 1975, 'The effects of doubling the CO_2 concentration on the climate of a general circulation model', *Journal of Atmospheric Sciences*, vol. 32, no. 1.

Schmidtko, S, Heywood, KJ, Thompson, AF & Aoki, S 2014, 'Multidecadal warming of Antarctic waters', *Science*, vol. 346, pp. 1227–1231.

Schmidt, G, GHCN V3 unadj dataset, viewed 4 March 2019, https://data.giss.nasa.gov/gistemp/stdata/.

Spence, P, Holmes, R, Hogg, A, Griffes, S, Stewart, K & England, M 2007, 'Localised rapid warming of West Antarctic subsurface waters by remote winds', *Nature Climate Change*, vol. 7, pp. 595–603.

8: No Evidence of Warming at Mawson, Antarctica

Australian Antarctic Division 2018, Australian Government, Department of Agriculture, Water and the Environment, 'Casey station: a brief history', viewed 13 March 2020, http://www.antarctica.gov.au/about-antarctica/history/stations/casey.

Bureau of Meteorology 2012, 'Australian Climate Observations Reference Network – Surface Air Temperature (ACORN-SAT): Observation practices', Technical report, Australian Government, viewed 17 May 2020, http://www.bom.gov.au/climate/data/acorn-sat/documents/ACORN-SAT_Observation_practices_WEB.pdf.

Bureau of Meteorology 2013, 'Australian Climate Observations Reference Network for Surface Air Temperature (Remote Australian Islands and Antarctica): Station catalogue', Technical report, Australian Government, viewed 17 May

2020, http://www.bom.gov.au/climate/data/acorn-sat/documents/ACORN-SAT-remote-Station-Catalogue-WEB.pdf.

Bureau of Meteorology 2018, 'Observation of air temperature', viewed 13 March 2020, http://www.bom.gov.au/climate/cdo/about/airtemp-measure.shtml.

Bureau of Meteorology 2019a, 'Climate Data Online', viewed 5 August 2019, http://www.bom.gov.au/climate/data/.

Bureau of Meteorology 2019b, 'Basic Climatological Station Metadata: Mawson', viewed 13 March 2020 (metadata compiled 28 July 2019), http://www.bom.gov.au/clim_data/cdio/metadata/pdf/siteinfo/IDCJMD0040.300001.SiteInfo.pdf.

IPCC 2013, *Climate Change 2013: The Physical Science Basis. Contribution of Working Group I to the Fifth Assessment Report of the Intergovernmental Panel on Climate Change*, Stocker, TF, Qin, D, Plattner, G-K, Tignor, M, Allen, SK, Boschung, J, Nauels, A, Xia, Y, Bex, V & Midgley, PM (eds.), Cambridge University Press, Cambridge, United Kingdom and New York, NY, USA, p. 1535.

James, FE 2006, *Statistical Methods in Experimental Physics*, World Scientific Publishing, Singapore, 2nd edition.

Jovanovic, B, Braganza, K, Collins, D & Jones, D 2012, 'Climate variations and change evident in high-quality climate data for Australia's Antarctic and remote island weather stations', *Australian Meteorological and Oceanographic Journal*, vol. 62, pp. 247–261.

Peterson, TC, Easterling, DR, Karl, TR, Groisman, P, Nicholls, N, Plummer, N, Torok, S, Auer, I, Boehm, R, Gullett, D, Vincent, L, Heino, R, Tuomenvirta, H, Mestre, O, Szentimrey, T, Salinger, J, Førland, EJ, Hanssen-Bauer, I, Alexandersson, H, Jones, P & Parker, D 1998, 'Homogeneity adjustments of in situ atmospheric climate data: a review', *International Journal of Climatology*, vol. 18, iss. 13, pp. 1493–1517.

Ribeiro, S, Caineta, J & Costa, AC 2016, 'Review and discussion of homogenisation methods for climate data', *Physics and Chemistry of the Earth*, vol. 94, pp. 167–179.

Trewin, B 2012, 'Techniques involved in developing the Australian Climate Observations Reference Network – Surface Air Temperature (ACORN-SAT) dataset', CAWCR Technical Report No. 049, https://cawcr.gov.au/technical-reports/CTR_049.pdf.

9: Deconstructing Two Thousand Years of Temperature Change at Antarctica

Abbot, J & Marohasy, J 2012, 'Application of artificial neural networks to rainfall forecasting in Queensland, Australia', *Advances in Atmospheric Science*, vol. 29(4), pp. 717–730.

Abbot, J & Marohasy, J 2013, 'The potential benefits of using artificial intelligence for monthly rainfall forecasting for the Bowen Basin, Queensland, Australia', *WIT Transactions on Ecology and the Environment: Water Resources Management VII*, vol. 171, pp. 287–297.

Abbot, J & Marohasy, J 2014, 'Input selection and optimisation for monthly rainfall forecasting in Queensland, Australia, using artificial neural networks', *Atmospheric Research*, vol. 138, pp. 166–178.

Abbot, J & Marohasy, J 2015a, 'Forecasting of Monthly Rainfall in the Murray Darling Basin, Australia: Miles as a Case Study', *River Basin Management*, vol. VIII, pp. 149–159.

Abbot, J & Marohasy, J 2015b, 'Improving Monthly Rainfall Forecasts Using Artificial Neural Networks and Single-month Optimisation: A Case Study of the Brisbane Catchment, Queensland, Australia', *Water Resources Management*, vol. VIII, pp. 3–13.

Abbot, J & Marohasy, J 2015c, 'Using artificial intelligence to forecast monthly rainfall under present and future climates for the Bowen Basin, Queensland, Australia', *International Journal of Sustainable Development and Planning*, vol. 10(1), pp. 66–75.

Abbot J & Marohasy, J 2015d, 'Using lagged and forecast climate indices with artificial intelligence to predict monthly rainfall in the Brisbane Catchment, Queensland, Australia', *International Journal of Sustainable Development and Planning*, vol. 10(1), pp. 29–41.

Abbot, J & Marohasy, J 2016a, 'Forecasting monthly rainfall in the Bowen Basin of Queensland, Australia, using neural networks with Niño Indices for El Niño–Southern Oscillation', Lecture Notes in Artificial Intelligence, vol. 9992, pp. 88–100.

Abbot, J & Marohasy, J 2016b, 'Forecasting Monthly Rainfall in the Western Australian Wheat-belt up to 18 months in Advance Using Artificial Neural networks', Lecture Notes in Artificial Intelligence, vol. 9992, pp. 71–87.

Abbot, J & Marohasy, J 2017a, 'The application of machine learning for evaluating anthropogenic versus natural climate change', *GeoResJ*, vol. 14, pp. 36–46.

Abbot, J & Marohasy, J 2017b, 'Skilful rainfall forecasts from artificial neural networks with long duration series and single-month optimization', *Atmospheric Research*, vol. 197, pp. 289–299.

Abbot, J & Marohasy, J 2017c, 'Forecasting Extreme Monthly Rainfall Events in Regions of Queensland, Australia Using Artificial Neural Networks', *International Journal of Sustainable Development and Planning*, vol. 12 (7), pp. 1117–1131.

Acemoglu & Robinson 2012, *Why Nations Fail – The Origins of Power, Prosperity and Poverty*, Profile Books, London.

Antoniades, D, Giralt, S, Geyer, A, Álvarez-Valero, AM, Pla-Rabes, S, Granados, I, Liu, EJ, Toro, M, Smellie, JL & Oliva, M 2018, 'The timing and widespread effects of the largest Holocene volcanic eruption in Antarctica', *Scientific Reports*, vol. 8, Article No. 17279.

Davis, WJ, Taylor, PJ & Davis, WB 2018, 'The Antarctic Centennial Oscillation: A Natural Paleoclimate Cycle in the Southern Hemisphere That Influences Global Temperature', *Climate*, vol. 6, iss. 1.

EPICA Members 2004, 'Eight glacial cycles from an Antarctic ice core', *Nature*, vol. 429, pp. 623–628.

Kuhn, TS 1962, *The Structure of Scientific Revolutions*, The University of Chicago Press.

Lüdecke, HJ, Hempelmann, A & Weiss, CO 2013, 'Multi-periodic climate dynamics: spectral analysis of long-term instrumental and proxy temperature records', *Climate of the Past*, vol. 9, pp. 447–452.

Marohasy, J 2018, 'Historical Temperature Reconstructions and Estimating the Contribution of the Industrial Revolution to 20th Century Warming', Jennifer Marohasy's blog, https://jennifermarohasy.com/temperatures/response-to-criticism-of-abbot-marohasy-2017-georesj/

Masson, V, Vimeux, F, Jouzel, J, Morgan, V, Delmotte, M, Ciais, P, Hammer, C, Johnsen, S, Lipenkov, VY, Moseley-Thompson, E, Petit, J-R, Steig, EJ, Stievenard, M & Vaikmae, R 2000, 'Holocene climate variability in Antarctica based on 11 ice-core isotopic records', *Quaternary Research*, vol. 54, pp. 348–358.

Münch, T & Laepple, T 2018, 'What climate signal is contained in decadal- to centennial-scale isotope variations from Antarctic ice cores?', *Climate of the Past*, vol. 14, pp. 2053–2070.

Petit, JR, Jouzel, J, Raynaud, D, Barkov, NI, Barnola, J-M, Basile, I, Bender, M, Chappellaz, J, Davis, M, Delaygue, G, Delmotte, M, Kotlyakov, M, Legrand, M, Lipenkov, VY, Lorius, C, Pépin, L, Ritz, E, Saltzman, E & Stievenard, M 1999, 'Climate and atmospheric history of the past 420,000 years from the Vostok ice core, Antarctica', *Nature*, vol. 399, pp. 429–436.

Stenni, B, Curran, MAJ, Abram, NJ, Orsi, A, Goursaud, S, Masson-Delmotte, V, Neukom, R, Goosse, H, Divine, D, van Ommen, T, Steif, EJ, Dixon, DA, Thomas, ER, Bertler, NA, Isaksson, E, Ekaykin, A, Werner, M & Frezzotti, M 2017, 'Antarctic climate variability on regional and continental scales over the last 2000 years', *Climate of the Past*, vol. 13, pp. 1609–1634.

Yiou, P, Fuhrer, K, Meeker, LD, Jouzel, J, Johnsen, S & Mayewski, PA 1997, 'Paleoclimatic variability inferred from the spectral analysis of Greenland and Antarctic ice-core data', *Journal of Geophysical Research-Oceans*, vol. 102, pp. 26441–26454.

Zhao, XH & Feng, XS 2014, 'Correlation between solar activity and the local temperature of Antarctica during the past 11,000 years', *Journal of Atmospheric and Solar Terrestrial Physics*, vol. 122, pp. 26–33.

10: Cosmoclimatology

Alroy, J, Aberhan, M, Bottjer, DJ, Foote, M, Fürsich, FT, Harries, PJ, Hendy, AJW, Holland, SM, Ivany, LC, Kiessling, W, Kosnik, MA, Marshall, CR, McGowan, AJ, Miller, AI, Olszewski, TD, Patzkowsky, ME, Peters, SE, Villier, L, Wagner, PJ, Bonuso, N, Borkow, PS, Brenneis, B, Clapham, ME, Fall, LM, Ferguson, CA, Hanson, VL, Krug, AZ, Layou, KM, Leckey, EH, Nürnberg, S, Powers, CM, Sessa, JA, Simpson, C, Tomašových, A & Visaggi, CC 2008, 'Phanerozoic Trends in the Global Diversity of Marine Invertebrates', *Science*, vol. 321, p. 97.

Eddy, JA 1976, 'The Maunder Minimum,' *Science*, vol. 192, pp. 1189–1202.

Haq, BU & Shutter, SR 2008, 'A Chronology of Paleozoic Sea-Level Changes', *Science*, vol. 322, pp. 64–68. <http://dx.doi.org/10.1126/science.1161648>.

Neff, U, Burns, SJ, Mangini, M, Mudelsee, D, Fleitmann, D & Matter, A 2001, 'Strong coherence between solar variability and the monsoon in Oman between 9 and 6 kyr ago', *Nature*, vol. 411, p. 290.

Pierce, JR & Adams, PJ 2009, 'Can cosmic rays affect cloud condensation nuclei by altering new particle formation rates?' *Geophysical Research Letters*, vol. 36, L09820, doi:10.1029/2009GL037946.

Shaviv, NJ 2003, 'The spiral structure of the Milky Way, cosmic rays, and ice age epochs on Earth', *New Astronomy*, vol. 8, pp. 39–77.

Shaviv, NJ 2008, 'Using the oceans as a calorimeter to quantify the solar radiative forcing', *J. Geophys. Res.*, 113, A11101, doi:10.1029/2007JA012989.

Svensmark, H & Friis-Christensen, E 1997, 'Variation of cosmic ray flux and global cloud coverage-a missing link in solar-climate relationships', *Journal of Atmospheric and Solar-Terrestrial Physics*, vol. 59, p. 1225.

Svensmark, H, Pedersen, OP, Marsh, ND, Enghoff MB & Uggerhøj, UI 2007, 'Experimental evidence for the role of ions in particle nucleation under atmospheric conditions', *Proceedings of the Royal Society A*, vol. 463, pp. 385–396 doi:10.1098/rspa.2006.1773.

Svensmark, H, Bondo, T & Svensmark, J 2009, 'Cosmic ray decreases affect atmospheric aerosols and clouds', *Geophysical Research Letters*, vol. 36, L15101, doi:10.1029/ 2009GL038429.

Svensmark, H 2012, 'Evidence of nearby supernovae affecting life on Earth', *Monthly Notices of the Royal Astronomical Society*, vol. 423, pp. 1234–1253 doi:10.1111/j.1365-2966.2012.20953.x.

Svensmark, J, Enghoff, MB, Shaviv, NJ & Svensmark, H 2016, 'The response of clouds and aerosols to cosmic ray decreases', *Journal of Geophysical Research:*

Space Physics, vol. 121, pp. 8152–8181, <http://dx.doi.org/10.1002/2016 JA022689>.

Svensmark, H, Enghoff, MB, Shaviv, NJ & Svensmark, J 2017, 'Increased ionization supports growth of aerosols into cloud condensation nuclei', *Nature Communications*, vol. 8, no. 2199, <https://doi.org/10.1038/ s41467-017-02082-2>.

Svensmark, H 2019, *Force Majeure: The Sun's Role in Climate Change*, The Global Warming Policy Foundation, report 33, ISBN 978-0-9931190-9-5.

Suggestion for further reading

Svensmark, H & Calder, N 2007, *The Chilling Stars: A New Theory of Climate Change*, Icon Books, London.

12: Tropical Convection: Cooling the Atmosphere

Manabe, S & Strickler, RF 1964, 'Thermal equilibrium of the atmosphere with a convective adjustment', *Journal of Atmospheric Sciences*, vol. 21, pp. 361–385.

Peixoto, JP & Oort, AH 1992, *Physics of Climate*, Springer.

Renno, NO and Ingersoll, AP 1996, Natural convection as a heat engine: A theory of CAPE, *American Meteorological society*, doi.org/10.1175/1520-0469(1996)053<0572:NCAAHE>2.0.CO;2

Riehl, H & Simpson, J 1979, 'The heat balance of the equatorial trough zone revisited', *Contributions to Atmospheric Physics*, vol. 52, pp. 287–305.

Thomas, GE & Stamnes, K 1999, *Radiative Transfer in the Atmosphere and Ocean* p. 446, in S Manabe & RF Strickler 1964, 'Thermal equilibrium of the atmosphere with a convective adjustment', *Journal of Atmospheric Sciences*, vol. 21, pp. 361–385.

13: Reflections on the Iris Effect

Cho, H, Ho, C-H & Choi, Y-S 2012, 'The observed variation in cloud-induced longwave radiation in response to sea surface temperature over the Pacific warm pool from MTSAT-1R imagery', *Geophysical Research Letters*, vol. 39, no. 18.

Choi, Y-S, Cho, H, Ho, C-H, Lindzen, RS, Park, SK & Yu, X 2014, 'Influence of non-feedback variations of radiation on the determination of climate feedback', *Theoretical and Applied Climatology*, vol. 15, no. 1–2, pp. 355–364.

Choi, Y-S & Ho, C-H 2006, 'Radiative effect of Cirrus with different optical properties over the tropics in MODIS and CERES observations', *Geophysical Research Letters*, vol. 33, no. 21.

Chou, M-D & Lindzen, RS 2005, 'Comments on "Examination of the Decadal Tropical Mean ERBS Nonscanner Radiation Data for the Iris Hypothesis"', *Journal of Climate,* vol. 18, no. 12, pp. 2123–2127.

Chou, M-D, Lindzen, RS & Hou, AY 2002a, 'Impact of Albedo Contrast between Cirrus and Boundary-Layer Clouds on Climate Sensitivity', *Atmospheric Chemistry and Physics*, vol. 2, pp. 99–101.

Chou, M-D, Lindzen, RS & Hou, AY 2002b, 'Comments on "The Iris hypothesis: A negative or positive cloud feedback?"', *Journal of Climate*, vol. 15, pp. 2713–2715.

Del Genio, AD & Kovari, W 2002, 'Climatic properties of tropical precipitating convection under varying environmental conditions', *Journal of Climate*, vol. 15, pp. 2597–2615.

Fu, Q, Baker, M & Hartmann, DL 2002, 'Tropical cirrus and water vapor: an effective Earth infrared iris feedback?', *Atmospheric Chemistry and Physics*, vol. 2, pp. 31–37.

Fyfe, JC, Gillett, NP & Zwiers, FW 2012, 'Overestimated global warming over the past 20 years', *Nature Climate Change*, vol. 3, pp. 767–769.

Hartmann, DL & Michelsen, ML 2002, 'No evidence for iris', *Bulletin of the American Meteorological Society*, vol. 83, pp. 249–254.

Horvath, A & Soden, BJ 2008, 'Lagrangian Diagnostics of Tropical Deep Convection and Its Effect upon Upper-Tropospheric Humidity', *Journal of Climate*, vol. 21, no. 5.

Lewis, N & Curry, JA 2014, 'The implications for climate sensitivity of AR5 forcing and heat uptake estimates', *Climate Dynamics*, vol. 45, no. 3–4, pp. 1009–1023.

Lin, B, Wielicki, B, Chambers, L, Hu, Y & Xu, K-M 2004, 'The iris hypothesis: A negative or positive cloud feedback?', *Journal of Climate*, vol. 15, no. 1, pp. 3–7.

Lindzen, RS 1990, 'Some coolness concerning global warming', *Bulletin of the American Meteorological Society*, vol. 71, pp. 288–299.

Lindzen, RS, Chou, M-D, Hou, A-Y 2001, 'Does the Earth have an adaptive infrared iris', *Bulletin of the American Meteorological Society*, vol. 82, no. 3.

Lindzen, RS 2004, https://www.dropbox.com/s/8i4l6e5iqup0ath/GSFC040519-a.pdf?dl=0.

Lindzen, RS 2012, 'Climate science: is it designed to answer questions', *Euresis Journal*, vol. 2, pp. 161–193.

Lindzen, RS 2014, 'Global warming, models and language', in A. Moran (ed), *Climate Change: The Facts 2014*, ch. 3, Stockade Books, Woodsville, New Hampshire.

Lindzen, RS 2018, 'An appropriate role for climate theory', in *Proceedings of the 51st International Seminar on Nuclear War and Planetary Emergencies*.

Lindzen, RS, & Choi, Y-S 2009, 'On the determination of climate feedbacks from ERBE data', *Geophysical Research Letters*, vol. 36, no. 16.

Lindzen, RS, & Choi, Y-S 2011, 'On the observational determination of climate sensitivity and its implications', *Asian Pacific Journal of Atmospheric Science*, vol. 47, pp. 377–390.

Lindzen, RS & Farrell, B 1980, 'The role of polar regions in global climate, and the parameterization of global heat transport', *Monthly Weather Review*, vol. 108, pp. 2064–2079.

Lindzen, RS, Chou, M-D & Hou, AY 2001, 'Does the Earth have an adaptive infrared iris?', *Bulletin of the American Meteorological Society*, vol. 82, pp. 417–432.

Lindzen, RS, Chou, M-D & Hou, AY 2002, 'Comments on "No evidence for iris"', *Bulletin of the American Meteorological Society*, vol. 83, pp. 1345–1348.

Lindzen, RS, Hou, AY & Farrell, BF 1982, 'The role of convective model choice in calculating the climate impact of doubling CO_2', *Journal of the Atmospheric Sciences*, vol. 39, pp. 1189–1205.

Manabe, S & Wetherald, RT 1975, 'The effects of doubling the CO_2 concentration on the climate of a general circulation model', *Journal of the Atmospheric Sciences*, vol. 32, pp. 3–15.

Mauritsen, T & Stevens, B 2015, 'Missing iris effect as a possible cause of muted hydrological change and high climate sensitivity in models', *Nature Geoscience*, vol. 8, pp. 346–351.

Roe, G & Baker, M 2007, 'Why Is Climate Sensitivity So Unpredictable?', *Science*, vol. 318, pp. 629–632.

Rondanelli, RF & Lindzen, RS 2008, 'Observed variations in convective precipitation fraction and stratiform area with SST', *Journal of Geophysical Research*, vol. 113, no. D16.

Rondanelli, RF & Lindzen, RS 2010, 'Can thin cirrus clouds in the tropics provide a solution to the faint young Sun paradox?', *Journal of Geophysical Research*, vol. 115, D2.

Sagan, C & Mullen, G 1972, 'Earth and Mars: Evolution of atmospheres and surface temperatures', *Science*, vol. 177, no. 4043, pp. 52–56.

Spencer, RW & Braswell, WD 1997, 'How dry is the tropical free troposphere? Implications for global warming theory', *Bulletin of the American Meteorological Society*, vol. 78, pp. 1097–1106.

Stott, P, Good, P, Jones, G, Gillett, N & Hawkins, E 2013, 'The upper end of climate model temperature projections is inconsistent with past warming', *Environmental Research Letters*, vol. 8, no. 1.

Sun, D-Z & Lindzen, RS 1993, 'Distribution of tropical tropospheric water vapor', *Journal of the Atmospheric Sciences* , vol. 50, pp. 1643–1660.

Trenberth, KE & Fasullo, JT 2009, 'Global warming due to increasing absorbed solar radiation', *Geophysical Research Letters*, vol. 36.

Udelhofen, PM & Hartmann, DL 1995, 'Influence of tropical cloud systems on the relative humidity in the upper troposphere', *Journal of Geophysical Research*, vol. 100, pp. 7423–7440.

14: Rainfall-Powered Aviation

Abbot, J 2019, 'Australia – A Land of Drought and Flooding Rain', in *Rainfall: Extremes, Distribution and Properties*, InTech Publishing.

Abbot, J & Marohasy, J 2012, 'Application of artificial neural networks to rainfall forecasting in Queensland, Australia', *Advances in Atmospheric Science*, vol. 29(4), pp. 717–730.

Abbot, J & Marohasy, J 2013, 'The potential benefits of using artificial intelligence for monthly rainfall forecasting for the Bowen Basin, Queensland, Australia', *WIT Transactions on Ecology and the Environment: Water Resources Management*, vol. 171, pp. 287–297.

Abbot, J & Marohasy, J 2014, 'Input selection and optimisation for monthly rainfall forecasting in Queensland, Australia, using artificial neural networks', *Atmospheric Research*, vol. 138, 166–178.

Abbot, J & Marohasy, J 2015a, 'Forecasting of Monthly Rainfall in the Murray Darling Basin, Australia: Miles as a Case Study', *River Basin Management*, vol. VIII, pp. 149–159.

Abbot, J & Marohasy, J 2015b, 'Improving monthly rainfall forecasts using artificial neural networks and single-month optimisation: a case study of the Brisbane Catchment, Queensland, Australia', *Water Resources Management*, vol. VIII, pp. 3–13.

Abbot, J & Marohasy, J 2015c, 'Using lagged and forecast climate indices with artificial intelligence to predict monthly rainfall in the Brisbane Catchment, Queensland, Australia', *International Journal of Sustainable Development and Planning*, vol. 10(1), pp. 29–41.

Abbot, J & Marohasy, J 2015d, 'Using artificial intelligence to forecast monthly rainfall under present and future climates for the Bowen Basin, Queensland, Australia', *International Journal of Sustainable Development and Planning*, vol. 10(1), pp. 66–75.

Abbot, J & Marohasy, J 2016a, 'Forecasting monthly rainfall in the Bowen Basin of Queensland, Australia, using neural networks with Niño Indices for El Niño-Southern Oscillation', *Lecture Notes in Artificial Intelligence*, vol. 9992, pp. 88–100.

Abbot, J & Marohasy, J 2016b, 'Forecasting monthly rainfall in the Western Australian wheat-belt up to 18 months in advance using artificial neural networks', *Lecture Notes in Artificial Intelligence*, vol. 9992, pp. 71–87.

Abbot, J & Marohasy, J 2017a, 'Forecasting extreme monthly rainfall events in regions of Queensland, Australia using artificial neural networks', *International Journal of Sustainable Development and Planning*, vol. 12(7), pp. 1117–1131.

Abbot, J & Marohasy J 2017b, 'Forecasting of Medium-term Rainfall Using Artificial Neural Networks: Case Studies from Eastern Australia', in *Engineering and Mathematical Topics in Rainfall,* InTech Publishing.

Barr, C, Tibby, J, Gell, PG, Tyler JJ, Zawadzki, A & Jacobsen, G 2014, 'Climatic variability in southeastern Australia over the last 1500 years inferred from the high resolution diatom records of two crater lakes', *Quarterly Science Reviews,* vol. 95, pp. 115–131.

Bureau of Meteorology 2012, Australian Water Resources Assessment, Tasmania, http://www.bom.gov.au/water/awra/2012/documents/tasmania-lr.pdf.

Cai, W, Purich, A, Cowan, T, van Rensch, P & Weller, E 2014, 'Did Climate Change-Induced Rainfall Trends Contribute to the Australian Millennium Drought?', *Journal of Climate,* vol. 27(9), pp. 3145–3168.

COAG 2019, Energy Council Hydrogen Working Group, Department of Industry, Science, Energy and Resources, *Australia's National Hydrogen Strategy,* https://www.industry.gov.au/sites/default/files/2019-11/australias-national-hydrogen-strategy.pdf.

Cook, BI, Palmer, JG, Cook, ER, Turney, CSM, Allen, K, Fenwick, P, O'Donnell, A, Lough, JM, Grierson, PF, Ho, M & Baker, PJ 2016, 'The paleoclimate context and future trajectory of extreme summer hydroclimate in eastern Australia', *Journal of Geophysical Research -Atmospheres,* vol. 121(21), pp. 12820–12838.

Department of Industry, Science, Energy and Resources 2020, 'Technology Investment Roadmap Discussion Paper: A framework to accelerate low emissions technologies', https://consult.industry.gov.au/climate-change/technology-investment roadmap/supporting_documents/technology investmentroadmapdiscussionpaper.pdf.

Donovan, D 2020, 'How The Airline Industry Will Transform Itself As It Comes Back From Coronavirus', *Forbes,* https://www.forbes.com/sites/deandonovan/2020/03/30/how-the-airline-industry-will-transform-itself-as-it-comes-back-from-cornonavirus/#2f37109d67b9.

Fowler, E 2020, 'Tasmania to invest $50m in "green hydrogen" export plan', *Australian Financial Review,* 5 March.

Gergis, J & Ashcroft, L 2013, 'Rainfall variations in South-Eastern Australia part 2: a comparison of documentary, early instrumental and palaeoclimate records, 1788–2008', *International Journal of Climatology,* vol. 33, pp. 2973–2987.

Hargreaves, S 2017, 'Using Neural Networks to Forecast Inflows to the Hydro Tasmania storages', email to Guy Barnett, Minister for Energy 28 November 2017 (supplied by the author).

Hawkins, AJ 2019, 'This company wants to fill the skies with hydrogen-powered planes by 2022', *The Verge,* https://www.theverge.com/2019/8/14/20804257/zeroavia-hydrogen-airplane-electric-flight.

Hawthorne, S, Wang, QJ, Schepen, A & Robertson, D 2013, 'Effective use of general circulation model outputs for forecasting monthly rainfalls to long lead times', *Water Resources Research*, vol. 49, pp. 5427–5436.

Hudson, D, Alves, O, Hendon, HH & Marshall, AG 2011, 'Bridging the gap between weather and seasonal forecasting: intraseasonal forecasting for Australia', *Quarterly Journal of the Royal Meteorological Society*, vol. 137, pp. 673–689.

Hydro Tasmania 2018, 'The Next Generation of hydropower', Annual Report 2018, https://www.hydro.com.au/about-us/our-governance/annual-report.

Hydro Tasmania 2019, 'Tasmania's "green hydrogen" opportunity. Tasmania's unique advantage as a "green hydrogen" development zone', *Hydro Tasmania White Paper*, https://www.hydro.com.au/docs/default-source/clean-energy/hydrogen/tasmanias-green-hydrogen-opportunity.pdf?sfvrsn=96539a28_2.

Kiem, AS & Verdon-Kidd, DC 2009, 'Towards understanding hydroclimatic change in Victoria, Australia – Why was the last decade so dry?' *Hydrology and Earth System Science Discussions*, vol. 6, pp. 6181–6206.

McBride, JL & Nichols N 1983, 'Seasonal Relationships between Australian Rainfall and the Southern Oscillation', *Monthly Weather Review*, vol. 111, pp. 1998–2004.

Moore, J 2020, 'Flying on hydrogen: California firm advancing fuel cell development overseas', AOPA, https://www.aopa.org/news-and-media/all-news/2020/march/05/flying-on-hydrogen.

Nayak, DR, Mahapatra, A & Mishra, P 2013, 'A survey on rainfall prediction using artificial neural network', *International Journal of Computer Applications*, vol. 72(16), pp. 32–40.

Nicholls, N 2010, 'Local and remote causes of the southern Australian autumn-winter rainfall decline, 1958–2007', *Climate Dynamics*, vol. 34(6), pp. 835–845.

Philip, NS, Joseph, KB 2013, 'A neural network tool for analyzing trends in rainfall', *Computational Geoscience*, vol. 29, no. 2, pp. 215–223.

Queensland Flood Commission of Enquiry, Final Report 2012, http://www.floodcommission.qld.gov.au/ data/assets/pdf_file/0007/11698/QFCI-Final-ReportMarch-2012.pdf.

Risbey, JS, Pook, MJ, McIntosh, PC, Wheeler, MC & Hendon, HH 2009, 'On the remote drivers of rainfall variability in Australia', *Monthly Weather Review*, vol. 137, pp. 3233–3253.

Royal Aeronautical Society 2020, 'High Time for Hydrogen', https://www.aerosociety.com/news/high-time-for-hydrogen/.

Schepen, A, Wang, QJ & Robertson, DE 2014, 'Seasonal Forecasts of Australian Rainfall through Calibration and Bridging of Coupled GCM Outputs', *Monthly Weather Review*, vol. 142(5), pp. 1758–1770.

Steffen W, Rice M, Hughes L & Dean, A 2018, Climate Change and Drought June 2018, Climate Council of Australia, https://www.climatecouncil.org.au/wp-content/uploads/2018/06/CC_MVSA0146-Fact-Sheet-Drought_V2-FA_High-Res_Single-Pages.pdf.

Tasmania Department of State Growth 2019, *Tasmanian Renewable Hydrogen Action Plan,* https://www.stategrowth.tas.gov.au/__data/assets/pdf_file/0003/207705/Draft_Tasmanian_Hydrogen_Action_Plan_-_November_2019.pdf

Tularam, GA 2010, 'Relationship between El Niño Southern Oscillation Index and rainfall (Queensland, Australia)', *International Journal of Sustainable Development and Planning,* vol. 5(4), pp. 378–391.

van den Honert, RC & McAneney, J 2011, 'The 2011 Brisbane floods: causes, impacts and implications', *Water,* vol. 3, pp. 1149–1173.

van Dijk, AIJ, Beck, HE, Crosbie, RS, de Jeu, RAM, Liu, YY, Podger, G, Timbal, B & Viney, NR 2013, 'The Millennium Drought in southeast Australia (2001-2009): Natural and human causes and implications for water resources, ecosystems, economy, and society', *Water Resources Research,* vol. 49(2), pp. 1040–1057.

Vance, TR, Roberts, JL, Plummer, CT, Kiem, AS & van Ommen, TD 2015, 'Interdecadal Pacific variability and eastern Australian mega-droughts over the last millennium', *Geophysical Research Letters,* vol. 42, pp. 129–137.

Wei, CC 2013, 'Soft computing techniques in ensemble precipitation nowcast', *Applied Soft Computing,* vol. 13(2), pp. 793–805.

Whitley, A 2020, 'How Coronavirus Will Forever Change Airlines and the Way We Fly', *Bloomberg,* https://www.bloomberg.com/news/features/2020-04-24/coronavirus-travel-covid-19-will-change-airlines-and-how-we-fly.

Wu, CL & Chau, KW 2013, 'Prediction of rainfall time series using modular soft computing methods', *Engineering Applications of Artificial Intelligence,* vol. 26, pp. 997–1007.

15: Wildfires in Australia: 1851 to 2020

Australian Government, Department of Agriculture, Fisheries and Forestry (Bureau of Rural Sciences) 2008, *Australia's State of the Forests Report,* viewed 24 January 2020, https://www.agriculture.gov.au/abares/forestsaustralia/sofr/sofr-2008.

Australian Government, Department of Agriculture (ABARES) 2013, *Australia's State of the Forests Report,* viewed 24 January 2020, https://www.agriculture.gov.au/abares/forestsaustralia/sofr/sofr-2013.

Australian Government, Department of Agriculture and Water Resources (ABARES) 2018, *Australia's State of the Forests Report,* viewed 24 January 2020, https://www.agriculture.gov.au/abares/forestsaustralia/sofr/sofr-2018.

Australian Government, Department of Agriculture 2020, *Australia's State of the Forests Report* series, viewed 24 January 2020, https://www.agriculture.gov.au/abares/forestsaustralia/sofr.

Bureau of Meteorology 2019, Basic Climatological Station Metadata, Rutherglen Research http://www.bom.gov.au/clim_data/cdio/metadata/pdf/siteinfo/IDCJMD0040.082039.SiteInfo.pdf.

Bowman, DMJS 1988, 'Stability amid turmoil?: towards an ecology of north Australian eucalyptus forests', *Proceedings of the Ecological Society of Australia*, vol. 15, pp. 149–158.

Bowman, DMJS & Panton, WJ 1993, 'Factors that control monsoon-rainforest seedling establishment and growth in north Australian Eucalyptus savanna', *Journal of Ecology*, vol. 81, pp. 297–304.

Bowman, DMJS & Panton, WJ 1995, 'Munmarlary revisited: Response of a north Australian *Eucalyptus tetrodonta* savanna protected from fire for 20 years', *Australian Journal of Ecology*, vol. 20, pp. 526–531.

Burrows, B 2016, *Inquiry into the Vegetation Management (Reinstatement) and other Legislation Amendment Bill 2016*, Queensland Parliament Agriculture and Environment Committee, SDNRAIDC Review Submission, Department of Environment and Science, Brisbane, https://www.parliament.qld.gov.au/documents/committees/AEC/2016/rpt19-11-VegetationMangt/submissions/214.pdf.

Cheney, NP 1995, 'Bushfires – an integral part of Australia's environment', 1301.0 – Year Book Australia, 1995, Australian Bureau of Statistics, viewed 14 January 2020.

Cooke, P 1998, 'Fire Management and Aboriginal Lands', *Summary Papers from the North Australia Fire Management Workshop*, Darwin, 24–25 March.

Duggin, JA 1976, *Bushfire History of the South Coast Study Area*, CSIRO Division of Land Use Research, Canberra, Technical Memorandum 76/13Ref.

Fensham, RJ & Bowman, DMJS 1992, 'Stand structure and the influences of overwood on regeneration in tropical Eucalypt forest on Melville Island', *Australian Journal of Botany*, vol. 40, pp. 335–352.

Fox, ID, Neldner, VJ, Wilson, GW & Bannink, PJ 2001, *The Vegetation of the Australian Tropical Savannas*, Environmental Protection Agency, Queensland Government Brisbane.

Gammage, B 2011, *The Biggest Estate on Earth, How Aborigines Made Australia*, Allen & Unwin, Crows Nest.

Lin, X & Hubbard, KG 2008, 'What are daily maximum and minimum temperatures in observed climatology', *International Journal of Climatology*, vol. 28, pp. 283–294.

Marohasy, J 2016, 'Temperature change at Rutherglen in south-east Australia', *New Climate*, http://dx.doi.org/10.22221/nc.2016.001.

Marohasy, J 2017, 'The Homogenisation of Rutherglen', in J Marohasy (ed), *Climate Change, The Facts 2017*, Institute of Public Affairs, Melbourne, Australia.

Marohasy, J 2018, 'Marohasy's Open Letter to Chief Scientist on BoM Failures', *IPA Today*, 16 May, https://ipa.org.au/ipa-today/marohasys-open-letter-to-chief-scientist-on-bom-failures.

Marohasy J 2020, 'After the Tragic Wildfires: History is Rewritten or Forgotten', https://jennifermarohasy.com/2020/01/after-the-tragic-wildfires-history-is-rewritten-or-forgotten/.

Milovanovic, S 2008, 'Plea for forest fuel burns', *Sydney Morning Herald*, https://www.smh.com.au/environment/plea-for-forest-fuel-burns-20081119-6 bm6.html.

Mulligan, J 2012, Submission to Senate Committee Investigating 'The effectiveness of threatened species and ecological communities protection in Australia', East Gippsland Wildfire Taskforce Inc.

Mulligan, J 2018, *Fire in East Gippsland*, South East Timber Association, https://volunteerfirefighters.org.au/fire-in-east-gippsland-recollections-of-john-mulligan.

Neldner, VJ, Butler, DW & Guymer, GP 2019, *Queensland's regional ecosystems. Building and maintaining a biodiversity inventory, planning framework and information system for Queensland, Version 2*, Queensland Herbarium, Queensland.

Trewin, B 2012, *Techniques involved in developing the Australian Climate Observations Reference Network – Surface Air Temperature (ACORN-SAT) dataset*, The Centre for Australian Weather and Climate Research, Technical Report No. 049.

Victoria. *Royal Commission to Inquire into the Causes of and Measures Taken to Prevent the Bush Fires of January 1939 & Stretton, Leonard Edward Bishop. 1939, Report of the Royal Commission to inquire into the causes of and measures taken to prevent the bush fires of January, 1939 and to protect life and property and the measures to be taken to prevent bush fires in Victoria and to protect life and property in the event of future bush fires*, Govt. Printer, Melbourne, viewed 18 May 2020, http://nla.gov.au/nla.obj-52798639.

Ward, DJ & Van Didden, G 2003, *A Partial Fire History of the Coolgardie District*, Consultant report to Department of Conservation and Land Management, Perth.

Ward, DJ 2010, *People, Fire, Forest and Water in Wungong: The Landscape Ecology of a West Australian Water Catchment*, PhD Thesis, Curtin University.

16: Rewriting Australia's Temperature History

Bureau of Meteorology 2012, 'Australian Climate Observations Reference Network – Surface Air Temperature (ACORN-SAT): Station catalogue', viewed 14 May 2017, http://www.bom.gov.au/climate/change/acorn-sat/documents/ACORN-SAT-Station-Catalogue-2012-WEB.pdf.

Lloyd, G 2019, 'Darwin warming claim triggers challenge to Bureau', *The Weekend Australian*, 25 February, https://www.theaustralian.com.au/national-affairs/climate/darwin-warming-claim-triggers-challenge-to-Bureau/news-story/bba138e1feb1c270b08b7e22c92f8659.

Marohasy, J 2016, 'Temperature change at Rutherglen in south-east Australia,' *New Climate*, http://dx.doi.org/10.22221/nc.2016.001.

McAneney, KJ, Salinger, MJ, Porteous, AS & Barber, RF 1990, 'Modification of an orchard climate with increasing shelter-belt height', *Agricultural and Forest Meteorology*, vol. 49, pp. 177–189.

Trewin, B 2018, 'The Australian Climate Observations Reference Network – Surface Air Temperature (ACORN-SAT) Version 2, Bureau Research Report 032', October, http://www.bom.gov.au/climate/change/acorn-sat/documents/BRR-032.pdf.

17: Perspectives on Sea Levels

Abbot, J & Marohasy, J 2012, 'Application of artificial neural networks to rainfall forecasting in Queensland, Australia', *Advances in Atmospheric Science*, vol. 29(4), pp. 717–730.

Belperio, A 1993, 'Land Subsidence and Sea-level Rise in the Port Adelaide Estuary: Implications for Monitoring the Greenhouse Effect', *Australian Journal of Earth Sciences*, vol. 40, pp. 359–368.

CSIRO 2017, 'Historical Sea Level Changes, Last few hundred years.' Online: http://www.cmar.csiro.au/sealevel/sl_hist_few_hundred.html

Denys, P & Hannah, J 1998, 'GPS and Sea Level Measurements in New Zealand', Department of Surveying, Otago University, NZ.

Han, S-C, Sauber, J, Pollitz, F & Ray, R 2019, 'Sea Level Rise in the Samoan Islands Escalated by Viscoelastic Relaxation After the 2009 Samoa-Tonga Earthquake', *Journal of Geophysical Research: Solid Earth*, vol. 124, pp. 4142–4156.

Hunter, J, Coleman, R & Pugh, D 2003, 'The Sea Level at Port Arthur, Tasmania, from 1841 to the Present', *Geophysical Research Letters*, vol. 30, pp. 54-1–54-4.

IPCC 2001, *Climate Change 2001: The Scientific Basis. Contribution of Working Group 1 to the Third Assessment Report of the Intergovernmental Panel on Climate Change*, Houghton, JT, Ding, Y, Griggs, DJ, Noguer, M, van der Linden, PJ, Dai, X, Maskell, K & Johnson, CA (eds.), Cambridge University Press, Cambridge, United Kingdom and New York, NY, USA, p. 881.

IPCC 2013, *Climate Change 2013: The Physical Science Basis. Contribution of Working Group I to the Fifth Assessment Report of the Intergovernmental Panel on Climate Change*, Stocker, TF, Qin, D, Plattner, G-K, Tignor, M, Allen, SK, Boschung, J, Nauels, A, Xia, Y, Bex, V & Midgley, PM (eds.), Cambridge University Press, Cambridge, United Kingdom and New York, NY, USA, p. 1535.

Lisitzin, E 1974, *Sea Level Changes*, Elsevier, New York.

Lord, R 1995, *The Isle of the Dead – Port Arthur*, 4th edn, Richard Lord and Partners, Taroona, Tasmania, pp. 24–28.

Marohasy, J & Abbot, J 2015, 'Assessing the quality of eight different maximum temperature time series as input when using artificial neural networks to forecast monthly rainfall at Cape Otway, Australia', *Atmospheric Research*, vol. 166, pp. 141–149.

Marohasy, J, Abbot, J, Stewart, K & Jensen, D 2014, 'Modelling Australian and Global Temperatures: What's Wrong? Bourke and Amberley as Case Studies', *The Sydney Papers Online*, iss. 26.

McCue, K 2015, *Historical Earthquakes in Tasmania – Revisited*, Australian Earthquake Engineering Society, aees.org.au/wp-content/uploads/2012/06/Historical-earthquakes-Tasmaniafinal.pdf.

Peltier, W & Tushingham, M 1991, 'ICE-3G: A New Global Model of Late Pleistocene Deglaciation based upon Geophysical Predictions of Post Glacial Relative Sea Level Change', *Journal of Geophysical Research*, vol. 96, p. 4497.

Permanent Service for Mean Sea Level (PSMSL) 2020, 'Tide Gauge Data', Retrieved 01 June 2020 from http://www.psmsl.org/data/obtaining/ Holgate, SJ, Matthews, A, Woodworth, PL, Rickards, LJ, Tamisiea, ME, Bradshaw, E, Foden, PR, Gordon, KM, Jevrejeva, S & Pugh J 2013, 'New Data Systems and Products at the Permanent Service for Mean Sea Level', *Journal of Coastal Research*, vol. 29, pp. 493–504.

Pugh, D, Hunter, J, Coleman, R & Watson, C 2002, 'A Comparison of Historical and Recent sea Level Measurements at Port Arthur, Tasmania', *International Hydrographic Review*, vol. 3, pp. 27–45.

Ross, Captain Sir JC 1847, *A Voyage of Discovery and Research in the Southern and Antarctic Regions, During the Years 1839–43*, John Murray, London, vol. 2, pp. 22–32.

Schneider, D 1998, 'The Rising Seas', *Scientific American*.

Shortt, Captain 1889, '*Notes on the Possible Oscillation of Levels of Land and Sea in Tasmania During Recent Years*', Royal Society of Tasmania papers, Hobart, pp. 18–20.

Quirk, T 2017, 'Taking Melbourne's Temperature', in J Marohasy (ed), *Climate Change, The Facts 2017,* Institute of Public Affairs, Melbourne, Australia.

Watts, A 2017, 'Creating a False Warming Signal in the US Temperature Record', in J Marohasy (ed), *Climate Change, The Facts 2017*, Institute of Public Affairs, Melbourne, Australia.

Vlok, JD 2019, 'Temperature reconstruction methods', University of Tasmania, Hobart, Australia. https://eprints.utas.edu.au/29788/.

18: Climate Science and Policy-based Evidence

Bates, J 2017, 'Climate scientists versus climate data', 4 February, viewed 8 May 2017, <https://judithcurry.com/2017/02/04/climate-scientists-versus-climate-data/>.

Boehmer-Christiansen, S 1994, 'Global climate protection policy: the limits of scientific advice: Part 2', *Global Environmental Change*, vol. 4, no. 3, pp. 185–200.

BBC News 2010, 'Q&A: Professor Phil Jones', 13 February, viewed 12 April 2017 <http://news.bbc.co.uk/2/hi/8511670.stm>.

Chan, D, Kent, EC, Berry, DI & Huybers, P 2019, 'Correcting datasets leads to more homogeneous early-twentieth-century sea surface warming', *Nature*, vol. 571, pp. 393–397.

Climate Audit 2009, 'A Peek Behind the Curtain', viewed 6 April 2017, <https://climateaudit.org/2009/03/04/a-peek-behind-the-curtain/>.

Climategate 2012, Archive of Climategate Files, available at: http://di2.nu/foia/ (accessed 20 February 2017).

Coady, D & Corry, R 2013, *The Climate Change Debate: An Epistemic and Ethical Enquiry*. Palgrave Macmillan, Basingstoke.

Cornwall, W & Voosen, P 2017, 'How a culture clash at NOAA led to a flap over a high-profile warming pause study.' Science Insider 8 February. <https://www.sciencemag.org/news/2017/02/how-culture-clash-noaa-led-flap-over-high-profile-warming-pause-study>

CRU (Climate Research Unit, University of East Anglia) 2016, CRU: Data – Temperature CRUTEM 2 (Updated January 2016), viewed 6 April 2017, <https://crudata.uea.ac.uk/cru/data/crutem2/>.

Daubert v. Merrel Dow Pharmaceuticals, Inc., 509 US 579, 113 S.Ct, 2786, 125 L.ed. 2d 469 (1993). On remand 43 F.3d 1311 (9th Cir. 1995), cert. denied, US, 116 S.Ct. 132 L.ed.2d 126 (1995).

Feyerabend, P 1975, *Against Method: Outline of an Anarchistic Theory of Knowledge*, New Left Books, London.

Funtowicz, SO, & Ravetz, JR 1993, 'Science for the post-normal age', *Futures*, vol. 25, no. 7, pp. 739–755.

Goldstein, L 2017, 'No, Santer et al. have not refuted Scott Pruitt', 25 May, *Watts Up With That?* viewed 20 June 2017, <https://wattsupwiththat.com/2017/05/25/no-santer-et-al-have-not-refuted-scott-pruitt/>.

Hughes, W 2005, 'WMO non respondo' email from Phil Jones to Warwick Hughes'. 21 February, viewed 17 January 2016, available at <http://www.warwickhughes.com/blog/?p=4203>.

Hilsenrath, J 2009, 'Cap-and-Trade's Unlikely Critics: Its Creators; Economists Behind Original Concept Question the System's Large-Scale Usefulness, and Recommend Emissions Taxes Instead', Wall Street Journal, 13 August, viewed

14 May 2017, <http://online.wsj.com/article/NA_WSJ_PUB:SB125011380 094927137.html>.

The Hockey Schtick 2018, <http://hockeyschtick.blogspot.com>

Ioannidis, JP 2005, 'Why most published research findings are false', *PLoS medicine*, vol. 2, no. 8, e124.

Janis, IL 1971, 'Groupthink' *Psychology Today*, vol. 5(6), pp. 84–90;

Jussim, L, Crawford, JT, Stevens, ST, Anglin, SM, & Duarte, JL 2016, 'Can high moral purposes undermine scientific integrity', in J Forgas, P van Lange & L Jussim (eds), *The social psychology of morality*, Psychology Press, London.

Karl, TR, Arguez, A, Huang, B, Lawrimore, JH, McMahon, JR, Menne, MJ, Peterson, TC, Vose, RS & Zhang, HM 2015, 'Possible artifacts of data biases in the recent global surface warming hiatus', *Science*, vol. 348, no. 6242, pp. 1469–1472.

Kellow, A 2007, *Science and Public Policy: The Virtuous Corruption of Virtual Environmental Science*, Edward Elgar, Cheltenham.

Kellow, Aynsley 2018, *Negotiating Climate Change: A Forensic Analysis*, Edward Elgar, Cheltenham.

Kent, EC, Woodruff, SD, & Berry, DI 2007, 'Metadata from WMO publication no. 47 and an assessment of voluntary observing ship observation heights in ICOADS.' *Journal of Atmospheric and Oceanic Technology*, 24(2): 214–234.

Kuhn, TS 1962, *The Structure of Scientific Revolutions*, University of Chicago Press, Chicago.

Laframboise, D 2011, *The Delinquent Teenager who was mistaken for the world's top climate expert*, Ivy Avenue Press, Toronto.

Lewin, B 2017, *Searching for the Catastrophe Signal: The Origins and Early History of The Intergovernmental Panel on Climate Change*, GWPF, London.

McLean, JD 2017, *An audit of uncertainties in the HadCRUT4 temperature anomaly dataset plus the investigation of three other contemporary climate issues*, PhD thesis, James Cook University.

Michaels, PJ, & Knappenberger, PC 1996, 'Human effect on global climate?' *Nature*, vol. 384, pp. 522–523.

Montford, AW 2010, *The Hockey Stick illusion: Climategate and the Corruption of Science*, Stacey International, London.

Montford, AW 2012, *Hiding the Decline: A History of the Climategate Affair*, Anglosphere Books, London.

NOAA 2019, 'Cold and Warm Episodes by Season', <https://origin.cpc.ncep.noaa.gov/products/analysis_monitoring/ensostuff/ONI_v5.php>.

Paltridge, G, Arking, A & Pook, M 2009, 'Trends in middle-and upper-level tropospheric humidity from NCEP reanalysis data', *Theoretical and Applied Climatology*, vol. 98, no. 3–4, pp. 351–359.

Popkin, G 2019, 'The forest question', *Nature*, vol. 565, pp. 280–282.

Popper, K 1963, *Conjectures and Refutations: The Growth of Scientific Knowledge*, Routledge, London.

Popper, K 1968, *The Logic of Scientific Discovery* (Rev ed), Hutchinson, London.

Santer, BD, Taylor, KE, Wigley, TML, Johns, TC, Jones, PD, Karoly, DJ, Mitchell, JFB, Oort, AH, Penner, JE, Ramaswamy, V & Schwarzkopf, MD 1996, 'A search for human influences on the thermal structure of the atmosphere', *Nature*, vol. 382, no. 6586, p. 39.

Santer, BD, Solomon, S, Wentz, FJ, Fu, Q, Pro-Chedley, S, Mears, C, Painter, JF & Bonfils, C 2017, 'Tropospheric Warming Over the Past Two Decades', *Scientific Reports*, no. 2336, doi:10.1038/s41598-017-02520-7.

Spencer, R 2017, 'Santer takes on Pruitt: the global warming pause and the devolution of climate science', 25 May, viewed 8 November 2017, <http://www.drroyspencer.com/2017/05/santer-takes-on-pruitt-the-global-warming-pause-and-the-devolution-of-climate-science/>.

Stocker, TF, Qin, D, Plattner, G-K, Alexander, LV, Allen, SK, Bindoff, NL, Bréon, F-M, Church, JA, Cubasch, U, Emori, S, Forster, P, Friedlingstein, P, Gillett, N, Gregory, JM, Hartmann, DL, Jansen, E, Kirtman, B, Knutti, R, Krishna Kumar, K, Lemke, P, Marotzke, J, Masson-Delmotte, V, Meehl, GA, Mokhov, II, Piao, S, Ramaswamy, V, Randall, D, Rhein, M, Rojas, M, Sabine, C, Shindell, D, Talley, LD, Vaughan, DG & Xie, S-P 2013, *Technical Summary. In: Climate Change 2013: The Physical Science Basis. Contribution of Working Group I to the Fifth Assessment Report of the Intergovernmental Panel on Climate Change*, Stocker, TF, Qin, D, Plattner, G-K, Tignor, M, Allen, SK, Boschung, J, Nauels, A, Xia, Y, Bex, V & Midgley, PM (eds.), Cambridge University Press, Cambridge, United Kingdom and New York, NY, USA, p. 115.

19: Of Blind Beetles and Shapeshifting Birds: Research Grants and Climate Catastrophism

Davidson, S 2005, 'Research shows it's time to axe ARC', *The Australian Financial Review*, August 10, 2005, p. 63.

Montford, A 2012, The Royal Society and Climate Change, Global Warming Policy Foundation. https://www.thegwpf.org/images/stories/gwpf-reports/montford-royal_society.pdf.

Nott, G 2018, 'Minister makes no vetoes in latest ARC funding round', www.computerworld.com, viewed December 5 2019, https://www.computerworld.com/article/3475303/minister-makes-no-vetoes-in-latest-arc-funding-round.html.

Tehan, D 2018, 'Strengthening public confidence in university research funding', viewed December 5 2019, https://ministers.education.gov.au/tehan/strengthening-public-confidence-university-research-funding.

20: A Descent into Sceptics' Hell

Alighieri, D 2015, *Divina Commedia*, Translated by Clive James, Pan Macmillan, London, United Kingdom.

Berg, C & Hargreaves, S 2019, '*Looking Forward Special Episode – Dr Peter Ridd*', Institute of Public Affairs, recorded 21 June, <https://ipa.org.au/ipa-today/ep-23-looking-forward-special-episode-dr-peter-ridd>.

Breheny, S 2017, 'Free speech and climate change', in J Marohasy (ed), *Climate Change, The Facts 2017*, Institute of Public Affairs, Melbourne, Australia.

Bureau of Meteorology 2012, Australian Climate Observations Reference Network – Surface Air Temperature (ACORN-SAT) Observation Practices, Commonwealth of Australia, Melbourne. http://www.bom.gov.au/climate/data/acorn-sat/documents/ACORN-SAT_Observation_practices_WEB.pdf

Bureau of Meteorology 2017, '*Review of Bureau of Meteorology Automatic Weather Stations*', Commonwealth of Australia, Melbourne. https://apo.org.au/sites/default/files/resource-files/2017-09/apo-nid106276.pdf

Cook, J, Nuccitelli, D, Green, SA, Richardson, M, Winkler, B, Painting, R, Way, R, Jacobs, P & Skuce, A 2013, 'Quantifying the consensus on anthropogenic global warming in the scientific literature', *Environmental Research Letters*, vol. 8, no. 2.

Costella, J 2010, 'Climategate emails', The Lavoisier Group, Melbourne, Australia. https://www.lavoisier.com.au/articles/greenhouse-science/climate-change/climategate-emails.pdf

Curry, J 2015, 'Assessments, meta-analyses, discussion and peer review', *Judith Curry*, 29 July 2015, viewed 4 May 2020, https://judithcurry.com/2015/07/29/assessments-meta-analyses-discussion-and-peer-review/

Curry, J 2017, 'Climate models for the layman', The Global Warming Policy Foundation, https://www.thegwpf.org/content/uploads/2017/02/Curry-2017.pdf

Ioannidis, JPA 2005, 'Why Most Published Research Findings Are False', *PLOS Medicine*, vol. 2, no. 8.

Kellow, A 2018, *Negotiating Climate Change: A Forensic Analysis*, Edward Elgar Publishing, Cheltenham, United Kingdom.

Lloyd, G 2017a, 'More BoM weather stations put on ice', *The Australian*, 3 August, viewed 4 May 2020, <https://www.theaustralian.com.au/nation/climate/more-bureau-of-meteorology-weather-stations-put-on-ice/news-story/0412a1aa5ab48d588fa71df90763aba2>.

Lloyd, G 2017b, 'BoM Opens Cold Case on Temperature Data', *The Australian*, 1 August, <https://www.theaustralian.com.au/nation/climate/bureau-of-meteorology-opens-cold-case-on-temperature-data/news-story/c3bac520af2e81fe05d106290028b783>.

Lloyd, G 2017c, 'Temperatures Plunge after BoM Orders Fix', *The Australian*, 4 August <https://www.theaustralian.com.au/nation/climate/temperatures-plunge-after-bureau-orders-weather-station-fix/news-story/9230dd914ac532fa735700ffc7799203>.

Lomborg, B 2017, 'The Impact and Cost of the 2015 Paris Climate Summit, with a Focus on US Policies', in J Marohasy (ed), *Climate Change, The Facts 2017*, Institute of Public Affairs, Melbourne, Australia.

Marohasy, J 2017a, 'Introduction' in J Marohasy (ed), *Climate Change: The Facts 2017*, Institute of Public Affairs, Melbourne, Australia.

Marohasy, J 2017b, 'The Homogenisation of Rutherglen', in J Marohasy (ed), *Climate Change, The Facts 2017*, Institute of Public Affairs, Melbourne, Australia.

Marohasy, J 2017c, 'John has Plus 10 Degrees, Bureau Loses Minus 10 Degrees', *Jennifer Marohasy*, September 17, viewed 4 May 2020, <https://jennifermarohasy.com/2017/09/john-plus-10-degrees-bureau-loses-minus-10-degrees/>.

Marohasy, J 2018, 'Marohasy's Open Letter to Chief Scientist on BoM Failures', *IPA Today*, 16 May, <https://ipa.org.au/ipa-today/marohasys-open-letter-to-chief-scientist-on-bom-failures>.

Marohasy, J & Abbot, J 2015, 'Assessing the quality of eight different maximum temperature time series as inputs when using artificial neural networks to forecast monthly rainfall at Cape Otway, Australia', *Atmospheric Research*, vol. 166, pp. 141–149.

Marohasy, J & Abbot, J 2016, 'Southeast Australian Maximum Temperature Trends, 1887–2013: An Evidence-Based Reappraisal', in D Easterbrook (ed) *Evidence-Based Climate Science*, 2nd edn, pp. 83–99.

Mitchell, MD & Boettke, PJ 2016, *Applied Mainline Economics, Bridging the Gap between Theory and Public Policy*, Mercatus Center, Virginia, USA.

Montford, A 2014, 'Fraud, Bias And Public Relations, The 97% 'consensus' and its critics', *The Global Warming Policy Foundation*, <https://www.thegwpf.org/content/uploads/2014/09/Warming-consensus-and-it-critics1.pdf>.

Nova, J 2017, 'Mysterious Revisions to Australia's Long Hot History', in J Marohasy (ed), *Climate Change, The Facts 2017*, Institute of Public Affairs, Melbourne, Australia.

Quirk, T 2017, 'Taking Melbourne's Temperature', in J Marohasy (ed), *Climate Change, The Facts 2017*, Institute of Public Affairs, Melbourne, Australia.

Roberts, R 2019, 'Bjorn Lomborg on the Costs and Benefits of Attacking Climate Change', *Econtalk*, recorded April 16, <http://www.econtalk.org/bjorn-lomborg-on-the-costs-and-benefits-of-attacking-climate-change/#audio-highlights>.

Rosling, H, Rosling, O & Rosling Rönnlund, A 2018, *Factfulness: Ten Reasons We're Wrong About the World – And Why Things Are Better Than You Think*, Hodder and Staunton, London, United Kingdom.

Rudd, K 2007, 'Climate change: the great moral challenge of our generation', Address to the National Climate Summit, 31 March, viewed 21 August, 2019, <https://youtu.be/CqZvpRjGtGM>.

Watts, A 2017, 'Creating a False Warming Signal in the US Temperature Record', in J Marohasy (ed), Climate Change, The Facts 2017, Institute of Public Affairs, Melbourne, Australia.

Wild, D 2018, 'Why Australia Must Withdraw From The Paris Climate Agreement', Institute of Public Affairs, <https://ipa.org.au/wp-content/uploads/2018/08/IPA-Report-Why-Australia-Must-Withdraw-from-Paris.pdf>.

Worrall, E 2016, 'Accused of sexual harassment crimes, former IPCC head Pachauri claims: "I was set up"', 30 March, viewed July 22, 2019, <https://wattsupwiththat.com/2016/03/30/accused-ipcc-pervert-pachauri-i-was-set-up/>.

Vlok, JD 2019, 'Temperature reconstruction methods', University of Tasmania, Hobart, Australia. <https://eprints.utas.edu.au/29788/>.

Acknowledgements

The Editor would like to acknowledge the efforts and expertise of Leonie Ryan (typesetting), Susan Prior (copy) and John Castle (graphics and illustrations), who kept going to get this book completed through the COVID-19 pandemic. The Editor also thanks Susan Ball, Ric Werme, Arthur Day and Jaco Vlok for their diligent proofreading, while accepting full responsibility for any remaining errors. The unwavering commitment of Bryant Macfie to honest and evidence-based scientific inquiry is a continuing source of inspiration.

Afterword

The previous book in this series *Climate Change: The Facts 2017*, began with a chapter by Peter Ridd about the Great Barrier Reef. This chapter was mentioned in the judgement of the Federal Court of Australia in James Cook University v Ridd [2020] FCAFC 123 – as this fourth book in the series is going to press. Specifically, that Peter Ridd's appearance on television, following the chapter's publication, contributed to his second censure by James Cook University (JCU) that resulted in his sacking by that university.

It is the case that *Climate Change: The Facts 2017* challenged some of the orthodoxies on Great Barrier Reef science and climate change more generally. Specifically, Dr Ridd raised concerns about the calculation of coral calcification rates and associated quality assurance issues.

Dr Ridd originally challenged his dismissal by James Cook University in the Australian Federal Circuit Court. In April 2019 the court found for Dr Ridd on all counts on the basis that he had been unlawfully dismissed by the university.

James Cook University subsequently appealed the case.

Now that the Federal Court of Australia has overturned the decision of the Federal Circuit Court, Dr Ridd has indicated he intends to appeal the decision of the Federal Court to the High Court of Australia.

The case is of enormous public importance, for free speech and also the traditions of the scientific method. As the Chairman of the IPA, Janet Albrechtsen, wrote in *The Australian* newspaper on 25 July 2020:

> Remember that Ridd wasn't querying the interpretation of Ovid's Meta-
> morphoses. He was raising questions, in one particular area of his expertise,

about the quality of climate change science. One of the fundamental challenges of our generation is to get the science right so we can settle on the right climate change policies. JCU told Ridd to keep quiet, then it sacked him. And a court has endorsed its actions.

JCU's conduct, and the court's decision, has sent intellectual inquiry down the gurgler in the 21st century at an institution fundamental to Western civilisation. Is that to be legacy of JCU's vice-chancellor Sandra Harding? And what oversight has JCU's governing council provided to this reputational damage, not to mention the waste of taxpayer dollars, in pursuing a distinguished scientist who was admired by his students?

Following this decision, no academic can assume that an Australian university will allow the kind of robust debate held at Oxford University in 1860 between the bishop of Oxford, Samuel Wilberforce, and Thomas Henry Huxley, a biologist and proponent of Darwin's theory of evolution.

The Historical Journal records how this legendary encounter unfolded: 'The Bishop rose, and in a light scoffing tone, florid and fluent he assured us there was nothing in the idea of evolution: rock-pigeons were what rock-pigeons have always been. Then, turning to his antagonist with a smiling insolence, he begged to know, was it through his grandfather or his grandmother that he claimed his descent from a monkey? On this Mr Huxley slowly and deliberately arose. A slight tall figure stern and pale, very quiet and very grave, he stood before us, and spoke those tremendous words … He was not ashamed to have a monkey for his ancestor, but he would be ashamed to be connected with a man who used his great gifts to obscure the truth.'

Not for nothing, Ridd's lawyers submitted this example of intellectual freedom during the first trial. In sacking Ridd, and to win in court, JCU had to argue against the means that seeks the truth – intellectual freedom.

In deciding whether to grant special leave for the appeal, the High Court will consider whether the case involves 'a question of law that is of public importance'. The Ridd matter easily meets this threshold. It would be the first time the High Court has been called upon to consider the meaning of 'academic and intellectual freedom', which is used in enterprise agreements covering staff at almost all Australian universities. The court's decision will therefore have very real consequences in terms of university governance, and the extent to which administrators tolerate controversial (and, often, commercially inconvenient) opinions from the professoriate.

Should 'intellectual freedom' be limited by the whims of university administrators, as the university is arguing? Or should it be wide enough to allow for the kind of controversial, but honestly held opinions for which Dr Ridd was ultimately sacked?

The Federal Court's answer to that question is deeply disturbing. In its judgement last week, the majority seemed to suggest that free speech on campus is past its use-by date.

'There is little to be gained in resorting to historical concepts of academic freedom,' claimed justices Griffiths and Derrington in the majority judgement. They were quoting from an academic textbook outlining 'a host of new challenges', like 'the rise of social media' and 'student demands for accommodations such as content warnings and safe spaces' as reasons for doing away with the concept of intellectual freedom.

While the IPA does not suggest the judges acted improperly, it is worrying that the boundaries of free speech should be defined in this way.

Intellectual freedom and free speech are not antiquated notions. They are ancient and important rights through which we may get closer to the truth – and that is ultimately the quest of this series of books: to get closer to the facts as they pertain to climate science.

For further background and current information, go to
ipa.org.au/peterridd

The IPA Gives Thanks

Climate Change: The Facts 2020 is a publication of the Institute of Public Affairs (IPA), an organisation dedicated to securing freedom for the next generation.

It is an independent, non-profit public policy think tank, founded in Australia in 1943 to advance social, economic and political freedom.

The IPA supports the free market of ideas, the free flow of capital, limited and efficient government, the rule of law, and representative democracy.

Its active program of research into climate science and related issues is led by Dr Jennifer Marohasy.

The IPA gratefully acknowledges the support of the 1417 people in Australia and across the world whose donations to our special fund-raising campaign made possible the publication of the important new research, reviews and commentary now published within these pages.

In particular, the IPA expresses its deep appreciation for the contributions made by the following 322 people, who gave consent for recognition of their support for free speech on climate change to appear in the following pages.

To learn more abut the IPA's climate change research program, view supplementary materials, and related publications, please go to

climatechangethefacts.org.au

To learn more about the benefits of becoming one of more than 6,000 IPA Members, go to

ipa.org.au/join

This book made possible through the generous contributions of:

CHRISTOPHER ABBOTT, ANDREW J ABERCROMBIE, ANTHONY ADAIR, KATE & PETER ADKINS, STEPHEN AINSWORTH, IAN AIREY, RICHARD W ANDERSEN, BRIAN ANDERSON, STEVE & JANE ANTONIO, STUART ASHTON, PETER ASSFALG, MICHAEL ASTEN, CHARLES BARNES, JOHN BARNES, MARCELLE & KEN BARNES, ESTIE BAV, STEPHEN BAXTER, MAX C S BECK, HAYDN L BEECK, RAYMOND BENHAM, DIANA BENNETT, PAUL BIDE, GEORGE & SYBIL BINDLEY, WILLIAM BINN, EDWARD BIRBARA, DON BISHOP, TERRY BIXLER, ROBERT BLACK, DAVID BLAKE, TOM BOSTOCK, GEOFF BROWN, ANDREW BROWNE, ROBERT BRYAN, GREG BUCHANAN, ELIZABETH BURNS, MICHAEL BURSTON, ANDY BUTTFIELD, JOHN BYKERK, ANTHONY CAIRNS, WALLACE & JOAN CAMERON, ALEX CAMPBELL, DOUGLAS CAMPBELL, KEDE CARBONI, LEN & WENDY CARLSON, ANTONY CARR, MYRON CAUSE, SCOTT CHALMERS, PETER CHAMPNESS, GREG CHAPMAN, PAUL CHAPMAN, ROD CHARLTON, GEORGE S CHOMLEY, GEORGE CHRISTENSEN, ANTHONY CLIFFORD, DAVID COAD, RICHARD COLEBATCH, OWEN COLTMAN, WALTER & CHRISTL COMMINS, PETER D CONDON, CHRIS CONNELLAN, JAMES COOMBE, CRAIG COOPER, RICHARD CORBETT, JOHN & JULIA CORDUKES, JOHN CORRIGAN, TERRY COUPLAND, HUGH COWLING, ALLAN M COX, ALF CRISTAUDO, JOHN CROMB, TIM & OLGA CROME, DIGBY CROZIER, FRANK CUFONE, FRANZIE CUMMINGS, ANDREW CUMMINS, DAVID CUNNINGHAM, DOUG CUSTANCE, MURRAY CUTBUSH, D'ANTOINE FAMILY, PETER DAVIES, PHILIP DAVIS, SANDY & JANE DAWSON, PIERS DAWSON-DAMER, EVERT DE BLOK, GEOFF DE ROSS, MIKE & GLENYS DEAM, KEITH DELACY, HOWARD DEWHIRST, JENNIFER DIXON, JOHN DOHERTY, CRAIG DONOHUE, JAMES DOOGUE, RUSSELL DOWLING, FRED & PHILOMENA DRAKE, AERT DRIESSEN, PETER DRYGALA, TIM DUNCAN, JAMES WALTER SYDNEY DURRANT, DAVID EDWARDS, JOHN EGAN, MALCOLM EGLINTON, CHRISTINE ELKINS, MIKE ELLIOTT, J K & A ELLIS, DAVID L ELSUM,

This book made possible through the generous contributions of:

PETER & VERONICA ENGLISH, RIHS ERMANNO, RONALD J EVANS, JIM EXCELL,
PETER FARRELL,   NEIL FEARIS,   PETER FRANCIS FENWICK,   SIMON FENWICK,
P C FERUGLIO, IAN FLANIGAN, JIM FLETCHER, CHRIS FLIPO, NICHOLAS FORD,
IAN FORSTER, JOLYON FORSYTH, PETER FRANKLIN, PAMELA FRASER, ROSS A FRICK,
MICHAEL FRY, FRANK FUNDER, CHRISTINE FUREDY, TONY GALL, ROGER GIBBONS,
ROGER & LESLEY GILLESPIE,   BRENDAN GILSENAN,   CHRISTOPHER GOLIS,
ADRIAN GOOD, RUPERT GOOD, IAN GRANT, CRAIG GREEN, JAMES GRESHAM,
BILL GRIFFITHS & ZANA SMITH,   THE GÜNTHER FAMILY,   ANDRZEJ GURBA,
MUTHURAJ GURUSWAMY, DOUG HALL, GRAHAM HAMILTON, BEN HARDING,
ANN HARE,   CHRISTOPHER HARMAN,   CHRIS HARRINGTON,   BOB HARRIS,
MICHAEL E HARRISON, GRAEME HAUSSMANN, GEOFF HEMM, VAUGHAN HENRY,
GORDON HERBERTSON,   JAMES HERCUS,   SHANE HIBBIRD,   IAN HILLIAR,
TONY HODGSON, MICHAEL HOGG, P HOLT, GEOFF HONE, JOHN HONEYCOMBE,
JEFF HUDSON,   BRIAN HURLOCK,   JOHN ILLINGWORTH,   HARRY IMBER,
ROY INGLETON, TIMOTHY IRELAND, ALAN IRVING, RICK JACKSON, PAT W JAMESON,
MURRAY JAMIESON,   COL JARVIS,   LYN JEFFRIES,   BJÖRN JOHANSSON,
DAVID B JOHNSTON, ALAN JONES, BRIAN JONES, GERARD A JOSEPH, PETER KALINA,
LILY KEANE, HUGH KENDRICK, SAM KENNARD, BRYAN KENSINGTON, BARRY KEOUGH,
MOSES KHOR, RAY KING, ALEXANDER KISS, PETER KNIGHT, STEFAN KOVIC,
GLEN LAMPERD, JOHN LANDER, STEFAN LANDHERR, KEN LANG, ALLAN LAWTON,
GUY LEBLANC SMITH, ALFRED LEDNER, JOHN LEVINY, CHRIS LIEBENBERG,
PHIL LOCKYER,   SANDY LONGWORTH,   PETER LYNN,   STUART MACDONALD,
WARWICK MACKAY,   ANDREW MACLAINE-CROSS,   MICHAEL MAHONY,
PETER MANGER, JAMES MARKE, IAN MARTIN, MIRRIE MASNJAK, CHRIS MATHER,
MARK MCCAULEY, JEFFREY MCCLOY, STUART MCDONALD, DAVID MCGARRY,
ROGER MCGHEE, GARY MCGILL, STUART MCGILL, TONY MCGILL, TONY MCLEAN,

This book made possible through the generous contributions of:

STEPHEN MCPHERSON, LAURIE & BERNADETTE MENTZ, PETER MILES, CRAIG MILLS, MICHAEL & SUE MINSHALL, GORDON MOCK, MAURICE MOLAN, PAUL MORGAN, RICHARD MORGAN, ELIZABETH MORRISON, DAVID MULLER, JOHN B MURRAY, GRAEME NEWING, GRAEME NICOL, PETER NIXON, EGIL NORDANG, BOB NORLIN, TERRY O'CONNOR, COLIN & HELEN O'NEIL, EDWARD O'NEIL, PETER O'REGAN, BERNARD O'SHEA, ALAN OXLEY, LES PARSONS, JOHN PASCOE, PETER PASZKIEWICZ, GERRY PAULUSZ, IAN & SANDRA PERROTT, BRIAN PFLAUM, ANDREW PIDGEON, FRASER POWER, FRANK POWNALL, PATRICK PURCELL, MARK JOHN RADKE, GEOFF RANKIN, GERHARD REDELINGHUYS, LACHLAN REEKS, ANTHONY REID, EVAN RICHARDS, JOHN RICHARDSON, BRIAN ROBSON, JOHN ROLFE, BRADLEY ROW, SAM & KATE SAIDI, MURRAY SANDLAND, STEPHEN SASSE, MAY SCOTT, GRAHAM SELLARS-JONES, JOE SELVAGGI, FRANCIS SEPPELT, PHILLIP SGHERZA, HARRY SHANN, GREG SHERIDAN, MICHAEL SHIRLEY, JOHN & JOAN SHRAPNEL, JEAN-MARIE SIMART, MARIO SIMIC, FRED SINGLETON, MARIAN SKWARNECKI, KATHIE SMORGON, ROSS SMYTH-KIRK, DOUGLAS SPENCE, EVE SPENCE, TREVOR & JUDITH ST BAKER, JOHN ST JULIAN, JOHN STAPLETON, MICHAEL STEYN, DALLAS STOKES, MARK STRETTON, PETER SWAN, TREVOR SYKES, STUART TAIT, JAMES THOMSON, TIM TIMERMANIS , PETER A TRAINE, IAN TRISTRAM, BRIAN TUCKER, EWEN TYLER, DAVID VAN GEND, PETER VANRENEN, ROGER & MARGARET VINES, COR VISSER-MARCHANT, D K VOSS, BARRY WAKELIN, MARK WALLAND, MIC WALSH, DICK WARBURTON, BRUCE WATSON, GRAEME WEBER, GORDON WHITE, JOHN WHITE, TERENCE WHITFIELD, KYLE WIGHTMAN, GARTH WILLIAMS, ROGER WILLIAMS, WESLEY WILLIAMS, ANTHONY WILLIAMSON, T ANDREW K WILSON, MICHAEL WOOD, JOHN WYLD, RANDALL WYNN, RAYMOND YEOW, SOPHIE YORK, IAN YOULES